Complex
Stochastic
Systems

MONOGRAPHS ON STATISTICS AND APPLIED PROBABILITY

General Editors

D.R. Cox, V. Isham, N. Keiding, T. Louis, N. Reid, R. Tibshirani, and H. Tong

Complex Stochastic Systems

EDITED BY

OLE E. BARNDORFF-NIELSEN

Department of Mathematical Sciences
University of Aarhus
Aarhus, Denmark

DAVID R. COX

Honorary Fellow
Nuffield College
Oxford, UK

AND

CLAUDIA KLÜPPELBERG

Center of Mathematical Sciences
Munich University of Technology
Munich, Germany

CRC Press
Taylor & Francis Group
Boca Raton London New York

CRC Press is an imprint of the
Taylor & Francis Group, an **informa** business
A CHAPMAN & HALL BOOK

Cover photographs courtesy of Michael C. Thomas

Senior Editor: John Sulzycki
Project Editor: Mimi Williams
Cover Design: Shayna Murry

CRC Press
Taylor & Francis Group
6000 Broken Sound Parkway NW, Suite 300
Boca Raton, FL 33487-2742

First issued in paperback 2019

© 2000 by Taylor & Francis Group, LLC
CRC Press is an imprint of Taylor & Francis Group, an Informa business

No claim to original U.S. Government works

ISBN-13: 978-1-58488-158-2 (hbk)
ISBN-13: 978-0-367-39828-6 (pbk)
Library of Congress Card Number 00-022172

Library of Congress Cataloging-in-Publication Data

Arnett, Ross H.
    American insects: a handbook of the insects of America north of Mexico/Ross H.
Arnett, Jr.--2nd ed.
        p. cm.
    Includes bibliographical references.
    ISBN 0-8493-0212-9 (alk. paper)
    1. Insects--United States--Classif cation. 2. Insects--Canada--Classif cation. I. Title.

QL474 .A76 2000
595.7'0973—dc21   00-022172

Visit the Taylor & Francis Web site at
http://www.taylorandfrancis.com

and the CRC Press Web site at
http://www.crcpress.com

Contents

2 Causal Inference from Graphical Models   **63**

by Steffen L. Lauritzen

3 State Space and Hidden Markov Models 109

by Hans R. Künsch

Contributors

Frank den Hollander
Mathematical Institute
University of Nijmegen
6525 ED Njimegen, The Netherlands
Email: denholla@sci.kun.nl

Peter J. Green
Department of Mathematics
University of Bristol
Bristol BS8 1TW, UK
Email: p.j.green@bristol.ac.uk
http://www.stats.bris.ac.uk/~peter/

Hans R. Künsch
Seminar für Statistik
ETH-Zürich
CH-8092 Zürich, Switzerland
Email: kuensch@stat.math.ethz.ch
http://stat.ethz.ch/kuensch.html

Steffen L. Lauritzen
Department of Mathematics
Aalborg University
DK-9220 Aalborg, Denmark
Email: steffen@math.auc.dk
http://www.math.auc.dk/~steffen/

Gesine Reinert
King's College Research Centre
and Statistical Laboratory
Cambridge CB2 1ST, UK
Email: g.reinert@statslab.cam.ac.uk
http://www.statslab.cam.ac.uk/Dept/People/reinert.html

Elizabeth A. Thompson
Department of Statistics
University of Washington
Seattle WA 98195, USA
Email: thompson@stat.washington.edu
http://www.stat.washington.edu/thompson/

Participants

DENIS ALLARD, Avignon (France), allard@avignon.inra.fr
Using (pseudo-) Bayes factors for selecting models in a spatial context.

CLAUDIA BECKER, Dortmund (Germany),
cbecker@amadeus.statistik.uni-dortmund.de
Identification of outliers in multivariate data.

MALGORZATA BOGDAN, Wroclaw (Poland), mbogdan@im.pwr.wroc.pl
Data driven version of Neyman's test for goodness-of-fit based on Bayesian rule.

PAOLO BORTOT, Padova (Italy), bortot@hal.stat.unipd.it

STEVE BROOKS, Bristol (United Kingdom), steve.brooks@bristol.ac.uk
Efficient construction of reversible jump MCMC proposal distributions.

ANGELIKA CAPUTO, München (Germany), caputo@stat.uni-muenchen.de
Definition and interpretation of properties of a class of multivariate distributions by means of graphical models.

JOSÉ ANDRÉS CHRISTEN GRACIA, Morelia-Guanajuato (Mexico),
jac@matem.unam.mx
A general Bayesian classification utility for heterogeneous data bases.

CATERINA CONIGLIANI, Roma (Italy), caterina@pow2.sta.uniroma1.it
Bayesian analysis of integrated autoregressive time series with change points.

RAMA CONT, Palaiseau (France), rama.cont@polytechnique.fr
Multiresolution analysis of financial time series: some probabilistic and pathwise approaches.

KLAUS DOHMEN, Berlin (Germany), dohmen@informatik.hu-berlin.de
Improved Bonferroni inequalities.

MICHAEL EICHLER, Heidelberg (Germany),
eichler@statlab.uni-heidelberg.de
Graphical models in time series analysis.

MARCO FERRANTE, Padova (Italy), ferrante@math.unipd.it
On the stochastic stability of simple threshold bilinear models.

FLORENCE FORBES, Montbonnot Saint Martin (France),
florence.forbes@inrialpes.fr
Mean field approximation principle and Markov random field model-
based image segmentation.

MICHAL FRIESL, Pilzen (Czech Republic), friesl@kma.zcu.cz
Bayesian estimation in exponential competing risks model.

A.J. GANESH, Bristol (United Kingdom), ajg@hplb.hpl.hp.com
Fixed points of single server queues and their large deviation behaviour.

MARTIN B. HANSEN, Aalborg (Denmark), mbh@math.auc.dk
Nonparametric Bayesian inference with a view towards inverse problems.

KATJA ICKSTADT, Darmstadt (Germany),
katja@mathematik.tu-darmstadt.de
Spatial Poisson regression for health and exposure data measured at
disparate resolutions.

MARTON ISPÁNY, Debrecen (Hungary), ispany@math.klte.hu
On state-space representation of some nonlinear time series models with
Gaussian innovation.

AGNIESZKA JURLEWICZ, Wroclaw (Poland), agniesz@im.pwr.wroc.pl
Limit theorems applied to model relaxation phenomena in disordered
physical systems.

JULIA KELSALL, Lancaster (United Kingdom), julia.kelsall@lancaster.ac.uk

NIKOLAI KOLESNIKOV, Freiberg (Germany),
kolesn@merkur.hrz.tu-freiberg.de
Identification of the spatial structure through the generalized
backfitting.

MICHAL KULICH, Praha (Czech Republic), kulich@karlin.mff.cuni.cz
Analysis of the case-cohort design by additive hazards model.

ERCAN KURUOGLU, Cambridge (United Kingdom),
kuruoglu@xrce.xerox.com
Signal processing with alpha-stable distributions.

VOLKMAR LIEBSCHER, Neuherberg (Germany), liebsche@gsf.de
Coherent states – a bridge between classical and noncommutative
probability.

JORGE MATEU, Castellon (Spain) mateu@mat.uji.es

ANTONIETTA MIRA, Varese (Italy), anto@ipvaim.unipv.it
Slice samplers.

FLORENCE MURI, Paris (France), muri@citi2.fr
Heterogeneity of DNA sequences and hidden Markov chains.

MARITA OLSSON, Gothenburg (Sweden), marita@math.chalmers.se
A parametric estimation procedure for relapse time distribution.

SANDRA PEREIRA, Western Australia (Australia),
spereira@maths.uwa.edu.au

ANTONIO PIEVATOLO, Milano (Italy), marco@iami.mi.cnr.it
Simple modelling of failures in large and complex repairable systems.

PIOTR POKAROWSKI, Warsaw (Poland), pokar@mimuw.edu.pl
The probability inequalities for relative error of sequential Monte Carlo
procedures.

SILVIA POLETTINI, Roma (Italy), silpol@pow2.sta.uniroma1.it
On a measure of association for bivariate failure times.

EVA RICCOMAGNO, Coventry West Midlands (United Kingdom),
emr@stats.warwick.ac.uk
Multi-strain species modelling via differential algebra reduction.

AILA SÄRKÄ, Gothenburg (Sweden), aila@math.chalmers.se
Modelling interactions between two ants' species.

MICHAEL SAHM, Heidelberg (Germany), msahm@statlab.uni-heidelberg.de

LUDMILA SAKHNO, Kiev (Ukraine), lms@mechmat.univ.kiev.ua
On the Kaplan-Meier estimator for long-range dependent data.

BARRY SCHOUTEN, Amsterdam (The Netherlands), schouten@cs.vu.nl
Bayesian inference for hidden Markov models of ion channel data.

MARTIN SEVERIN, München (Germany), severin@ma.tum.de
IBNR (incurred but not reported) claims in insurance.

NICO STOLLENWERK, Cambridge (United Kingdom), nks22@cam.ac.uk
Testing nonlinear stochastic models on empirical phytoplankton biomass
time series.

DIRK TASCHE, München (Germany), tasche@ma.tum.de
Credit decisions based on various notions of risk.

ELKE THÖNNES, Göteborg (Sweden), thonnes@math.chalmers.se
Perfect simulation in stochastic geometry.

PAOLO VIDONI, Udine (Italy), vidoni@dss.uniud.it
Some new examples of finite dimensional filters in discrete-time.

NAFTALI ZILVERBERG, Ashdod (Israel), zilber@is.elta.co.il

Preface

The chapters of this volume represent the revised versions of the main papers given at the fourth Séminaire Européen de Statistique on "Complex Stochastic Systems", held at Eurandom, Eindhoven (The Netherlands), March 15-20, 1999. The aim of the Séminaire Européen de Statistique is to provide talented young researchers with an opportunity to get quickly to the forefront of knowledge and research in areas of statistical science which are of major current interest. This volume is tutorial as are the books based on the first three seminars in the series entitled

- Networks and Chaos – Statistical and Probabilistic Aspects,

- Time Series Models in Econometrics, Finance and Other Fields,

- Stochastic Geometry: Likelihood and Computation.

In the present Séminaire more than 40 young scientists, most from European countries, participated. They all presented their recent work, partly in short contributed talks, partly in a poster session; a list of the attendants of the Séminaire and the titles of their presentations is given on pp. ix–xi.

Complex stochastic systems is a vast area ranging from modelling aspects of specific applications and the associated probabilistic analysis to statistical model fitting, estimation procedures, simulation and computing issues. In particular, statistical analysis has been revolutionized by the constantly increasing computer power during the last decades. The Séminaire and this corresponding book mostly concentrate on statistical aspects, but show in very concrete applications the progress which has been made as a result of the recent development in statistical computing. The book is organised as follows, some details about the contents of the chapters being given below: Chapters 1 and 3 concentrate on sophisticated statistical methods for analysing complex systems. Chapter 2 deals with a very fundamental and much discussed problem in statistics, namely causality. Chapter 4 then treats some applications from genetics, showing convincingly how models like hidden Markov models and computational methods like MCMC contribute to advance research in this important area. Many statistical models have their roots in probabilistic models; in this spirit, Chapters 5 and 6 deal with probabilistic modelling in physics and biology.

The first chapter of the book, written by Peter Green, is a primer on Markov chain Monte Carlo (MCMC). This has become an all-pervading technique in statistical computation when modelling and analyzing complex stochastic systems. As Peter Green writes in his introduction: "The chapter is not primarily intended for those who wish to make use of standard MCMC methods as implemented in a package, and to make sense of the output; however, it should be of some use to those wishing to apply standard methods to some new application by means of their own code. The focus is on understanding the principles underlying the methods, and the main ideas in evaluating their performance." Consequently, the chapter offers an ideal mixture of the mathematical and statistical ideas, enriched with concrete examples and more than one hundred references.

The second chapter, written by Steffen Lauritzen, presents causal concepts in connection with modelling complex stochastic systems. His main aspect is the prediction of the effect of interventions in a given system. Consequently, he distinguishes conditioning by *intervention* and conditioning by *observation*. As the base for a language he employs directed acyclic graphs and their Markov properties; causal Markov fields and associated intervention probabilities are introduced. He also gives ample references to very diverse fields of applications such as clinical trials, genetics, social sciences, economics and expert systems.

The third chapter on state space and hidden Markov models has been written by Hans R. Künsch. These are models for observations which are incomplete and noisy functions of some underlying unobservable process with simple Markov dynamics. The chapter shows the variety of applications of this concept to time series in engineering, biology, finance and geophysics. Recently developed Monte Carlo methods for fitting such models are presented as well as prediction, filtering and smoothing methods. Again the long list of references bears witness of the recent growth of this important field.

The fourth chapter by Elizabeth Thompson is on Monte Carlo methods for some genetic structures. Here special complex systems are investigated, one important goal being to detect and localize the genes contributing to genetically complex human diseases. The tools used include MCMC as well as hidden Markov models and Monte Carlo methods to compute probabilities and likelihoods. This chapter offers not only a concise introduction to the relevant biological methodology, but also some insight into in the complexity of the problems involved.

The two short chapters 5 and 6 cover some aspects of probabilistic modelling of complex stochastic systems. Frank den Hollander has reviewed some recent results on the large space-time behaviour of infinite systems of interacting diffusions, a topic from statistical physics. And, finally, Gesine Reinert investigates the mean field behaviour of a general stochastic epidemic with explicit bounds for the rate of convergence depending on the time length that the epidemic is observed and on the total population size.

The Séminaire Européen de Statistique is an activity of the *European Regional Committee* of the Bernoulli Society for Mathematical Statistics and Probability. The scientific organization of the fourth Séminaire Européen de Statistique was in the hands of Ole E. Barndorff-Nielsen, Aarhus University; Muriel Casalis, Université Paul Sabatier Toulouse; Rainer Dahlhaus, University of Heidelberg; Wilfrid Kendall, University of Warwick; Claudia Klüppelberg, Munich University of Technology; Hans R. Künsch, ETH-Zürich; Peter Spreij, Free University Amsterdam. The Séminaire was sponsored by Eurandom, also the local organisation was conducted by the staff of Eurandom. We are grateful to the Scientific Director Willem van Zwet for this generous support; without this support the Séminaire might not have taken place. We also thank the Managing Director Wim Senden and the staff for the smooth running of the Séminaire.

On behalf of the Organizers and Editors
Claudia Klüppelberg
Munich

CHAPTER 1

A Primer on
Markov Chain Monte Carlo

Peter J. Green
University of Bristol

1.1 Introduction

Markov chain Monte Carlo is probably about 50 years old, and has been both developed and extensively used in physics for the last four decades. However, the most spectacular increase in its impact and influence in statistics and probability has come since the late '80s.

It has now come to be an all-pervading technique in statistical computation, in particular for Bayesian inference, and especially in complex stochastic systems. A huge research effort is being expended, in devising new generic techniques, in extending the application of existing techniques, and in investigating the mathematical properties of the methods.

The target audience for the *Sémstat* lectures is European post-doctoral researchers in probability and statistics, and the present chapter is both the written version of these lectures and a primer for others seeking to get started in some aspect of MCMC research. By 'MCMC research' I mean *both* research into the mathematical properties of MCMC algorithms, and research that aims to develop new classes of algorithms for new and challenging problems; in both cases, I am thinking primarily but not quite exclusively of ultimate application in Bayesian statistics. Thus the chapter is not primarily intended for those who wish to make use of standard MCMC methods as implemented in a package, and to make sense of the output; however, it should be of some use to those wishing to apply standard methods to some new application by means of their own code. The focus is on understanding the principles underlying the methods, and the main ideas in evaluating their performance. With that objective, I will begin with some very basic examples, covered in detail, which are aimed at those who are complete novices. Those with a basic understanding of Bayesian analysis and the Gibbs sampler may not need this motivation, and can skip Section 1.2.

The selection of material is necessarily a personal one — the subject is by now too big for the 4 or 5 hours allocated to the lectures, and indeed I would not claim expertise over all of the potential coverage of a lecture series of this kind. To save space, some sections have been reduced to just a few key references.

I have decided not to try to cover any very substantial applications, although plenty of reference is made to such work. I do make use of a running example — on point processes with change points, exemplified by a Bayesian analysis of some data on cyclones — that is intended to provide continuity as I cover the main topics.

1.2 Getting started: Bayesian inference and the Gibbs sampler

1.2.1 Bayes theorem and inference

The recent great impetus to research in MCMC has been the widespread realisation of its important application in Bayesian inference, following the work of Besag and York (1989) and Gelfand and Smith (1990), building on the 'Gibbs sampler' (popularly ascribed to Geman and Geman (1984)). The book of Gilks, Richardson and Spiegelhalter (1996), comprising articles contributed by 32 authors, provides an excellent introduction and overview to the theory, implementation and application of Bayesian MCMC.

Let us start with the simplest basic set-up, a model relating data Y and parameters $\boldsymbol{\theta} = (\theta_1, \theta_2, \ldots, \theta_p)$. We need two probabilistic models: a data model specifying the likelihood: $p(Y|\boldsymbol{\theta})$, and a prior model, specifying the prior distribution $p(\boldsymbol{\theta})$.

In the Bayesian approach, inference is based on the *joint posterior*

$$p(\boldsymbol{\theta}|Y) = \frac{p(\boldsymbol{\theta})p(Y|\boldsymbol{\theta})}{\int p(\boldsymbol{\theta})p(Y|\boldsymbol{\theta})d\boldsymbol{\theta}}$$

$$\propto p(\boldsymbol{\theta})p(Y|\boldsymbol{\theta})$$

i.e.   *Posterior* $\propto$ *Prior* $\times$ *Likelihood*

For a proper account of Bayesian theory, the reader is referred to Bernardo and Smith (1994) or O'Hagan (1994).

1.2.2 Cyclones example: point processes and change points

We are going to illustrate the ideas of MCMC with a running example: the observations are a point process of *events* at times $y_1, y_2, \ldots, y_N$ in an observation interval $[0, L)$. For simplicity, we suppose the events occur *at random* — that is, as a Poisson process — but at a possibly non-uniform rate: say rate $x(t)$ per unit time, at time t. The objective is to make inference about $x(t)$. We will work up through a series of models, ultimately allowing

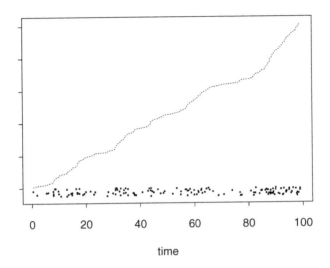

time

Figure 1.1 *Cyclones data, as a jittered dot plot, and their cumulative counting process.*

an unknown number of change points, unknown hyperparameters, and a parametric periodic component.

The models and the respective algorithms and inferences will be illustrated by an analysis of a data set of the times of cyclones hitting the Bay of Bengal; there were 141 cyclones over a period of 101 years (Mooley, 1981). The data are plotted, both as a jittered dot plot, and by means of their cumulative counting process, in Figure 1.1.

Model 1: constant rate

First suppose that $x(t) \equiv x$ for all t.

Then the times of the events are immaterial: we observe N events in a time interval of length L; the obvious estimate of x is

$$\widehat{x} = \frac{N}{L}.$$

This is the *maximum likelihood estimator* of x under the assumption (implied by the 'randomness' assumption above), that N has a Poisson distribution, with mean xL:

$$p(N|x) = e^{-xL} \frac{(xL)^N}{N!}.$$

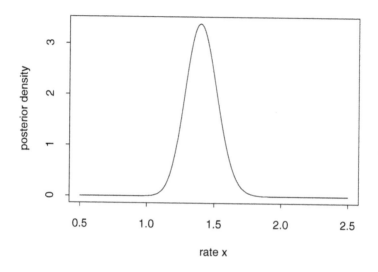

Figure 1.2 *Cyclones data: posterior for x in model 2.*

Model 2: constant rate, the Bayesian way

To take a Bayesian approach to this example, suppose that we have prior information about x (from previous studies, for example). Let us suppose we can model this by saying

$$x \sim \Gamma(\alpha, \beta),$$

a Gamma distribution (with mean α/β and variance α/β^2).

Then since

$$p(x|N) \propto p(x)p(N|x),$$

we find that

$$
\begin{aligned}
p(x|N) \quad &\propto \quad \frac{\beta^\alpha x^{\alpha-1} e^{-\beta x}}{\Gamma(\alpha)} e^{-xL} \frac{(xL)^N}{N!} \\
&\propto \quad x^{\alpha+N-1} \exp(-(\beta + L)x)
\end{aligned}
$$

or in other words

$$x|N \sim \Gamma(\alpha + N, \beta + L).$$

So x has a Gamma distribution with mean $(\alpha + N)/(\beta + L)$, or approximately N/L if N and L are large compared with α and β. Thus *with a lot of data*, the Bayesian posterior mean is close to the maximum likelihood estimator. The posterior distribution of x for model 2 fitted to the cyclones data is shown in Figure 1.2; we used $\alpha = \beta = 1$ here.

There is no need for MCMC in this model: you can calculate the posterior exactly, and recognise it as a standard distribution. It would not have worked out like this for any other prior; this choice is called *conjugate*.

1.2.3 The Gibbs sampler for a Normal random sample

Before we elaborate the cyclones example to a point where exact calculation is no longer practicable, let alone formally introduce Markov chain Monte Carlo methods, let us consider an even simpler, and completely familiar, example, but following an elementary Bayesian approach.

Our data are a random sample of size n from $N(\mu, \sigma^2)$. We place independent priors on μ and σ:

$$\mu \sim N(\xi, \kappa^{-1})$$
$$\sigma^{-2} \sim \Gamma(\alpha, \beta),$$

and it is easy to see that the resulting joint posterior is

$$p(\mu, \sigma^{-2}|Y) \propto (\sigma^{-2})^{\alpha + n/2 - 1}$$
$$\times \exp\left\{-\frac{\beta}{\sigma^2} - \frac{\kappa(\mu - \xi)^2}{2} - \frac{\sum(Y_i - \mu)^2}{2\sigma^2}\right\}. \quad (1.1)$$

This is somewhat awkward to handle; the parameters are dependent *a posteriori*, although they were independent *a priori*. However, the *full conditionals* — the conditional distributions of each parameter given the other parameter(s) and the data — are easily found:

$$\mu|\sigma, Y \sim N\left(\frac{\sigma^{-2}\sum Y_i + \kappa\xi}{\sigma^{-2}n + \kappa}, \frac{1}{\sigma^{-2}n + \kappa}\right)$$
$$\sigma^{-2}|\mu, Y \sim \Gamma(\alpha + n/2, \beta + \sum(Y_i - \mu)^2/2).$$

What happens if we generate a sample of (μ, σ) pairs by alternately drawing μ and σ^{-2} from these distributions? The beginning of this process is illustrated in Figure 1.3, using the (improper) uninformative prior setting $\xi = \kappa = \alpha = \beta = 0$.

This is a simple example of a *Gibbs sampler*. The alternating updates of one variable conditioned on the other induces Markov dependence: the successively sampled pairs form a Markov chain (on the uncountable state space $\mathcal{R} \times \mathcal{R}^+$), and it is readily shown that the joint posterior (1.1) is the (unique) invariant distribution of the chain. Standard theorems, quoted in Section 1.3.1 below, imply that the chain converges to this invariant distribution in several useful senses, so that we can treat the realised values as a sample from the posterior. A sample of 1000 pairs is shown in Figure 1.4, and the shape of the joint distribution can now be discerned. Examples of

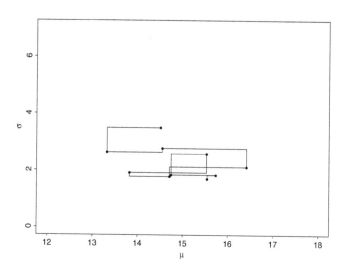

Figure 1.3 *First 10 samples from a Gibbs sampler of* (μ, σ) *from Normal random sample with* $n = 10$, $\overline{Y} = 15$, $s_Y^2 = 4$. *Uninformative prior.*

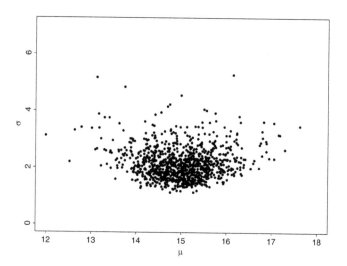

Figure 1.4 *Posterior sample of* (μ, σ) *from Normal random sample with* $n = 10$, $\overline{Y} = 15$, $s_Y^2 = 4$. *Uninformative prior.*

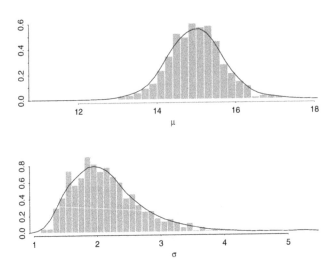

Figure 1.5 *Posterior distributions of μ and σ from Normal random sample with $n = 10$, $\overline{Y} = 15$, $s_Y^2 = 4$. Uninformative prior.*

possible outputs of interest are the marginal distributions shown in Figure 1.5.

However, we need not be confined to pictorial displays of marginal posteriors. One of the great liberating influences of MCMC in Bayesian inference has been the flexibility of inference afforded by sample-based computation. For example, consider prediction: we can calculate $P\{Y_{n+1} > 19\}$ by averaging $1 - \Phi(\{19 - \mu\}/\sigma)$:

$$\frac{1}{N} \sum_{t=1}^{N} \left[1 - \Phi(\{19 - \mu^{(t)}\}/\sigma^{(t)}) \right] \approx 0.045$$

for the sample of Figure 1.4. Incidentally, it is interesting that this is more than twice the value (0.0175) that a frequentist would obtain by plugging the maximum likelihood estimates into $1 - \Phi(\{19 - \mu\}/\sigma)$. (Of course, this, like any other inference based on this model, is influenced by the prior setting used.)

1.2.4 Cyclones example, continued

For a more interesting and substantial application, let us return to the cyclones example, and consider some elaborations of the basic model 2.

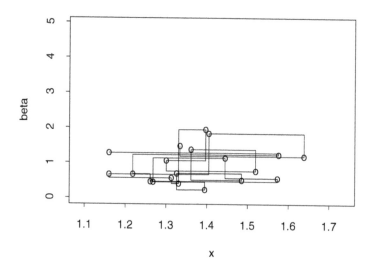

Figure 1.6 *First few moves of the Gibbs sampler for the cyclones data, model 3.*

Model 3: constant rate, with hyperparameter

Suppose you are reluctant to specify your prior fully: you are happy to say

$$x \sim \Gamma(\alpha, \beta)$$

and can specify α but not β, and want to state only

$$\beta \sim \Gamma(e, f)$$

for fixed e and f. (This formulation actually makes rather more sense in our next formulation, model 4).

Now $p(x|N, \alpha, e, f)$ is no longer available: it does not have an explicit form. But $p(x|N, \alpha, \beta, e, f)$ and $p(\beta|x, N, \alpha, e, f)$ are simple:

$$x|N, \alpha, \beta, e, f \sim \Gamma(\alpha + N, \beta + L)$$

as before, and

$$\beta|x, N, \alpha, e, f \sim \Gamma(e + \alpha, f + x).$$

So we can use the Gibbs sampler, and sample from these distributions in turn, updating x and β alternately. This creates a Markov chain with states (x, β), the unknown parameters in this model.

Figure 1.6 shows the first few moves of a Gibbs sampler applied to model 3 on the cyclones data; we took $e = 1$ and $f = N/L = 1.396$, and kept

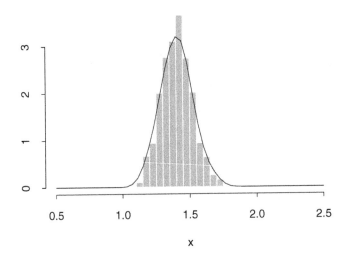

Figure 1.7 *Marginal distribution for x for the cyclones data, model 3.*

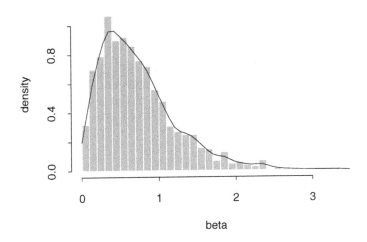

Figure 1.8 *Marginal distribution for β for the cyclones data, model 3.*

$\alpha = 1$. The marginal distributions for x and β, as accumulated from the first 1000 sweeps of this Gibbs sampler are displayed in Figures 1.7 and 1.8.

Model 4: constant rate, with change point

Now let us allow $x(t)$ to vary, but in a particular way.

Suppose $x(t)$ is piecewise constant, that is, a step function. This might be a suitable model if we postulate one or more *change points*; the process is completely random, but the rate switches between levels, perhaps as part of an underlying process, perhaps due to the recording mechanism.

Let us first take one change point, at *known* time $T \in (0, L)$, so that

$$x(t) = \begin{cases} x_0 & \text{if } 0 \le t < T \\ x_1 & \text{if } T \le t < L \end{cases}.$$

Suppose that x_0 and x_1 are *a priori* independently drawn from Gamma distributions, as before:

$$x_j \sim \Gamma(\alpha, \beta).$$

Then if N_0 and N_1 are the numbers of events before and after T, the above method extends to sampling in turn from

$$x_0 | \cdots \sim \Gamma(\alpha + N_0, \beta + T),$$

$$x_1 | \cdots \sim \Gamma(\alpha + N_1, \beta + (L - T)),$$

and

$$\beta | \cdots \sim \Gamma(e + 2\alpha, f + x_0 + x_1),$$

forming a Markov chain with a three-dimensional state space $\{(x_0, x_1, \beta)\}$. Note that for the sake of clarity and compactness we write '$|\cdots$' to mean 'given all other variables' — including the data.

The hierarchical model using random β makes more sense now: the effect is to 'borrow strength' in estimation from both halves of the data together: x_0 and x_1 are conditionally independent given β, but are *un*conditionally *dependent*. In inference their values will be shrunk together.

Model 5: multiple change points

If there are k change points $T_1, T_2, \ldots, T_k$ with

$$x(t) = \begin{cases} x_0 & \text{if } 0 \le t < T_1 \\ x_1 & \text{if } T_1 \le t < T_2 \\ \cdots & \cdots \\ x_k & \text{if } T_k \le t < L \end{cases},$$

then everything is extended in a very similar way, giving a Markov chain with states $(x_0, x_1, \ldots, x_k, \beta)$.

1.2.5 Other approaches to Bayesian computation

Do we have to resort to Gibbs sampling for this application, and examples like it? Under the posterior distribution in a Bayesian formulation, the parameters $\boldsymbol{\theta}$ are generally *dependent*, so we have to compute with a multivariate distribution, often in a high number of dimensions, with arbitrarily complex patterns of dependence. Here, "compute with" could mean almost anything; examples would be to calculate a marginal (posterior) density or make a probabilistic prediction. See Bernardo and Smith (1994) and O'Hagan (1994).

There are various possible approaches to Bayesian computation:

- Exact analytic integration: this is usually only available when we make use of conjugate priors, which is in itself often an unreasonable restriction, and in any case is usually restricted to very simple formulations.

- Asymptotic analytic approximations (e.g. Laplace; see, for example, Kass *et al.*, 1988): these are somewhat awkward to set up, and can be unreliable.

- Conventional numerical methods: these require expertise and careful design to set up, and are only efficient in a low number of dimensions.

- Ordinary ("static") simulation: this is always available in principle, since any posterior distribution can be factorised as

$$p(\boldsymbol{\theta}|Y) = p(\theta_1, \theta_2, \ldots, \theta_p|Y)$$
$$= p(\theta_1|Y)p(\theta_2|\theta_1, Y)\ldots p(\theta_p|\theta_1, \ldots, \theta_{p-1}, Y)$$

but the univariate distributions on the right-hand side are rarely all available for simulation purposes (even after re-ordering).

Markov chain Monte Carlo (MCMC, also sometimes known as iterative or dynamic simulation) works even where static simulation does not, essentially because

- All simulation methods rely on the Law of Large Numbers, and this remains true (in the guise of the Ergodic theorem) when you have a Markov chain instead of an independent, identically distributed sequence.

- If you can tolerate Markov dependence, then you can update the parameters $\theta_1, \theta_2, \ldots, \theta_p$ one-by-one (or in small groups).

The result of combining these two simple points is very far-reaching indeed!

1.3 MCMC — the general idea and the main limit theorems

Having motivated the idea of MCMC by use of the Gibbs sampler in two very basic problems, we are now in a position to discuss the subject from a rather more general perspective.

Our object of interest is the *target distribution* π of a random quantity

$x \in \mathcal{X}$. In Bayesian statistics, x are the unknowns (parameters, latent variables, missing values, future data) in a statistical experiment, and π is the posterior distribution of these variables given the data Y:

$$\pi(A) = p(x \in A|Y)$$

Henceforth in this chapter, we shall use x, π in this generic way, and reserve the $p(\cdot|\cdot)$ notation for discussion of specific models. One of many advantages of the generic notation is that it helps us not lose sight of other, non-Bayesian, applications of MCMC. (Although by far the greatest impact of MCMC in statistics has been in Bayesian analysis, because of the ubiquitous need there for integration, it has also found application in other contexts where variables are integrated out, for example in latent variables models, contingency tables and in models with complicated conditional likelihoods.)

The objective now is to construct a time-homogeneous discrete time Markov chain whose state space is $\mathcal{X}$ (the parameter space in Bayesian statistics), and whose limiting distribution is the specified target. That is, we want a transition kernel P such that

$$P\{x^{(t)} \in A|x^{(0)}\} \to \pi(A) \quad \text{as } t \to \infty, \forall\, x^{(0)}.$$

Having constructed such a Markov chain, in the sense of devising a transition kernel with this limiting property, we then construct it in another sense — we form a realisation of the chain $\{x^{(1)}, x^{(2)}, \dots, x^{(N)}\}$ and treat this *as if it was a random sample* from π.

Of course, in fact we should not be so naive as to ignore completely the fact that this is *not* a simple random sample! However, in practice we will routinely make displays (histograms, density estimates) of the empirical distribution as estimating the target, estimate moments of the target from those of the sample

$$E_\pi(g) = \int g(x)\pi(dx) \approx \frac{1}{N}\sum_{t=1}^{N} g(x^{(t)}) \tag{1.2}$$

(for suitable functions g), and compute probabilities under the target distribution by empirical frequencies

$$\pi(A) \approx \frac{1}{N}\sum_{t=1}^{N} I[x^{(t)} \in A]. \tag{1.3}$$

All such computations are justified by the limit theory of Markov chains; in order to handle the countless real applications where the space $\mathcal{X}$ is not discrete, we need these limit theorems for chains in a general state space.

1.3.1 The basic limit theorems

Our treatment of the limit theory for Markov chains given here is not at all complete, but will at least review the main concepts and results that are important to MCMC. A fuller treatment with the same objective can be found in Tierney (1994) and Tierney (1996), and the complete story is in Meyn and Tweedie (1993). This treatment borrows heavily from these sources.

The most important theorem in practice concerns convergence of sample means, and justifies (1.2) and (1.3) above. It requires the concepts of invariance and irreducibility. A probability distribution π is *invariant* for a transition kernel P if $\int P(x, A)\pi(dx) = \pi(A)$. The kernel P is *irreducible* if there exists a probability distribution, ψ say, on $\mathcal{X}$ such that $\psi(A) > 0 \Rightarrow P(\tau_A < \infty | x^{(0)} = x) = 1$ for all π-almost all $x \in \mathcal{X}$, where τ_A is the hitting time $\min\{t : x^{(t)} \in A\}$. Any such ψ is called an irreducibility distribution for P.

If $\{x^{(t)}\}$ is an irreducible Markov chain with transition kernel P and invariant distribution π, and g is a real valued function with $\int |g(x)|\pi(dx) < \infty$, then

$$\frac{1}{N} \sum_{t=1}^{N} g(x^{(t)}) \rightarrow \int g(x)\pi(dx) \qquad (1.4)$$

almost surely, for π-almost all $x^{(0)}$.

Sometimes, it is useful to say a little more — that the distribution of $x^{(t)}$ converges to π. As in the simple discrete case, this requires the additional assumption that the chain is not periodic.

An m-cycle for an irreducible chain with kernel P is a collection of subsets $\{E_0, E_1, \ldots, E_{m-1}\}$ such that $P(x, E_{i+1 \bmod m}) = 1$ for all $x \in E_i$ and all i; the period d is the largest m for which an m-cycle exists, and the chain is aperiodic if $d = 1$.

If the chain is aperiodic, the t-step transition kernel converges:

$$||P^t(x^{(0)}, \cdot) - \pi(\cdot)|| \rightarrow 0 \qquad (1.5)$$

as $t \rightarrow \infty$, for π-almost all $x^{(0)}$. Here, the norm is the total variation distance between two probability measures, defined by $||\nu_1 - \nu_2|| = 2\sup_A |\nu_1(A) - \nu_2(A)|$.

1.3.2 Harris recurrence

The assumptions of invariance and irreducibility are usually rather easy to check for a given transition kernel, so the results of the previous subsection are then available. However, when used to justify a simulation computation, they are subject to a crucial *caveat*. Both of these limit theorems apply only to π-almost all starting values $x^{(0)}$. For routine purposes, this restriction is of little concern, but in simulation, we really need to know that we were not unlucky enough to be running our chain from an initial state in the probability-zero exceptional set!

We say that an irreducible kernel P is *Harris recurrent* if, for any irreducibility distribution ψ and any A such that $\psi(A) > 0$, we have $P\{x^{(t)} \in A \text{ i.o.}|x^{(0)} = x\} = 1$ for all x (where 'i.o.' means 'infinitely often').

If the chain is Harris recurrent, then (1.4) holds for all $x^{(0)}$, as does (1.5) if it is also aperiodic.

1.3.3 Rates of convergence

Knowing that the chain converges is not the same as knowing that it converges quickly enough to be useful. It is therefore important to try to study rates of convergence. This is a challenge for practically useful chains in general state spaces.

Only in very rare cases can numerical bounds be found for rates of convergence, and when they can, they are often very discouraging. However, there have been several successful approaches to the qualitative study of convergence.

The chain is *geometrically ergodic* if

$$||P^t(x^{(0)}, \cdot) - \pi(\cdot)|| \leq M(x^{(0)})\rho^t$$

for finite $M(x)$, $\rho < 1$.

It is *uniformly ergodic* if for all $x^{(0)}$,

$$||P^t(x^{(0)}, \cdot) - \pi(\cdot)|| \leq M\rho^t.$$

Various conditions are known to imply uniform ergodicity, for example Doeblin's condition: there exists a probability measure ϕ and constants $\varepsilon < 1$, $\delta > 0$, t such that

$$\phi(A) > \varepsilon \Rightarrow P^t(x, A) \geq \delta \quad \text{for all} \quad x.$$

There are both positive and negative results about uniform or geometric ergodicity of popular MCMC recipes. For example, see Mengersen and Tweedie (1996), Roberts and Tweedie (1996), Roberts and Rosenthal (1999), and Mira, Møller and Roberts (1999).

A rather different approach to assessing speed of convergence is via computational complexity; for example, there are recent interesting results by Frigessi, Martinelli and Stander (1997).

1.4 Recipes for constructing MCMC methods

One might think initially that to construct a Markov chain with a specified target as its limiting distribution would be a complicated matter. Fortunately, several standard 'recipes' are available to automate this task.

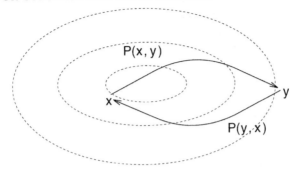

Figure 1.9 *Illustrating the idea of detailed balance. The transitions described by P are neutral with respect to the contours of probability of π.*

In this section, introducing the main recipes for MCMC methods, we assume the state space of our chain is countable, and work with a notation in which the target distribution π and the transition kernel P are expressed as densities with respect to counting measure, that is, as probability mass functions. Modifications to deal with other dominating measures, such as Lebesgue measure, are straightforward.

The key idea in most practical approaches to constructing MCMC methods is reversibility or detailed balance. The target π is invariant for P if we have detailed balance (time-reversibility):

$$\pi(\boldsymbol{x})P(\boldsymbol{x},\boldsymbol{y}) = \pi(\boldsymbol{y})P(\boldsymbol{y},\boldsymbol{x})$$

for all $\boldsymbol{x}, \boldsymbol{y} \in \mathcal{X}$. Detailed balance is sufficient but not necessary for invariance; however it is far easier to work with. You can think of reversibility as requiring a balance in the flow of probability; see Figure 1.9.

We will ignore the issues of irreducibility and aperiodicity for the moment.

1.4.1 The Gibbs sampler

In the Gibbs sampler, the basic step is simple: discard the current value of a single component $\boldsymbol{x}_i$, and replace it by a value $\boldsymbol{y}_i$ drawn from the *full conditional* distribution induced by π:

$$\pi(\boldsymbol{x}_i|\boldsymbol{x}_{-i}),$$

keeping the current values of other variables: $\boldsymbol{y}_{-i} = \boldsymbol{x}_{-i}$ (where "$-i$" stands for $\{j : j \neq i\}$). Then we are using the kernel

$$P(\boldsymbol{x},\boldsymbol{y}) = \pi(\boldsymbol{y}_i|\boldsymbol{x}_{-i})I[\boldsymbol{x}_{-i} = \boldsymbol{y}_{-i}],$$

and detailed balance holds because given $\boldsymbol{x}_{-i}$, $\boldsymbol{x}_i$ and $\boldsymbol{y}_i$ are independent, and identically distributed as $\pi(\boldsymbol{x}_i|\boldsymbol{x}_{-i})$.

This recipe was named the Gibbs sampler by Geman and Geman (1984), whose work brought the idea to the attention of spatial statisticians. However, it is earlier than that: it was well known as the 'heat bath' by statistical physicists, see for example, Creutz (1979), but the earliest appearance I know of is in statistics, in a Finnish Ph.D. thesis by Suomela (1976).

1.4.2 The Metropolis method

In the Metropolis method, we find a candidate new value (or "proposal") y by drawing y_i from an arbitrary density $q_i(y_i; x)$ parameterised by x, and setting $y_{-i} = x_{-i}$. We write $q_i(y_i; x) = q_i(x, y)$, and impose the symmetry requirement $q_i(x, y) = q_i(y, x)$. (Note the deliberate reversal of the order of arguments: $q_i(y_i; x)$ is a density in y_i parameterised by x, while $q_i(x, y)$ is a transition kernel, and so the arguments are used in the conventional time-oriented order.)

This proposal is accepted as the next state of the chain with probability

$$\alpha(x, y) = \min\left\{1, \frac{\pi(y)}{\pi(x)}\right\} = \min\left\{1, \frac{\pi(y_i | x_{-i})}{\pi(x_i | x_{-i})}\right\}, \qquad (1.6)$$

and otherwise x is left unchanged.

This recipe is due to Metropolis, *et al.* (1953). Note that the target density π is only needed up to proportionality, and then only at two values, the current and proposed next states.

1.4.3 The Metropolis-Hastings sampler

In a paper astonishingly overlooked by statisticians for nearly 20 years, Hastings (1970) introduced an important generalisation of Metropolis, in which symmetry of q is not needed; the acceptance probability becomes:

$$\alpha(x, y) = \min\left\{1, \frac{\pi(y)q_i(y, x)}{\pi(x)q_i(x, y)}\right\} = \min\left\{1, \frac{\pi(y_i | y_{-i})q_i(x_i; y)}{\pi(x_i | x_{-i})q_i(y_i; x)}\right\}. \qquad (1.7)$$

The optimality in some senses of this particular choice of $\alpha(x, y)$ over any other choice preserving detailed balance is demonstrated by Peskun (1973).

Note that Metropolis is the special case where q is symmetric, and Gibbs the special case where the proposal density $q_i(y_i; x)$ is just the full conditional $\pi(y_i | x_{-i}) = \pi(y_i | y_{-i})$, so that the acceptance probability is 1.

1.4.4 Proof of detailed balance

The proof of correctness of each is the same: the choice of acceptance probability simply ensures that detailed balance is satisfied.

For $x \neq y$,

$$\begin{aligned} \pi(x)P(x,y) &= \pi(x_{-i})\pi(x_i|x_{-i})q_i(y_i;x)\alpha(x,y) \\ &= \pi(x_{-i})\min\{R(x,y), R(y,x)\}, \end{aligned}$$

from (1.7), where $R(x,y) = \pi(x_i|x_{-i})q_i(y_i;x)$. The term R and hence the whole expression above is symmetric in x and y (recall that $x_{-i} = y_{-i}$). So detailed balance holds. (Note that we have only used the fact that

$$\frac{\alpha(x,y)}{\alpha(y,x)} = \frac{\pi(y)}{\pi(x)}\frac{q_i(y,x)}{q_i(x,y)}.$$

The argument for the particular choice of $\alpha(x,y)$ in (1.7) will be made in Section 1.6.2.)

This argument for the Hastings method obviously covers Gibbs and Metropolis *a fortiori*.

1.4.5 Updating several variables at once

Each of the Gibbs, Metropolis and Hastings methods is equally valid if a group of variables $x_A = \{x_j : j \in A\}$ is updated simultaneously; each uses the full conditional $\pi(x_A|x_{-A})$. You could update *all* variables at once in Metropolis or Hastings. (It is a subtle question whether it is a good idea to update many variables.)

An important special case arises where the variables in x_A are *conditionally independent* (under the full conditional). They can then be updated in parallel.

1.4.6 The role of the full conditionals

All of the basic methods use the full conditionals $\pi(x_A|x_{-A})$, where A indexes the variables being updated. In Gibbs, you have to *draw* from this distribution; in Metropolis and Hastings, you only have to *evaluate* it (up to a multiplicative constant) at the old and new values.

1.4.7 Combining kernels to make an ergodic sampler

All of the methods above satisfy detailed balance, and hence preserve the equilibrium distribution: if

$$x \sim \pi$$

before the transition, then so it will afterwards.

To ensure that this is also the limiting distribution of the chain (ergodicity), we must combine such kernels to make a Markov chain transition mechanism that is irreducible (and aperiodic).

To do that, scan over the available kernels (indexed by i or A) either systematically or randomly, or in various other ways that are valid, provided you visit each variable often enough. You can use different recipes (Gibbs, Metropolis,...) for different A. The most common strategies for combining kernels $P_1, P_2, \ldots, P_m$ are the systematic *cyclic* combination giving an overall kernel

$$P = P_1 P_2 \cdots P_m$$

or the equally-weighted random or *mixture* kernel

$$P = \frac{1}{m} \sum_{i=1}^{m} P_i.$$

Time in a MCMC simulation is usually measured in *sweeps*, the smallest period such that the chain is time-homogeneous, for example, after m individual transitions if the cyclic kernel is being used.

Note that the mixture kernel preserves detailed balance, while the cyclic one does not, so that reversibility at the sweep time scale is lost; of course π remains invariant for both combinations.

1.4.8 Common choices for proposal distribution

The user has a completely free choice of proposal distribution; there is no need even to worry about dividing by zero in (1.7), since y_i with $q_i(y_i; x) = 0$ will (almost surely) not get proposed! Nevertheless, typically, one of a small number of standard specifications is very often used.

Independence Metropolis-Hastings. If the proposed new state y is independent of the current x (so in particular we are proposing to update all components of the state simultaneously), then $q(x, y) = q(y)$, say, and the acceptance probability simplifies to

$$\alpha(x, y) = \min \left\{ 1, \frac{w(y)}{w(x)} \right\},$$

where $w(x) = \pi(x)/q(x)$.

This choice is of little use in practical terms (except perhaps in splitting, see Section 1.6.2) but often yields kernels amenable to theoretical investigation.

Random walk Metropolis. If $q_i(x, y) = q_i(y_i - x_i)$ where $q_i(\cdot)$ is a density function symmetric about 0, then

$$\frac{q_i(y, x)}{q_i(x, y)} = 1$$

so the acceptance probability simplifies; the proposal amounts to adding a random walk increment $\sim q_i$ to the current x_i.

Random walk Metropolis on the log scale. When a component x_i of the state vector is necessarily positive, it may be convenient to only propose changes to its value that leave it positive, in which case a multiplicative rather than additive update is suggested. If the proposed increment to $\log x_i$ has any distribution symmetric about 0, then we find

$$\frac{q_i(\boldsymbol{y}, \boldsymbol{x})}{q_i(\boldsymbol{x}, \boldsymbol{y})} = \frac{y_i}{x_i}.$$

1.4.9 Comparing Metropolis-Hastings to rejection sampling

There is a superficial resemblance of Metropolis-Hastings to ordinary rejection sampling, which may cause confusion. Recall that in rejection sampling, to sample from π, we first draw $\boldsymbol{y}$ from a density q, and then accept this value with probability $\pi(\boldsymbol{y})/(Mq(\boldsymbol{y}))$, where M is any constant such that $M \geq \sup_{\boldsymbol{y}} \pi(\boldsymbol{y})/q(\boldsymbol{y})$. If the generated $\boldsymbol{y}$ is not accepted, this procedure is repeated until it is. As with Metropolis-Hastings, π and q are needed only up to proportionality. The crucial differences are that in Metropolis-Hastings: (a) π/q need not be bounded, (b) you do *not* repeat if the proposal is rejected, and (c) you end up with a Markov chain, not an independent sequence.

1.4.10 Example: Weibull/Gamma experiment

Let us consider a different but still very simple example, where Gibbs sampling would not be straightforward. Our data will be a random sample, possibly censored, from the Weibull(ρ, κ) distribution:

$$p(Y|\rho, \kappa) = \kappa^m \rho^{m\kappa} \prod_U Y_i^{\kappa-1} \exp\left(-\rho^\kappa \sum Y_i^\kappa\right)$$

where m and $\prod_U$ are the number of and product over uncensored observations. We place independent Gamma priors on ρ and κ:

$$p(\rho, \kappa) \propto \rho^{\alpha-1} e^{-\beta\rho} \kappa^{\gamma-1} e^{-\delta\kappa}$$

The resulting posterior is

$$p(\rho, \kappa|Y) \quad \propto \quad \kappa^m \rho^{m\kappa} \prod_U Y_i^{\kappa-1} \exp\left(-\rho^\kappa \sum Y_i^\kappa\right)$$
$$\rho^{\alpha-1} e^{-\beta\rho} \kappa^{\gamma-1} e^{-\delta\kappa}$$

which is not a standard distribution.

Let us define a Markov chain with states $\boldsymbol{x} = (\rho, \kappa)$ and limiting distribution $\pi(\boldsymbol{x}) = p(\rho, \kappa|Y)$. The full conditionals for the two parameters are

$$p(\rho|\kappa) \quad \propto \quad \rho^{m\kappa} \exp\left(-\rho^\kappa \sum Y_i^\kappa\right) \rho^{\alpha-1} e^{-\beta\rho}$$
$$p(\kappa|\rho) \quad \propto \quad \kappa^m \rho^{m\kappa} \prod_U Y_i^{\kappa-1} \exp\left(-\rho^\kappa \sum Y_i^\kappa\right) \kappa^{\gamma-1} e^{-\delta\kappa}.$$

This is hardly of standard form, so Gibbs is problematical, but the full conditionals are easily evaluated for a Metropolis or Hastings algorithm.

An easily implemented Metropolis method for this setting would consist of the following ingredients:

1. alternate between updating ρ and κ,

2. propose a new value for the parameter from a distribution symmetric about its present value,

3. reject the update if the result is negative,

4. otherwise, accept it with probability (e.g.) $\min\{1, p(\rho'|\kappa)/p(\rho|\kappa)\}$.

1.4.11 Cyclones example, continued

Model 6: another hyperparameter

Let's now suppose α is also unknown, with, *a priori*,

$$\alpha \sim \Gamma(c, d)$$

for fixed constants c and d. (In our analysis of the cyclones data, we took $c = d = 2$.) This last change means that Gibbs sampling is not enough. In a Markov chain with states $x = (x_0, x_1, \ldots, x_k, \alpha, \beta)$, we can update α using a random walk Metropolis move, on the $\log(\alpha)$ scale: the acceptance ratio is

$$\min\left\{1, \frac{p(\log \alpha'|\cdots)}{p(\log \alpha|\cdots)}\right\}$$

which simplifies to

$$\min\left\{1, \left(\frac{\Gamma(\alpha)}{\Gamma(\alpha')}\right)^{k+1} \left(\frac{\alpha'}{\alpha}\right)^c \left(e^{-d}\beta^{k+1} \prod x_j\right)^{\alpha'-\alpha}\right\}$$

Model 7: unknown change points

If $x_0, x_1, \ldots, x_k$ are unknown, so probably are the times of the change points $T_1 < T_2 < \cdots < T_k$. The state vector is now $x = (x_0, x_1, \ldots, x_k, T_1, T_2, \ldots, T_k, \alpha, \beta)$.

Let us assume *a priori*

$$p(T_1, T_2, \ldots, T_k) \propto T_1(T_2 - T_1) \ldots (T_k - T_{k-1})(L - T_k),$$

a joint density providing a gentle preference against two changes occurring too closely in succession (this is actually the joint distribution of the even-numbered order statistics for a sample of size $2k + 1$ from $U(0, L)$).

The posterior marginal or joint conditional distributions are quite complex, for this or any prior, so Metropolis-Hastings is needed. The details are a little messy but straightforward. For a proposal drawing T_j' uniformly from $[T_{j-1}, T_{j+1}]$, the acceptance probability is

$$\min\left\{1, (\text{likelihood ratio})\frac{(T_j' - T_{j-1})(T_{j+1} - T_j')}{(T_j - T_{j-1})(T_{j+1} - T_j)}\right\}.$$

A sample of step functions drawn from the resulting MCMC sample is shown in Figure 1.10.

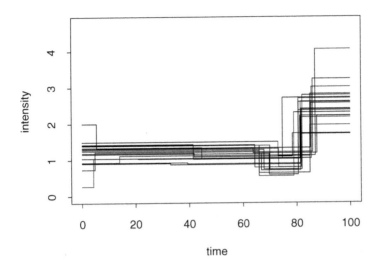

Figure 1.10 *Posterior sample of step functions $x(t)$ for model 7 with $k = 2$, applied to cyclones data.*

1.5 The role of graphical models

Graphical modelling provides a powerful language for specifying and understanding statistical models.

Graphs consist of vertices representing variables, and edges (directed or otherwise) that express conditional dependence properties. For a full treatment of the theory, see Lauritzen (1996).

1.5.1 Directed acyclic graphs

The DAG (directed acyclic graph) — a graph in which all edges are directed, and there are no directed loops — expresses the natural factorisation of a joint distribution into factors each giving the joint distribution of a variable x_v given the values of its *parents* $x_{\mathrm{pa}(v)}$; for example, in Figure 1.11,

$$\pi(a, b, c, d) = \pi(a)\pi(b)\pi(c|a, b)\pi(d|c)$$

In general, we can write

$$\pi(x) = \prod_{v \in V} \pi(x_v | x_{\mathrm{pa}(v)}) \tag{1.8}$$

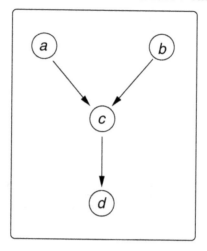

Figure 1.11 *A simple directed acyclic graph on four variables.*

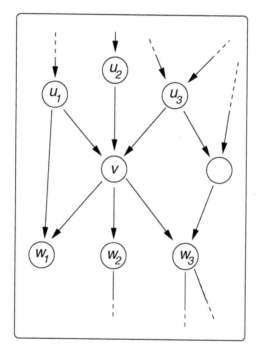

Figure 1.12 *A larger directed acyclic graph: the vertices labelled $\{u_i\}$ are the parents of v, and $\{w_i\}$ are its children.*

(see Figure 1.12), which in turn implies a *Markov property*, that variables are conditionally independent of their non-descendants, given their parents.

From the perspective of setting up MCMC methods, the graphical structure assists in identifying which terms need be included in a full conditional. Equation (1.8) implies

$$\pi(\boldsymbol{x}_v | \boldsymbol{x}_{-v}) \propto \pi(\boldsymbol{x}_v | \boldsymbol{x}_{\mathrm{pa}(v)}) \prod_{w:v \in \mathrm{pa}(w)} \pi(\boldsymbol{x}_w | \boldsymbol{x}_{\mathrm{pa}(w)})$$

where the right-hand side has one term for the variable of interest itself, and one for each of its children.

Graphical modelling, the construction of MCMC methods through full conditional distributions, and good practice in statistical model building all exploit the same modular structure.

A concrete example of this modularity has already been seen implicitly; in Section 1.4.7, we discussed how an ergodic kernel might be assembled (by cycling or mixing) from a collection of kernels $P_1, P_2, \ldots$ that were individually in detailed balance but not irreducible.

1.5.2 Undirected graphs, and spatial modelling

Directed acyclic graphs are a natural representation of the way we usually *specify* a statistical model (directionally, disease $\rightarrow$ symptom, past $\rightarrow$ future, parameters $\rightarrow$ data), but

- sometimes (e.g. spatial models) there is *no natural direction*;

- in *understanding associations* between variables implied by a model, however specified, directions can confuse; and

- these associations represent the full conditionals needed in *setting up MCMC* methods.

To form the conditional independence graph for a multivariate distribution, draw an (undirected) edge between variables α and β if they are **not** conditionally independent given all other variables.

Markov properties

The Markov property is familiar from temporal stochastic processes, where we learn that it may be expressed in several equivalent ways. For variables located on an arbitrary graph, the situation is more subtle: we can distinguish four related properties, each capturing an aspect of Markovness.

P: Pairwise   Non-adjacent pairs of variables are conditionally independent given the rest (see definition of graph).

L: Local   Conditional only on adjacent variables (neighbours), each variable is independent of all others (so that full conditionals are simplified).

G: Global Any two subsets of variables separated by a third are conditionally independent given the values of the third subset.

F: Factorisation The joint distribution factorises as a product of functions
on cliques (that is, maximal complete subgraphs).

The four properties are illustrated in Figures 1.13 and 1.14.

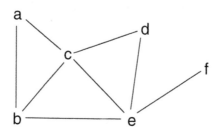

Figure 1.13 *Illustrating the pairwise, local and factorisation Markov proper-*
ties: $P : c \perp f|(a,b,d,e)$, $L : d \perp (a,b,f)|(c,e)$ *and* $F : \pi(a,b,c,d,e,f) = \psi_1(a,b,c)\psi_2(b,c,e)\psi_3(c,d,e)\psi_4(e,f)$.

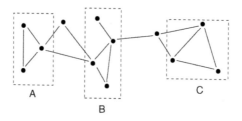

Figure 1.14 *Illustrating the global Markov property:* $G : A \perp C|B$.

It is always true that $F \Rightarrow G \Rightarrow L \Rightarrow P$, but these four Markov properties are in general different (there are easy counter-examples for each of the
reverse implications). However, in many statistical contexts, the four properties are the same; a sufficient but not necessary condition is that the joint
distribution has the positivity property ("any values realisable individually
are realisable jointly"). This result includes the Clifford-Hammersley theorem (Markov random field = Gibbs distribution, $L = F$). See, for example,
Besag (1974), Clifford (1990). A typical context in which the Markov properties may not coincide is where there are logical implications between some
subsets of variables.

For directed acyclic graphs, the situation is simpler: the directed local
Markov property is *always* equivalent to the directed graph factorisation
criterion: $DL = DF$ (subject to existence of a dominating product measure).

Modelling directly with an undirected graph

With a DAG, because of the acyclicity, *any* set of conditional distributions $\pi(\boldsymbol{x}_v|\boldsymbol{x}_{\mathrm{pa}(v)})$ combine to form a consistent joint distribution.

In an undirected graph, however, we need consistency conditions on the full conditionals $\pi(\boldsymbol{x}_v|\boldsymbol{x}_{-v})$ (using L, this is equal to $\pi(\boldsymbol{x}_v|\boldsymbol{x}_{\partial v})$, where ∂v denotes the neighbours of v). The only safe strategy is to use property F, to model the joint distribution as a product of functions on cliques

$$\pi(\boldsymbol{x}) = \prod_C \psi_C(\boldsymbol{x}_C)$$

We can then use property L, the local Markov property, to read off the full conditionals needed to set up MCMC:

$$\pi(\boldsymbol{x}_v|\boldsymbol{x}_{-v}) = \prod_{C:v\in C} \psi_C(\boldsymbol{x}_C) = \pi(\boldsymbol{x}_v|\boldsymbol{x}_{\partial v}).$$

Most of the applications in Besag, *et al.* (1995) have a spatial flavour, and provide illustrations of this style of modelling.

1.5.3 Chain graphs

In hierarchical spatial models, we need a hybrid modelling strategy: there will be some directed and some undirected edges. If there are no one-way cycles, the graph can be arranged to form a DAG with composite nodes called *chain components* Δ_t, that are the connected subgraphs remaining when all directed edges are removed: we call this a *chain graph*.

Model specification uses an appropriate combination of the two approaches; this builds a joint distribution

$$
\begin{aligned}
\pi(\boldsymbol{x}) &= \prod_t \pi(\boldsymbol{x}_{\Delta_t}|\boldsymbol{x}_{\mathrm{pa}(\Delta_t)}) \\
&= \prod_t \prod_{C\in\mathcal{C}_t} \psi_C(\boldsymbol{x}_C)
\end{aligned}
$$

where $\mathcal{C}_t$ are the cliques in an undirected graph with nodes $(\Delta_t, \mathrm{pa}(\Delta_t))$ and undirected edges consisting of (a) those already in Δ_t, (b) the links between Δ_t and parents, with directions dropped, and (c) links between all members of $\mathrm{pa}(\Delta_t)$.

1.6 Performance of MCMC methods

There are two main issues to consider when evaluating the performance of a Markov chain used for Monte Carlo calculations, for example in choosing between alternative chains for a particular target, or in assessing if a particular run of a particular chain is adequate for its purpose:

- Convergence (how quickly does the distribution of $x^{(t)}$ approach $\pi(x)$?);

- Efficiency (how well are functionals of $\pi(x)$ estimated from $\{x^{(t)}\}$?) .

In both cases, performance will be measured in relation to the computing effort expended, and of course this effort should be measured in seconds, not sweeps, although this does beg questions about whether for example, two rival methods have been coded comparably efficiently.

In this section, we will review some of the issues involved in these assessments, and some of the methods proposed. However, we should not lose sight of a third factor:

- Simplicity (how convenient is the method to code reliably and to use?)

We return to some issues of implementation in Section 1.11.

Contrary to a popular misconception, it should not be supposed that Gibbs is necessarily superior to other methods on *any* of these three criteria, so it does not provide a gold standard for comparison.

1.6.1 Monitoring convergence

An active and important subfield of MCMC research has aimed at investigating and developing methods for analysis of a Markov chain realisation, to determine empirically whether the chain can safely be said to 'have converged', and to provide a reliable basis for estimation of aspects of the target distribution.

It is undoubtedly important in practice to obtain some reassurance on these issues, and grossly irresponsible, for example, to accept at face value a statistical analysis of an important real-world problem, where this analysis is computed by a MCMC sampler whose performance on the model in question is unknown. However, there is a limit to the degree of reassurance that can be obtained from an empirical analysis, and this should always be supplemented by a sound understanding of the qualitative form of the target distribution, with an eye to the possible presence of features that the chosen MCMC sampler may have difficulty with.

Attempts to place the activity of convergence monitoring on a firm logical footing seem unconvincing. Apart from some contrived exceptional cases, no finite segments of Markov chain path are truly in equilibrium, so the question is not a deterministic decision problem. But it is also wrong to regard the issue as one of hypothesis testing. We *know* the sample is not 'in equilibrium', so the logic of testing that is aimed at detection of departures from the null hypothesis is not relevant. Whether the sample is large enough to enable such detection is inevitably bound up with the quantity – the closeness of the approximation to equilibrium – that is being measured.

Finally, of course, there can never be any protection based on an empirical analysis alone against the possibility that immediately after monitoring ceases, the chain jumps into a part of the parameter space that it has not previously visited!

Notwithstanding all these caveats, diagnostic techniques, intelligently used, are valuable, and the reader is referred to Brooks and Gelman (1998) for a thorough guide to the topic.

Some researchers have expressed optimism in the last year or two that perfect (or exact) simulation – the organisation of a MCMC simulation so that it delivers a sample guaranteed to be an exact draw from the target – will make reliance on diagnostics redundant. This may or may not happen, but it is still in the future! For an introduction to the role of *coupling from the past* in perfect simulation, see Section 1.9.

1.6.2 Monte Carlo standard errors

Since any Monte Carlo method is used to provide numerical estimates of *deterministic* quantities, even if these quantities arise in a stochastic model, it is important to be aware of, and in general to evaluate, the *Monte Carlo standard error* of estimated quantities, which should not of course be confused with the standard deviation of the posterior!

Because of Markov dependence, this is not quite straightforward, even though we (mostly) just use empirical averages as estimates.

Consider a Markov chain in equilibrium. Estimating $E_\pi(g) = \int g(x)\pi(dx)$ by $N^{-1} \sum_{t=1}^{N} g(x^{(t)}) = \bar{g}_N$, we find

$$
\begin{aligned}
\mathrm{var}(\bar{g}_N) &= N^{-2} \sum_{t=-N+1}^{N-1} (N - |t|)\gamma_t \\
&\sim N^{-1} \sum_{t=-\infty}^{\infty} \gamma_t
\end{aligned}
$$

where $\gamma_t = \mathrm{cov}_{\pi,P}\{g(x^{(s)}), g(x^{(s+t)})\}$ (note that unlike the equilibrium mean and variance, which depend only on π, the autocovariances depend also on the kernel P). This quantity is equivalently written

$$
N^{-1}\mathrm{var}_\pi(g) \sum_{t=-\infty}^{\infty} \rho_t = N^{-1}\mathrm{var}_\pi(g)\tau(g) = N^{-1}v(g,\pi,P)
$$

where ρ_t is the corresponding autocorrelation at lag t. The factor $\tau(g)$ by which the variance of the sample mean exceeds the value obtained in independent sampling is sometimes called the integrated autocorrelation time; it depends on π, g and the transition kernel P.

Several possibilities for estimating $\mathrm{var}(\bar{g}_N)$, $\tau(g)$ or $v(g,\pi,P)$ have been proposed, most of which are in common use:

- Blocking (also known as batching) (Hastings, 1970)

- Time-series methods (e.g. Sokal, 1989)
- Initial series estimates (Geyer, 1992)
- Regeneration (Mykland, Tierney and Yu, 1995)

There are also Central Limit theorems for Markov chain averages, of the form

$$\sqrt{N}\,(\bar{g}_N - E_\pi(g)) \xrightarrow{D} N(0, v(g, \pi, P)).$$

The theorems take various forms, but broadly speaking, we need ergodicity of the Markov chain, a finite variance of g and sufficiently good mixing that $v(g, \pi, P)$ is finite. Kipnis and Varadhan (1986) give such a result assuming reversibility, while Gordon and Lifšic (1978) do not need this condition, but make stronger assumptions elsewhere.

There are results comparing $v(g, \pi, P)$ for different kernels P. The best known is due to Peskun (1973), proved for a general state space setting by Tierney (1998); this states that if P and Q are two kernels with the same invariant distribution π, with P dominating Q off the diagonal — that is, $P(x, B) \geq Q(x, B)$ for all B not containing x — then $v(g, \pi, P) \leq v(g, \pi, Q)$ for all $g \in L_2(\pi)$, so that P is preferable. In particular, among all Metropolis-Hastings methods for a given π and proposal mechanism, that maximising the acceptance probability $\alpha(\boldsymbol{x}, \boldsymbol{y})$ is best: this explains the almost universal use of the acceptance probability formula (1.7). There are other recent interesting results on ordering Markov chains in Mira and Geyer (1999).

Blocking (or batching)

After satisfying ourselves that our Markov chain is in equilibrium, we divide a run of length N into b blocks of k consecutive samples. Then if k is large, so that block means are approximately independent, and b is also large, so that between-block variability can be estimated adequately, we have

$$\mathrm{var}(\bar{g}_N) \approx \{b(b-1)\}^{-1} \sum_{i=1}^{b} \{\bar{g}_{k,i} - \bar{g}_{N,1}\}^2$$

where

$$\bar{g}_{k,i} = k^{-1} \sum_{j=(i-1)k+1}^{ik} g(\boldsymbol{x}^{(j)})$$

is the mean of the ith block of length k.

This extends to nonlinear functionals of expectations; see Aykroyd and Green (1991).

Using empirical covariances

As is well-known from the time-series literature, we cannot estimate $\sum_{-\infty}^{\infty} \gamma_t$ consistently by $\sum_{-\infty}^{\infty} \widehat{\gamma}_t$, where $\widehat{\gamma}_t$ is the lag-t product-moment autocovariance of $g(\boldsymbol{x}^{(t)})$: we should, for example, use some kind of windowed estimate $\sum_{-\infty}^{\infty} w(t)\hat{\gamma}_t$ instead. Since $\sum_{-\infty}^{\infty} \gamma_t$ is proportional to the spectral density function evaluated at 0, this is a well-studied problem. See, for example, Priestley (1981, p. 225). A convenient estimator of $\tau(g)$ in practice is the truncated periodogram estimator of Sokal (1989): $\hat{\tau} = \sum_{t=-M}^{M} \hat{\gamma}_t/\hat{\gamma}_0$, where M is the smallest integer $\geq 3\hat{\tau}$.

Initial series estimators

Geyer (1992) observes that, for a reversible ergodic chain, $\gamma_{2t} + \gamma_{2t+1}$ is non-negative, decreasing and convex in t. This suggests a class of estimators obtained by truncating $\sum_{t:|t|<M} \hat{\gamma}_t$ when one or other of these properties is first violated.

Regeneration

Regeneration points in the Markov chain path are times $\{\tau_i, i = 1, 2, \ldots\}$ such that the *tours* $(\boldsymbol{x}^{(\tau_{i-1}+1)}, \boldsymbol{x}^{(\tau_{i-1}+2)}, \ldots, \boldsymbol{x}^{(\tau_i)})$ are independent and identically distributed for $i = 1, 2, \ldots$. If we can find such times, then renewal theory and ratio estimation give estimates of posterior expectations, and simulation standard errors that are valid without quantifying Markov dependence.

More specifically, let

$$L_i = (\tau_i - \tau_{i-1}), \qquad G_i = \sum_{t=\tau_{i-1}+1}^{\tau_i} g(\boldsymbol{x}^{(t)})$$

be the length of the ith tour, and the total of a function g evaluated at the states visited in the tour, then (L_i, G_i) are i.i.d. pairs, and

$$\frac{\sum_{i=1}^{n} G_i}{\sum_{i=1}^{n} L_i} \overset{a.s.}{\to} E_\pi(g) \quad \text{as } n \to \infty$$

by the renewal theorem.

Finding such regeneration times is easy in a discrete state space chain, since the chain regenerates at visits to any specified state. For general state space chains, the process of finding regeneration points is facilitated by use of Nummelin's splitting technique.

Regeneration using Nummelin's splitting

Suppose the transition kernel $P(\boldsymbol{x}, A)$ satisfies

$$P(\boldsymbol{x}, A) \geq s(\boldsymbol{x})\nu(A)$$

where ν is a probability measure, and s is a non-negative function such that $\int s(\boldsymbol{x})\pi(d\boldsymbol{x}) > 0$.

Let $r(\boldsymbol{x}, \boldsymbol{y})$ denote the Radon-Nikodym derivative

$$r(\boldsymbol{x}, \boldsymbol{y}) = \frac{s(\boldsymbol{x})\nu(d\boldsymbol{y})}{P(\boldsymbol{x}, d\boldsymbol{y})} \leq 1.$$

Now, given a realisation $\boldsymbol{x}^{(0)}, \boldsymbol{x}^{(1)}, \ldots$ from P, construct conditionally independent 0/1 random variables $S^{(0)}, S^{(1)}, \ldots$ with

$$P(S^{(t)} = 1 | \ldots) = r(\boldsymbol{x}^{(t)}, \boldsymbol{x}^{(t+1)})$$

Then by simple probability calculus we find

$$P(S^{(t)} = 1 | \boldsymbol{x}^{(\leq t)}, S^{(<t)}) = s(\boldsymbol{x}^{(t)})$$

and

$$P(\boldsymbol{x}^{(t+1)} \in A | \boldsymbol{x}^{(\leq t)}, S^{(<t)}, S^{(t)} = 1) = \nu(A)$$

that is, we can post-process the chain stochastically to generate binary 'splitting variables'. Whenever $S^{(t)} = 1$, the next state $\boldsymbol{x}^{(t+1)}$ is drawn from ν, independently of the past! The chain regenerates.

The problem with using the technique in practice is that in the Markov chains we tend to create for Bayesian computation, $P(\boldsymbol{x}, A)$ is difficult to handle algebraically, and/or impossible to bound below by $s(\boldsymbol{x})\nu(A)$ as required. Mykland, Tierney and Yu (1995) examine the possibilities of exploiting splitting in practical MCMC. Their perspective introduces another role for naive MCMC methods such as Independence Metropolis-Hastings (see Section 1.4.8), which although of limited efficiency may be amenable to algebraic manipulation to discover the required bounds.

1.7 Reversible jump methods: Metropolis-Hastings in a more general setting

The formulation of Metropolis-Hastings given in Subsection 1.4.3 is the standard one, and close to the original specification of Hastings (1970). It is already fairly general in that the densities $\pi(\boldsymbol{x})$ and $q(\boldsymbol{x}, \boldsymbol{y})$ appearing there may be with respect to an arbitrary measure on $\mathcal{X}$, so that both discrete and continuous distributions in any finite number of dimensions are covered. However, the formulation is a little restrictive when we come to consider MCMC samplers for certain new tasks, most notably problems where the dimension of the parameter varies, so that there is no elementary dominating measure for the target distribution.

The more general Metropolis-Hastings method we define here addresses this wider range of problems, but also offers a new perspective on the standard formulation, one that has certain pedagogical merits, and also may sometimes be more straightforward to implement. This reversible jump approach is based on Green (1995); see also Tierney (1998).

The detailed balance condition for a general transition kernel P and its invariant distribution π is written in integral form as

$$\int_{(\boldsymbol{x},\boldsymbol{y})\in A\times B} \pi(d\boldsymbol{x})P(\boldsymbol{x},d\boldsymbol{y}) = \int_{(\boldsymbol{x},\boldsymbol{y})\in A\times B} \pi(d\boldsymbol{y})P(\boldsymbol{y},d\boldsymbol{x}) \qquad (1.9)$$

for all Borel sets $A, B \subset \mathcal{X}$. If P is constructed in two steps, according to the Metropolis-Hastings paradigm, we make a transition by first drawing a proposed new state $\boldsymbol{y}$ from the proposal measure $q(\boldsymbol{x}, d\boldsymbol{y})$ and then accepting it with probability $\alpha(\boldsymbol{x}, \boldsymbol{y})$. If we reject, we stay in the current state, so that $P(\boldsymbol{x}, d\boldsymbol{y})$ has an atom at $\boldsymbol{x}$. This makes an equal contribution to each side of equation(1.9), so can be neglected, and we are left with the requirement

$$\int_{(\boldsymbol{x},\boldsymbol{y})\in A\times B} \pi(d\boldsymbol{x})\alpha(\boldsymbol{x},\boldsymbol{y})q(\boldsymbol{x},d\boldsymbol{y}) = \int_{(\boldsymbol{x},\boldsymbol{y})\in A\times B} \pi(d\boldsymbol{y})\alpha(\boldsymbol{y},\boldsymbol{x})q(\boldsymbol{y},d\boldsymbol{x}).$$
$$(1.10)$$

When can we 'solve' this collection of equations of measures to give an explicit equation for the function $\alpha(\boldsymbol{x}, \boldsymbol{y})$? Suppose that $\pi(d\boldsymbol{x})q(\boldsymbol{x}, d\boldsymbol{y})$ is dominated by a *symmetric* measure μ on $\mathcal{X} \times \mathcal{X}$, and has density (Radon-Nikodym density) f with respect to this μ. Then (1.10) becomes

$$\int_{(\boldsymbol{x},\boldsymbol{y})\in A\times B} \alpha(\boldsymbol{x},\boldsymbol{y})f(\boldsymbol{x},\boldsymbol{y})\mu(\boldsymbol{x},d\boldsymbol{y}) = \int_{(\boldsymbol{x},\boldsymbol{y})\in A\times B} \alpha(\boldsymbol{y},\boldsymbol{x})f(\boldsymbol{y},\boldsymbol{x})\mu(\boldsymbol{y},d\boldsymbol{x}),$$

and, using the symmetry of μ, this is clearly satisfied for all appropriate A, B if

$$\alpha(\boldsymbol{x},\boldsymbol{y})f(\boldsymbol{x},\boldsymbol{y}) = \alpha(\boldsymbol{y},\boldsymbol{x})f(\boldsymbol{y},\boldsymbol{x}).$$

As with the standard Metropolis-Hastings method, we usually take the acceptance probabilities as large as possible subject to detailed balance, so

$$\alpha(\boldsymbol{x},\boldsymbol{y}) = \min\left\{1, \frac{f(\boldsymbol{y},\boldsymbol{x})}{f(\boldsymbol{x},\boldsymbol{y})}\right\}. \qquad (1.11)$$

If we wrote this rather more informally as

$$\alpha(\boldsymbol{x},\boldsymbol{y}) = \min\left\{1, \frac{\pi(d\boldsymbol{y})q(\boldsymbol{y},d\boldsymbol{x})}{\pi(d\boldsymbol{x})q(\boldsymbol{x},d\boldsymbol{y})}\right\} \qquad (1.12)$$

then the similarity with the usual expression using densities (1.7) is apparent, but we must not forget that the meaning of the ratio of measures derives from equation (1.11), and assumes the existence and symmetry of μ.

The formulation in this section applies to a completely general state space Markov chain. For the particular context of spatial point processes, a very similar development was given by Geyer and Møller (1994), providing an alternative to the usual spatial birth-and-death process approach. In this setting, although the dominating measure for the target distribution is not as familiar as Lebesgue, it is perfectly explicit: models are expressed via their densities with respect to a unit rate Poisson process. Detailed balance can therefore be established directly. In other situations, the dominating measure is much less explicit, and the constructions of the following two subsections very often prove useful.

1.7.1 Explicit representation using random numbers

The general Metropolis-Hastings method of the preceding subsection hardly lives up to the claim that it offers advantages in implementation, as it seems rather abstract. Fortunately, in many cases the dominating measure and Radon-Nikodym derivatives can be generated almost automatically.

To see this, imagine how the transition will actually be implemented. Take the case where $X \subset R^d$, and suppose π has a density (also denoted π) with respect to d-dimensional Lebesgue measure ν_d. At the current state x, the program-writer will generate, say, r random numbers u from a known density g, and then form the proposed new state as some suitable deterministic function of the current state and the random numbers: $y = y(x, u)$. The left-hand side of (1.10) becomes:

$$\int_{(x,y)\in A\times B} \pi(x)g(u)\alpha(x,y)\nu_d(dx)\nu_r(du)$$

Now consider how the reverse transition from y to x would be made, with the aid of random numbers $u' \sim g$ giving $x = x(y, u')$. If the transformation from (x, u) to (y, u') is a bijection, and if both it and its inverse are differentiable, then by the standard change-of-variable formula, the $(d+r)$-dimensional integral equality (1.10) holds if

$$\pi(x)g(u)\alpha(x,y) = \pi(y)g(u')\alpha(y,x)\left|\frac{\partial(y,u')}{\partial(x,u)}\right|,$$

whence a valid choice for α is

$$\alpha(x,y) = \min\left\{1, \frac{\pi(y)g(u')}{\pi(x)g(u)}\left|\frac{\partial(y,u')}{\partial(x,u)}\right|\right\}. \tag{1.13}$$

It is often easier to work with this expression than the usual one, equation (1.7).

1.7.2 MCMC for variable dimension problems

What if the number of things you don't know is one of the things you don't know?

There is a huge variety of statistical problems of this kind, where the parameter dimension is not fixed, and itself subject to inference. Examples range from classical statistical tasks such as variable selection, mixture estimation, change-point analysis, and model determination in general, through to the kinds of problems raised in modern applications of stochastic modelling to gene-mapping, analysis of ion channel data, image segmentation and object recognition.

For a fully Bayesian analysis based on a single simulation run, we need a MCMC sampler that jumps between parameter subspaces of differing dimensions: given the reversible jump framework of the previous subsection, this is now a fairly modest generalisation. Our state variable x now lives in a union of spaces of differing dimension: $\mathcal{X} = \cup_k \mathcal{X}_k$.

We will use a range of *move types* m, each providing a transition kernel P_m, and insist on detailed balance for each:

$$\int_{x \in A} \pi(dx) P_m(x, B) = \int_{y \in B} \pi(dy) P_m(y, A)$$

for all Borel sets $A, B \subset \mathcal{X}$. The idea of a family of move types is implicit even in the simplest formulation of Metropolis-Hastings, where we have a different proposal density q_i for each component i, but compute the acceptance probability using the joint target distribution (equation (1.7)). In the present more elaborate context, there may be a richer variety of move types, recognising that different approaches may be needed to enable transitions between different pairs of spaces $\mathcal{X}_k, \mathcal{X}_{k'}$.

The Metropolis-Hastings idea still works, but you need to work a bit harder to make the acceptance ratio make sense. The proposal measure q is now the *joint* distribution of move type m and proposed destination y, so for each $x \in \mathcal{X}$, $\sum_m \int_{y \in \mathcal{X}} q_m(x, dy) \leq 1$ (allowing a positive probability of not attempting a move, if required). The detailed balance condition (see (1.10)) becomes

$$\int_{(x,y) \in A \times B} \pi(dx) \alpha_m(x, y) q_m(x, dy) = \int_{(x,y) \in A \times B} \pi(dy) \alpha_m(y, x) q_m(y, dx).$$

for all m, A, B. This leads to the formal solution

$$\alpha_m(x, y) = \min\left\{1, \frac{\pi(dy) q_m(y, dx)}{\pi(dx) q_m(x, dy)}\right\}$$

as in (1.12).

Apart from the addition of the subscript m, this is just a special case of the earlier general Metropolis-Hastings method, and the ratio of measures makes sense subject to the existence of a symmetric dominating measure μ_m for $\pi(dy) q_m(y, dx)$.

Again, this is most easily understood in the concrete terms of the preceding subsection: we need a differentiable bijection between (x, u) and (y, u'), where u, u' are the vectors of random numbers used to go between x and y in each direction. Suppose these have densities $g_m(u; x)$ and $g_m(u'; y)$. In the variable dimension context, move type m might use transitions between $\mathcal{X}_{k_1}$ and $\mathcal{X}_{k_2}$; if these spaces have dimensions d_1 and d_2, and π is absolutely continuous with respect to ν_{d_1} and ν_{d_2} in the respective spaces, then the dimensions of u and u', r_1 and r_2 say, must satisfy the dimension-balancing condition

$$d_1 + r_1 = d_2 + r_2.$$

We can then write

$$\alpha_m(x, y) = \min \left\{ 1, \frac{\pi(y)g_m(u'; y)}{\pi(x)g_m(u; x)} \left| \frac{\partial(y, u')}{\partial(x, u)} \right|, \right\} \qquad (1.14)$$

The ratio is of joint densities with the same degrees of freedom, together with the Jacobian needed to account for the change of variable.

Apart from the illustrative applications in Green (1995), this methodology has been widely implemented for problems with a variable-dimension parameter, for example Richardson and Green (1997), Uimari and Hoeschele (1997), Denison, Mallick and Smith (1998), Heikkinen and Arjas (1998), Holmes and Mallick (1998), Pievatolo and Green (1998), Green and Richardson (1999), Hodgson (1999), Hodgson and Green (1999), and Rue and Hurn (1999).

1.7.3 Example: step functions

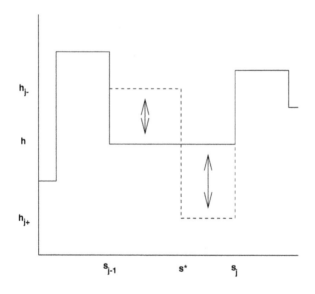

Figure 1.15 *Split and merge of a step.*

Let us illustrate the methods of the preceding subsection in what is almost the simplest setting possible, by studying the situation where the state variable x represents a step function, as might be part of the parameterisation of a model for change-point analysis in regression or point processes. We can readily evaluate each of the factors in (1.14), and end up with a useful sampler that — with a little modification — will find application in the next subsection.

A simple prior model for a step function on $[0, L)$ would parameterise the function in terms of its number of steps k, the positions $\{s_1 < s_2 < \cdots < s_k\}$ of those steps, and the heights $\{h_0, h_1, \ldots, h_k\}$, h_j being the value of the function on the interval $[s_{j-1}, s_j)$. For illustration here, we assume that the number of steps is drawn from an arbitrary $p(k)$, and that given k, the step heights are i.i.d. from some density $f_H(\cdot)$, and that the step positions are drawn as the order statistics from a uniform distribution on the observation interval:

$$p(s_1, s_2, \ldots, s_k) = \frac{k! \, I \, [0 < s_1 < s_2 < \cdots < s_k < L]}{L^k}.$$

As with ordinary Metropolis-Hastings, you have freedom to use intuition in designing proposals; validity is ensured by using the correct acceptance probability (1.14).

Consider a move which allows the number of steps to change, by 'birth' and 'death'. When x has k steps, we propose birth with probability b_k, draw two random numbers u_1 and u_2 from $g(u_1, u_2)$, and use them to split an existing step interval into two. Let the new step position be $s^* = u_1$, located between s_{j^*-1} and s_{j^*} say, and use u_2 to divide the current step height h_{j^*} into two values with weighted average h_{j^*}: $h_{j-} = h_{j^*} + u_2/w_-$ and $h_{j+} = h_{j^*} - u_2/w_+$, where $w_- = s^* - s_{j^*-1}$ and $w_+ = s_{j^*} - s^*$. Turning now to death of a step: this is proposed with probability d_k, and we choose a step at random to delete; if step $j^\dagger$ is deleted, then the new step height for the interval $[s_{j^\dagger-1}, s_{j^\dagger+1})$ is the weighted average $\{(s_{j^\dagger} - s_{j^\dagger-1})h_{j^\dagger} + (s_{j^\dagger+1} - s_{j^\dagger})h_{j^\dagger+1}\}/(s_{j^\dagger+1} - s_{j^\dagger-1})$. This precisely reverses the effect of the birth just described.

Note that this formulation has the dimension balance we require: when there are k steps, there are k positions and $k+1$ heights, making $d_1 = 2k+1$ variables in all. With a birth, we are proposing a move to $d_2 = 2k+3$. The dimensions of the random numbers u, u' are $r_1 = 2$ and $r_2 = 0$ respectively, and indeed $d_1 + r_1 = d_2 + r_2$. (An alternative choice, equally valid, would be $r_1 = 3, r_2 = 1$, which would be obtained if we had dropped the requirement of preserving the weighted average on birth and death, and generating new random heights as needed, for example independently of the current state.)

Suppose we let x, y denote the states of the chain before the split is proposed, and the state as modified by the birth proposal. Then

$$\pi(x) \quad \propto \quad p(Y|x)p(k) \prod_{j=0}^{k} f_H(h_j) \frac{k! \, I \, [0 < s_1 < s_2 < \cdots < s_k < L]}{L^k}$$

$$\pi(y) \quad \propto \quad p(Y|y)p(k+1) \prod_{j \neq j^*} f_H(h_j) f_H(h_{j-}) f_H(h_{j+})$$

$$\frac{(k+1)!\, I\,[0<s_1<s_2<\cdots<s_{j^*-1}<s^*<s_{j^*}<\cdots<s_k<L]}{L^{k+1}},$$

the constant of proportionality being the same in each case. The proposal terms are

$$g_m(u; x) = b_k g(u_1, u_2)$$
$$g_m(u'; y) = \frac{d_{k+1}}{k+1},$$

reflecting the described mechanism for choosing to propose birth or death, the drawing of (u_1, u_2), and the random choice of a step to delete. Finally, the Jacobian we need is an order $2k+3$ determinant, but with many of the components of the state vector unaltered by the transformation, it reduces to

$$\left| \frac{\partial(h_{j-}, h_{j+}, s^*)}{\partial(h_{j^*}, u_1, u_2)} \right| = \frac{w_- + w_+}{w_- w_+}.$$

We can now compute the acceptance probability for a birth from (1.14):

$$\alpha = \min\left\{ 1, \Lambda \frac{p(k+1)}{p(k)} \frac{(k+1)}{L} \frac{f_H(h_{j-}) f_H(h_{j+})}{f_H(h_{j^*})} \right.$$
$$\left. \frac{d_{k+1}/(k+1)}{b_k g(u_1, u_2)} \frac{w_- + w_+}{w_- w_+} \right\}$$

where Λ is the likelihood ratio $p(Y|y)/p(Y|x)$.

1.7.4 Cyclones example, continued

Model 8: unknown number of change points

What if the *number* of change points, k, is also unknown? We might place a prior on k, say Poisson(λ):

$$p(k) = e^{-\lambda} \frac{\lambda^k}{k!}$$

and then make Bayesian inference about all unknowns: $x = (k, \alpha, \beta, T_1, \ldots, T_k, x_0, \ldots, x_k)$. There are $2k+4$ parameters: the number of things you don't know is one of the things you don't know!

For a MCMC solution, the only additional ingredient we need over model 7 is a birth/death move to allow a variable number of steps. This follows closely the setup of the preceding subsection, except that since the step function represents an intensity and is necessarily non-negative, we arranged to preserve the weighted *geometric* mean:

$$h_{j-}^{w_-} h_{j+}^{w_+} = h_{j^*}^{w_- + w_+}.$$

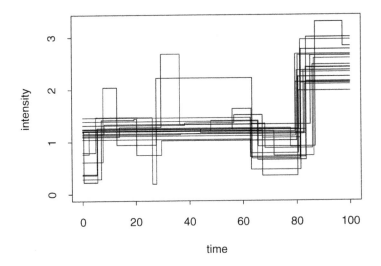

Figure 1.16 *Posterior sample of step functions $x(t)$ for model 8, applied to cyclones data.*

Also the joint density of the step positions is now $\propto \prod_j (s_j - s_{j-1})$ with a corresponding change to the acceptance ratio. The moves for this sampler are described in detail in Green (1995), except that here we have slightly extended the model to include variable hyperparameters α and β, as seen in Sections 1.2.4 and 1.4.11.

We applied this model to the cyclones data, using $\lambda = 3$; a small sample from the posterior distribution of the step function $x(\cdot)$ is shown in Figure 1.16. Various aspects of the posterior distribution can be summarised by appropriate analysis and display of much larger MCMC samples; see, for example, Figures 1.17, 1.18 and 1.19.

Model 9: with a cyclic component

Finally, here, as further illustration of the flexibility in modelling allowed by the approach, we include another ingredient, that will be justified in many real time-series point process problems: periodicity.

This could be handled in various ways — parametric, nonparametric, with known and unknown period(s) — but the simplest is to take a simple sinusoid, and assume that the data are generated from a Poisson process

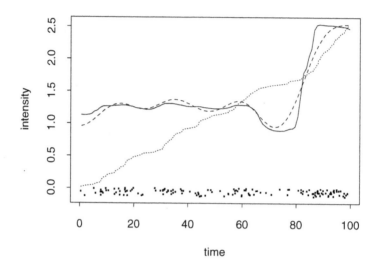

Figure 1.17 *Posterior mean of step function $x(t)$ for model 8 (solid line) and kernel estimate of $x(t)$ (broken line), for cyclones data.*

with instantaneous rate

$$x(t)\left\{1 + \gamma\cos(2\pi ft) + \delta\sin(2\pi ft)\right\},$$

where $x(t)$ is the step function defined above, and f denotes the assumed (known) frequency).

If *a priori* (γ, δ) are taken uniform on the unit disc, then a simultaneous Metropolis update is easily implemented.

A small sample from the posterior for the cyclic component is shown in Figure 1.20.

1.7.5 Bayesian model determination

It is wrong to behave as if the statistical model for our data was not subject to question.

Suppose we have a (countable) collection of models that we wish to entertain: $M_1, M_2, \ldots, M_k, \ldots$. *A priori*, we assign probabilities to these: $p(k)$.

For each model, there is a parameter vector $\theta = \theta_k \in \mathcal{R}^{n_k}$ say, with a prior: $p(\theta_k|k)$, and a likelihood for the observed data Y: $p(Y|k, \theta_k)$. The

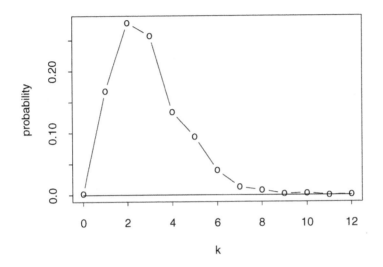

Figure 1.18 *Posterior distribution of number of change points k for model 8, applied to cyclones data.*

joint distribution of all variables is

$$p(k, \theta_k, Y) = p(k)p(\theta_k|k)p(Y|k, \theta_k).$$

(Note that in this section, the subscript k on θ indicates the model to which θ_k belongs, not the kth element of a vector θ.)

Observing Y provides information about both the model indicator k and the corresponding parameter vector θ_k, through their posterior distributions:

$$p(k|Y) = \frac{\int p(k, \theta_k, Y)d\theta_k}{\sum_k \int p(k, \theta_k, Y)d\theta_k}$$

and

$$p(\theta_k|Y, k) = \frac{p(k, \theta_k, Y)}{\int p(k, \theta_k, Y)d\theta_k}$$

involving integrals that as usual seem to need MCMC! There are two main approaches: within-model and across-model simulation.

Within-model simulation

Here we treat each model M_k separately.

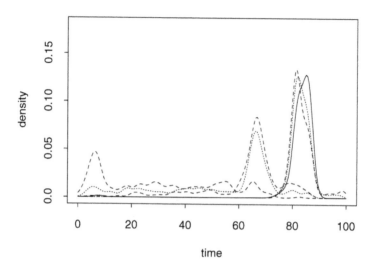

Figure 1.19 *Estimates of posterior density for change point positions for model 8, applied to cyclones data:* $k = 1$ *(solid line),* $k = 2$ *(dotted lines) and* $k = 3$ *(broken lines).*

The posterior for the parameters θ_k is in any case a within-model notion:

$$p(\theta_k | Y, k) = \frac{p(\theta_k | k) p(Y | k, \theta_k)}{\int p(\theta_k | k) p(Y | k, \theta_k) d\theta_k}$$

As for the posterior model probabilities, since

$$\frac{p(k_1 | Y)}{p(k_0 | Y)} = \frac{p(k_1)}{p(k_0)} \frac{p(Y | k_1)}{p(Y | k_0)}$$

(the *Bayes factor* for model M_{k_1} vs. M_{k_0}), it is sufficient to estimate the *marginal likelihoods*

$$p(Y | k) = \int p(\theta_k, Y | k) d\theta_k$$

separately for each k, using individual MCMC runs.

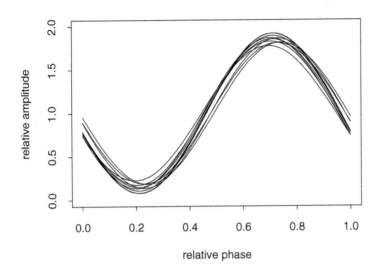

Figure 1.20 *Posterior samples from the harmonic component, model 9, applied to cyclones data.*

Estimating the marginal likelihood

There are many possible estimates based on importance sampling, some of which are well-studied, for example

$$\widehat{p}_1(Y|k) = N \left/ \sum_{t=1}^{N} \left\{ p(Y|k, \theta_k^{(t)}) \right\}^{-1} \right.$$

based on a MCMC sample $\theta_k^{(1)}, \theta_k^{(2)}, \ldots$ from the posterior $p(\theta_k|Y, k)$, or

$$\widehat{p}_2(Y|k) = N^{-1} \sum_{t=1}^{N} p(Y|k, \theta_k^{(t)})$$

based on a sample from the *prior* $p(\theta_k|k)$. Both of these have their faults, and composite estimates can perform better. See, for example, Newton and Raftery (1994).

Across-model simulation

Here we conduct a *single* simulation that traverses the entire (k, θ_k) space. Since the dimension n_k of θ_k typically varies with k, this requires a MCMC sampler that works in more general spaces than $\mathcal{R}^d$. The reversible jump sampler of Section 1.7.2 is an obvious candidate. Applications of this approach include Giudici and Green (1999), and Nobile and Green (2000).

See also Madigan and Raftery (1994), Carlin and Chib (1995), Phillips and Smith (1996) and George and McCulloch (1997) for other recent approaches to Bayesian computation for model determination.

1.8 Some tools for improving performance

1.8.1 Tuning a MCMC simulation

Having implemented an MCMC sampler, there are various quite simple techniques available to amend the algorithm to try to improve performance.

Most Metropolis-Hastings methods involve proposal distributions with freely chosen parameters – the spread of the perturbation distribution in a random-walk Metropolis method, for example. As the parameter is varied, different acceptance rates will be obtained. Of course, 100% acceptance is not necessarily desirable; in random-walk Metropolis it is only achieved in the limit as the spread goes to 0. As this is approached, the sample path will exhibit increasingly high autocorrelation. In the 'bold' opposite to this 'timid' strategy, the steps taken will be large, but they will be taken rarely. The right balance, where convergence may be faster and autocorrelation less, will be in the middle.

There is an interesting theoretical study of optimal acceptance rates for random walk Metropolis in Gelman, Roberts and Gilks (1996), based on the artificial case of multivariate normal target and proposal distributions; this has been quite influential in establishing a 'rule of thumb' advising aiming for 20-40% acceptance generally, but this study is perhaps a rather narrow basis for such a sweeping conclusion.

Metropolis-Hastings methods can be designed for updating single variables, or groups of any size. Larger groups offer the possibility of allowing the sampler to beat the restrictions on performance imposed by high correlations between variables in the target distribution, but may carry a burden in cumbersome tuning of a multivariate proposal distribution. A careful study of the merits of grouping variables is one of the themes of Roberts and Sahu (1997); see also Besag et al (1995, p. 10).

Another opportunity for reducing correlation between variables is to consider re-parameterisation; a hierarchical centring formalism in the linear mixed models context is introduced by Gelfand, Sahu and Carlin (1995), and there is a broader discussion in Gilks and Roberts (1996).

1.8.2 Antithetic variables and over-relaxation

The essential idea of antithetic variables in ordinary static Monte Carlo is one of the classic ideas for variance reduction: to aim to introduce negative correlation among some of the summands in an empirical average $N^{-1} \sum_1^N g(\boldsymbol{x}^{(t)})$ by using coupled pairs $(u, 1-u)$ of uniform random numbers in generating pairs of state vectors $\boldsymbol{x}$. Of course, the success of the method requires some monotonicity in both the mapping from u to $\boldsymbol{x}$, and in the function g.

As applied to MCMC, the aim would be to choose an update of $\boldsymbol{x}^{(t)}$ that has detailed balance, as usual, but also introduces negative *serial* autocorrelation in the process $g(\boldsymbol{x}^{(t)})$, or at least reduces the value of a positive autocorrelation.

Barone and Frigessi (1989) studied the effect of antithetic variables on the convergence of samplers for Gaussian processes. The full conditional for a single variable x_i in a multivariate Gaussian distribution is of course a normal distribution, $N(\mu_i, \sigma_i^2)$, say. It is easy to check that drawing the updated variable x_i' from $N((1+\theta)\mu_i - \theta x_i, (1-\theta^2)\sigma_i^2)$ is in detailed balance for any $\theta \in (-1, 1)$; $\theta = 0$ gives the Gibbs sampler, and if $\theta > 0$ then x_i and x_i' are conditionally negatively correlated. They show that in the case of entirely positive association between the variables, the spectral radius of the corresponding Markov chain is a decreasing function of θ at $\theta = 0$; thus convergence is improved by using the dynamic version of antithetic variables, $\theta > 0$.

Green and Han (1992) (see also Besag and Green, 1993) examine the effect of this antithetic modification on the autocorrelation time, and show that it is reduced by a factor $(1 - \theta)/(1 + \theta)$. They also propose using antithetically-modified Gaussian approximations to full conditionals as proposal distributions for Metropolis-Hastings for non-Gaussian targets, although the empirical evidence assessing this idea suggests that convergence is not always improved. Barone, Sebastiani and Stander (1998) have developed the idea further. Neal (1998) has proposed a related method, based on order statistics, that seems much more widely applicable.

This whole topic has close parallels with the theory of over-relaxation in the iterative solution of simultaneous equations in numerical analysis.

1.8.3 Augmenting the state space

Perhaps counter-intuitively, it is sometimes possible to improve MCMC performance by augmenting the state vector to include additional components. Two particularly successful recipes are those in which the original model appears as a *conditional* distribution in an augmented model (simulated tempering) and in which it appears as a *marginal* (auxiliary variables); these approaches are described in the next two subsections.

Two other devices might also be bracketed under the heading of augmentation. In multigrid methods, spatial problems are treated on a variety of spatial scales, sometimes by coupling together several different models, sometimes merely by using a family of MCMC samplers that update groups of variables together, the sizes of the groups varying with sweep. In hybrid MCMC, additional variables are introduced, bearing a relationship to the original ones analogous to that between momentum and position variables in dynamics. The MCMC updates maintain this physical analogy.

1.8.4 Simulated tempering

The approach here is to combat slow mixing by embedding the desired model in a family of models, indexed say by α, and treat α now as an additional dynamic variable. Thus the target is changed from $\pi(\boldsymbol{x})$ to $\pi^\star(\boldsymbol{x}, \alpha_0)$. The family $\{\pi^\star(\boldsymbol{x}, \alpha)\}$ is designed so that for some α, a much better-mixing chain can be found than for the original target. We run MCMC on $\pi^\star(\boldsymbol{x}, \alpha)$, and condition on $\alpha = \alpha_0$ by selecting from the output.

This 'serial' approach can be compared with the 'parallel' one of Metropolis-coupled MCMC (Geyer, 1991; see also Gilks and Roberts, 1996).

Simulated tempering, by changing the temperature

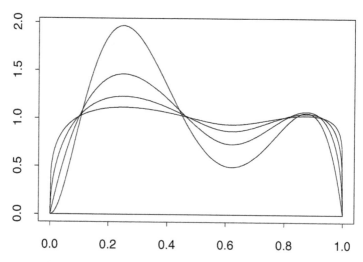

Figure 1.21 *The effect of tempering on a univariate full conditional: the beta mixture density $0.7Be(3,7) + 0.3Be(8,2)$, and the results after raising to the powers $\alpha = 0.5, 0.25, 0.125$ and renormalising.*

This was the original idea of Marinari and Parisi (1992), independently derived by Geyer and Thompson (1995); we set

$$\pi^\star(\theta, \alpha) \propto \{\pi(\theta)\}^\alpha$$

where $\alpha = \alpha_0 = 1$ corresponds to the original model, and $\alpha \to 0$ makes the probability surface 'flatter', or in physical terms, 'warmer'. A graphical illustration of the effect of the α power on a univariate density can be seen in Figure 1.21.

The full conditionals change in the same way as the joint distribution:

$$\pi^\star(\theta_i|\theta_{-i}, \alpha) \propto \{\pi(\theta_i|\theta_{-i})\}^\alpha$$

so implementation is very easy.

We normally place a (discrete) artificial prior on α so that the *marginal* for α is approximately uniform.

Simulated tempering, by inventing models

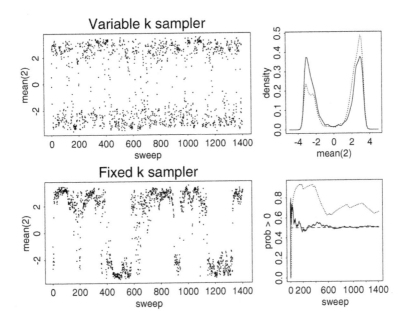

Figure 1.22 *Better mixing with variable dimensions, illustrated by a mixture analysis application (from Richardson and Green (1997)).*

A more general perspective on what tempering achieves and how it works can be obtained by envisaging it as embedding the target into a bigger model space, and there may be many ways to do that. For example, a model indicator k may be allowed to vary, although in truth its value is known, or at least fixed. An example from mixture analysis is shown in Figure 1.22. The left-hand panels show the sample paths for one component of the parameter vector, which has a strongly bimodal distribution under the target; two samplers are compared, one (bottom) in which the model indicator k (in this case the number of mixture components) is held fixed, the other (top) in which it varies, but we condition on its value by

selecting from the output. In the right-hand side panels we see (top) the
resulting estimates of the marginal density of this parameter and (bottom)
the evolution of the ergodic average estimating the probability that the parameter is positive; from the symmetry of the setup of the experiment, this
is known to be 0.5. Results for the variable-k sampler are shown in solid
lines, those for fixed k are dotted. Allowing the number of components to
vary can give much better mixing. See Richardson and Green (1997) for
details.

1.8.5 Auxiliary variables

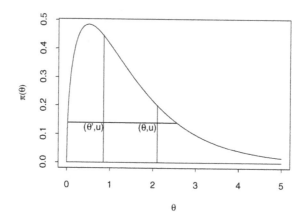

Figure 1.23 *The slice sampler.*

Edwards and Sokal (1988) proposed a way to improve mixing by augmenting the state space so that the original target appears as the *marginal*
equilibrium distribution. The following interpretation of their approach in
statistical language can be found in Besag and Green (1993).

Starting from $\pi(\boldsymbol{x})$, introduce some additional variables u, with $\pi(u|\boldsymbol{x})$
arbitrarily chosen. Then the joint is $\pi(\boldsymbol{x}, u) = \pi(\boldsymbol{x})\pi(u|\boldsymbol{x})$, for which $\pi(\boldsymbol{x})$
is certainly the marginal for $\boldsymbol{x}$.

We could now run a MCMC method for the *joint* target $\pi(\boldsymbol{x}, u)$ (usually a
method that updates $\boldsymbol{x}$ and u alternately), and simply ignore the u variables
in extracting information from the simulation.

When might this idea be useful? Suppose $\pi(\boldsymbol{x})$ factorises as:

$$\pi(\boldsymbol{x}) = \pi_0(\boldsymbol{x})b(\boldsymbol{x})$$

where $\pi_0(\boldsymbol{x})$ is a (possibly unnormalised) distribution that is easy to simulate from, and $b(\boldsymbol{x})$ is the awkward part, often representing the 'interactions' between variables that are slowing down the chain.

Then take a one-dimensional u with $u|\boldsymbol{x} \sim U[0, b(\boldsymbol{x})]$: we find

$$\pi(\boldsymbol{x}, u) = \pi(\boldsymbol{x})\pi(u|\boldsymbol{x}) = \pi_0(\boldsymbol{x})b(\boldsymbol{x})\frac{I[0 \le u \le b(\boldsymbol{x})]}{b(\boldsymbol{x})}$$

so that

$$\pi(\boldsymbol{x}|u) \propto \pi_0(\boldsymbol{x})$$

restricted to (conditional on) the event $\{\boldsymbol{x} : b(\boldsymbol{x}) \ge u\}$. At least when this $\pi(\boldsymbol{x}|u)$ can be sampled without rejection, we can easily implement a Gibbs sampler, drawing u and $\boldsymbol{x}$ in turn.

This method has recently been popularised under the name of the 'slice sampler', a picturesque but otherwise unnecessary name, reflecting the fact that if $\pi_0(\boldsymbol{x}) = $ constant, $\pi(\boldsymbol{x}|u)$ is a uniform distribution, corresponding to a horizontal slice through the graph of $\pi(\boldsymbol{x})$. For statistical applications of the idea, see Neal (1997) and Damien, Wakefield and Walker (1999), and for a detailed analysis of the method, see Roberts and Rosenthal (1999).

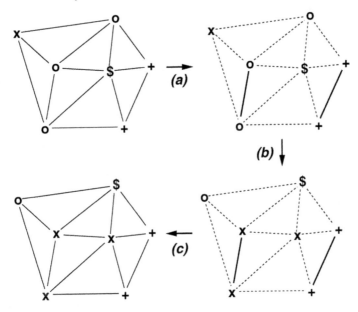

Figure 1.24 *Illustrating the Swendsen-Wang algorithm: (a) bond variables between like-coloured nodes are 'on' with probability $1 - e^{-\beta}$, always 'off' between unlike-coloured ones; (b) clusters formed by 'on' bonds are re-coloured uniformly at random; (c) the new colouring.*

The original applications of auxiliary variable methods were to statistical physics problems, where in particular the Swendsen-Wang method (Swendsen and Wang, 1987) has had a profound influence; see also Edwards and Sokal (1988) and Sokal (1989).

The Swendsen-Wang method is a MCMC method for the Potts model on an arbitrary graph (V, E), the target distribution

$$\pi(\boldsymbol{x}) \propto \exp \left\{ -\beta \sum_{(v,w) \in E} I[\boldsymbol{x}_v \neq \boldsymbol{x}_w] \right\} = \prod_{e \in E} b_e(\boldsymbol{x}),$$

say. We define one auxiliary variable u_e for each edge e, conditionally independent given $\boldsymbol{x}$, with $u_e | \boldsymbol{x} \sim U(0, b_e(\boldsymbol{x}))$. If $u_e > e^{-\beta}$ we say the edge e is 'on', otherwise 'off'. It is easy to see that in drawing $\boldsymbol{u}$ given $\boldsymbol{x}$, edges are on with probability $1 - e^{-\beta}$ if $\boldsymbol{x}_v = \boldsymbol{x}_w$, always off if $\boldsymbol{x}_v \neq \boldsymbol{x}_w$. Simple manipulation shows that $\pi(\boldsymbol{x}|\boldsymbol{u})$ is a random uniform colouring on the clusters determined by the on bonds.

Figure 1.24 illustrates one sweep of the Swendsen-Wang algorithm, applied to the Potts model on a small graph.

1.9 Coupling from the Past (CFTP)

Coupling from the Past (CFTP) is a beautiful idea due to Propp and Wilson (1996): it provides a way of organising a Markov chain simulation so that after a finite but random amount of work, it *exactly* delivers a sample from the target distribution! (Another such protocol, based on an elaborate form of rejection sampling was given by Fill (1998).)

Since the CFTP idea first appeared in preprint form, it has generated much excitement among MCMC researchers, keen both to understand and generalise the basic formulation, and to discover the practical potential for computation in stochastic processes and statistical applications.

For an example of Propp and Wilson's construction, consider the partial simulation of a symmetric random walk with reflecting barriers shown in Figure 1.25. To appreciate the message of this figure, it is not necessary to know anything about the order in which the displayed steps were generated, nor anything at all about any steps not displayed. All that we need is that the successive steps along each partially-drawn path are independent, and have the correct law: equally probably ± 1, except where steps attempting to go outside the interval $[1, 5]$ are suppressed.

One can see that *for the random numbers used to make this simulation*, and regarding the figure as part of a conceptual simulation of paths from *all* initial states at *all* initial times < 0, all paths of the chain starting at time $-\infty$ have the same state (*viz.*, 3) at time 0. This state, $\boldsymbol{x}^{(0)}$, must be drawn from π!

Generally, imagine multiple coupled paths of a Markov chain run from all initial states in the indefinite past, and look at the state at time 0, $\boldsymbol{x}^{(0)}$.

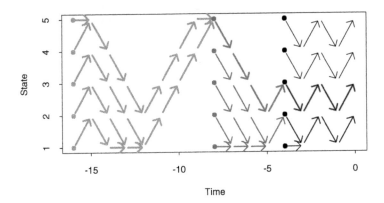

Figure 1.25 *Monotone CFTP for a simple random walk.*

If this is unique, then $x^{(0)} \sim \pi$. For this to be of any practical consequence in computing, we must be able to conduct this conceptually infinite amount of simulation in a finite time. But, we can see that Figure 1.25 was indeed constructed with a finite amount of work — fewer than 60 steps are shown.

Generalising from this example, if there exists a (random) initial time $-T$ such that for all initial states x_{-T}, $x^{(0)}$ is the same, then $x^{(0)} \sim \pi$. We do not even need to find T exactly, since coalescence occurs from all initial times $< -T$. So we can just try a decreasing sequence of initial times $-1, -2, -4, -8, \ldots$ until we discover coalescence.

1.9.1 Is CFTP of any use in statistics?

There have been some spectacular successes in finding CFTP implementations for certain models in statistical physics and spatial processes possessing a lot of symmetry, even with huge numbers of variables (4 million in one case).

But it seems much harder to make it work for even quite low-dimensional continuous distributions without symmetry.

1.9.2 The Rejection Coupler

Here is a simple approach to CFTP for a continuous state space, namely the unit interval, from Murdoch and Green (1998). It is more of a 'proof of existence' (of a CFTP method in a continuous state space) than a practical

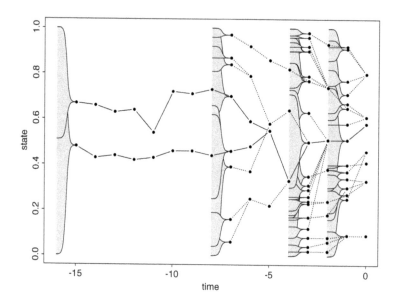

Figure 1.26 *Example realisation of rejection coupler for* $f(y|x) = (\delta + 1)\min(\{y/x\}^{\delta}, \{(1-y)/(1-x)\}^{\delta})$ *with* $\delta = 6$.

method, for we have to suppose we know $f(y|x)$, where

$$P\{X_{t+1} \leq y | X_t = x\} \propto \int_{-\infty}^{y} f(u|x)du,$$

and that the (not necessarily normalised) densities $f(y|x)$ are bounded above by an integrable $h(y)$. We cannot expect the transition density to be available for a practically-useful MCMC method.

Recall the familiar rejection sampler, expressed in geometrical terms. To sample from the (not necessarily normalised) $f(\cdot|x)$, we repeatedly draw (Y, Z) uniformly under the graph of h until $Z < f(Y|x)$. The rejection coupler generalises this scheme. To sample from $f(\cdot|x)$ for all x, again we repeatedly draw (Y_i, Z_i) uniformly under the graph of h. Let $A_i = \{x : Z_i < f(Y_i|x)\}$; then Y_i is a valid update for all $x \in A_i$. We continue until $\cup_i A_i = \chi$, obtaining a random-length list $\{Y_i\}$.

When incorporated into the CFTP protocol, this procedure gives partial realisations of a continuum of coupled paths exemplified by the simulation shown in Figure 1.26. This shows a single realisation of CFTP using rejection coupling, for the kernel density $f(y|x) = (1 + \delta)\min(\{y/x\}^{\delta}, \{(1 - y)/(1 - x)\}^{\delta})$ with $\delta = 6$, which is bounded above by the envelope function

$h(y) = 1 + \delta$ for $0 < y < 1$. The solid lines indicate the paths ultimately followed by all realisations starting from the indefinite past.

1.9.3 Towards generic methods for Bayesian statistics

In contrast to the rejection coupler, a practical technique for Bayesian CFTP should be based only on the target distribution, and created by some generic recipe, just as is the case for standard MCMC.

Evidence that this will become possible is still quite unconvincing, although this is an extremely active research area, and success may be obtained soon. Some experiments in this direction are the random walk Metropolis coupler of Green and Murdoch (1998), the methods using Gibbs sampling and (anti-) monotonicity of Møller (1999), perfect slice sampling (Mira, Møller and Roberts, 1999) and the perfect simulated tempering approach of Møller and Nicholls (1999).

One general reason for pessimism about the future of CFTP in Bayesian statistics is found by noting that much of the success of ordinary MCMC in this field is based on its modularity: as a model is elaborated, the parameter vector is augmented, and the current sampler is supplemented by new moves for the new components. Existing methods for perfect simulation are not modular.

1.10 Miscellaneous topics

1.10.1 Diffusion methods

A number of MCMC methods have been developed recently, inspired by the Langevin stochastic differential equation

$$dx_t = dB_t + \frac{1}{2}\nabla \log \pi(x_t)dt$$

where B_t denotes Brownian motion on $\mathcal{X}$. Here we describe only the case where $\mathcal{X} = \mathcal{R}$. This diffusion has invariant distribution π, and suggests use of the discrete-time chain

$$x^{(t+\varepsilon)}|x^{(\leq t)} \sim N\left(x^{(t)} + \frac{1}{2}\varepsilon\nabla \log \pi(x^{(t)}), \varepsilon\right), \qquad (1.15)$$

where the time increment is ε, not 1.

Unfortunately, this simple discretisation is too crude: not only does it only, at best, deliver an approximation to π as its invariant distribution, it can actually create a transient chain!

However, that can be fixed by using the *Metropolis-adjusted* Langevin algorithm (Besag, 1994), in which (1.15) is used simply as a proposal distribution, with acceptance determined as usual by (1.7).

Among examples of practical methodology using diffusion-based MCMC are the jump-diffusion methods of Grenander and Miller (1994), combining (unadjusted) Langevin diffusion with dimension-jumping moves to address variable-dimension problems, and the work of Phillips and Smith (1996) applying this approach in various statistical settings.

The Metropolis-adjusted Langevin method is known to fail to be geometrically ergodic for heavy-tailed targets, a problem addressed by the richer class of 'self-targetting' Metropolis-adjusted Langevin algorithms due to Stramer and Tweedie (1998), in which the proposal distribution is

$$\boldsymbol{x}^{(t+\varepsilon)}|\boldsymbol{x}^{(\leq t)} \sim N\left(\boldsymbol{x}^{(t)} + \varepsilon\mu(\boldsymbol{x}^{(t)}), \varepsilon\sigma^2(\boldsymbol{x}^{(t)})\right)$$

where

$$\mu(\boldsymbol{x}) = \frac{1}{2}\sigma^2(\boldsymbol{x})\nabla\log\pi(\boldsymbol{x}) + \sigma(\boldsymbol{x})\nabla\sigma(\boldsymbol{x}).$$

This is derived from the diffusion generated by

$$d\boldsymbol{x}_t = \sigma(\boldsymbol{x}_t)d\boldsymbol{B}_t + \mu(\boldsymbol{x}_t)dt.$$

Stramer and Tweedie (1998) discuss the extent to which their theory for these methods can be extended to the practically important cases where $X = \mathcal{R}^d, d > 1$.

1.10.2 Sensitivity analysis via MCMC

In responsible Bayesian inference, it is important to assess the effect on the posterior of changes to the model, especially variations in the prior. Suppose that, having completed a MCMC-based analysis using a prior $\pi_0(\boldsymbol{\theta})$ and likelihood $f(Y|\boldsymbol{\theta})$, we wish to entertain an alternative model built from $\pi_0^*(\boldsymbol{\theta})$ and $f^*(Y|\boldsymbol{\theta})$.

We could just repeat MCMC computation on the new model: note that even where the base model is rather tractable (for example, $\pi_0(\boldsymbol{\theta})$ conjugate to $f(Y|\boldsymbol{\theta})$), we should consider alternatives that are not. Thus MCMC may be needed in sensitivity analysis even where exact analytic calculation handles the standard model, or we may need Metropolis where Gibbs was sufficient in the standard case.

As an alternative to treating the revised model as a completely fresh problem, we may be able to make use of importance sampling to assess sensitivity using only the original simulation. This uses the importance sampling identity

$$E_{\pi^*}(g) = E_\pi\left(\frac{\pi^*}{\pi}g\right)$$

showing that expectations under π^* can be estimated from an MCMC run aimed at π, by

$$\frac{\sum_{t=1}^N w(\boldsymbol{x}^{(t)})g(\boldsymbol{x}^{(t)})}{\sum_{t=1}^N w(\boldsymbol{x}^{(t)})}$$

where

$$w(\boldsymbol{x}) \propto \frac{\pi^*(\boldsymbol{x})}{\pi(\boldsymbol{x})}.$$

There are several practical examples of MCMC-based sensitivity analysis in Besag, et al. (1995).

One problem of the importance sampling approach is that, except in very low dimensional problems or where π and π^* are very similar, $w(\boldsymbol{x}^{(t)})/\sum_t w(\cdot)$ is effectively concentrated on very few samples, implying very poor efficiency. This can sometimes be mitigated by considering infinitesimal perturbations instead: $\pi_\varepsilon(\boldsymbol{x}) \propto (\pi(\boldsymbol{x}))^{(1-\varepsilon)}(\pi^*(\boldsymbol{x}))^\varepsilon$, or, of course, by running another chain.

1.10.3 Bayes with a loss function

We have seen the tremendous advantages that MCMC offers to the practising Bayesian through the opportunities it gives for computing posterior distributions. However, the complete Bayesian agenda for statistical analysis does not stop at computing posteriors — in the full decision theoretic framework, a loss function is introduced, and optimal Bayes estimates and decisions determined by minimising the expectation of the loss under the posterior distribution.

Writing the posterior $p(\boldsymbol{x} \in \cdot | Y)$ in the generic $\pi(\cdot)$ notation, we wish to choose an action $\boldsymbol{z}$ to minimise

$$E(L(\boldsymbol{x}, \boldsymbol{z})|Y) = \int L(\boldsymbol{x}, \boldsymbol{z})\pi(d\boldsymbol{x}),$$

where $L(\boldsymbol{x}, \boldsymbol{z})$ is the loss incurred through taking action $\boldsymbol{z}$ when the true state of nature is $\boldsymbol{x}$.

When the posterior is computed using MCMC, the expectation is replaced by an empirical average over the realisation $\boldsymbol{x}^{(1)}, \boldsymbol{x}^{(2)}, \ldots$:

$$E(L(\boldsymbol{x}, \boldsymbol{z})|Y) \approx \frac{1}{N} \sum_{t=1}^{N} L(\boldsymbol{x}^{(t)}, \boldsymbol{z}).$$

The difficulty with this approach lies in the interplay between the averaging and the optimisation with respect to $\boldsymbol{z}$.

One class of loss functions that can be easily handled is that of finite sums of separable loss functions, where

$$L(\boldsymbol{x}, \boldsymbol{z}) = \sum_r a_r(\boldsymbol{x})b_r(\boldsymbol{z}). \tag{1.16}$$

Then the MCMC computation and the optimisation separate, since

$$E(L(\boldsymbol{x}, \boldsymbol{z})|Y) \approx \sum_r A_r b_r(\boldsymbol{z}) \quad \text{where} \quad A_r = \frac{1}{N} \sum_{t=1}^{N} a_r(\boldsymbol{x}^{(t)}).$$

The optimisation can even sometimes be done analytically, as for example in the elementary case of squared-error loss for some functional $g(\boldsymbol{x})$: $L(\boldsymbol{x}, z) = (g(\boldsymbol{x}) - z)^2$. Then $A_r, r = 0, 1, 2$ are the 0th, 1st and 2nd empirical moments of g, and $b_r(z) = z^2, -2z, 1$ for $r = 0, 1, 2$, leading of course to the optimal $z = A_1/A_0$, the MCMC estimate of the posterior mean.

More commonly, numerical optimisation is necessary. Several research papers recently (for example, Rue and Hurn (1999), Rue and Syversveen (1998)) have used simulated annealing, and have exploited the representation (1.16). The A_r are first computed by MCMC, and then a second, annealing, simulation, set up for the artificial probability distributions

$$p_T(z) \propto \exp\left(\frac{-1}{T} \sum_r A_r b_r(z)\right),$$

where the temperature T is sent to 0 on some suitable schedule.

1.11 Some notes on programming MCMC

1.11.1 The BUGS software

The only software I am aware of that provides MCMC computation for a wide range of statistical models, without requiring the user to code the sampling algorithms is BUGS (Gilks, Thomas and Spiegelhalter, 1994). The model is specified in a high-level specification language – or, if using the WINBUGS version, a graphical interface – a few options controlling the simulation are entered, and the system does the rest. Particularly for rather well-understood standard models, hierarchical versions of generalised linear models, for example, the facilities are easy to use, and the system extremely effective. The suite of implemented and documented examples distributed as part of the software release demonstrates the remarkable flexibility of the system, and some very complex models can be handled. However, facilities for using anything other than single-variable Gibbs sampling are very limited, so for some models, BUGS may be inefficient or even completely incapable. The authors are careful to stress that setting up a MCMC sampler and interpreting the output, even with BUGS, requires knowledge of the user beyond appreciation of the statistical model itself.

BUGS is very useful for many practitioners, but may be too limited for most MCMC researchers.

1.11.2 Your own code

Programming your own MCMC method from scratch is much less daunting than it might first appear, and provides flexibility and (run-time) efficiency that cannot really be matched by package software. The basic recipes are simple in structure, and may be coded following the algebraic notation almost exactly. I do not usually bother with Gibbs sampling unless the full conditionals are entirely standard distributions for which I have a random number generator. (The adaptive rejection sampling method of Gilks and Wild (1992) provides a means of extending Gibbs sampling to a wider range of full conditionals.) In Metropolis-Hastings algorithms, it is necessary to take some care with floating point arithmetic, as in complex models, there may be many multiplicative factors entering the acceptance ratio, with a wide numerical range. I find it convenient to accumulate the sum of the logarithms of the factors, and then truncate onto a safe range before exponentiating.

High and low level languages

The poor performance of looping code in most high-level statistical languages such as S precludes their use in coding all but the smallest problems. I always use Fortran or C. On the other hand, the flexibility of control and the availability of a wide range of statistical and graphical procedures in S, and similar systems, is absolutely invaluable in analysis of MCMC in a research environment. My usual strategy is to dump large quantities of raw MCMC output into a collection of files, with structured filenames, and then employ a suite of S functions to read, display and analyse these.

Validating your code

It is absolutely essential to check and double-check MCMC code. The very nature of the output of the computation — simulation results in a context where other numerical methods are not available for cross-checking — makes this problematical, especially in Bayesian statistical contexts, where problems are one-off, and data subject to sampling variation. Testing on simulated data-sets with known parameter values does not tell you very much.

I find two particular tricks extremely useful. First, I always use restartable random number streams, so that I can conveniently and reliably duplicate a run, with additional diagnostic output, if I suspect a bug. This is also often useful to compare results before and after a minor edit. Second, in programs implementing posterior simulation for a Bayesian model in which the variables are organised as a directed acyclic graph, and in which the observed data have no 'children', I always include a 'prior' option, which ignores the data and the likelihood terms. The posterior simulation program, largely unaltered, is then actually simulating from the prior distribution, and typically many marginal and conditional aspects thereof may be directly checked as the true distributions are known.

Many useful hints on the practical details of algorithm design, including matters such as thinning and burn-in, will be found in Geyer (1992) and Gilks and Roberts (1996).

1.12 Conclusions

1.12.1 Some strengths of MCMC

MCMC is evidently a very powerful and flexible tool for computation with complex multivariate distributions. Its availability has transformed practical Bayesian statistics, and it is making an important if less dramatic impact on other areas of computational statistics.

In the Bayesian context, its power derives from the two kinds of flexibility it offers. First there is flexibility in modelling, permitting the analyst to get much closer to his or her understanding of the reality of the process generating the data, and liberating the modelling process from the constraints only imposed for the sake of tractability. A desirable by-product is the encouragement to model builders to think in graphical terms, as MCMC is particularly well-adapted for models defined on sparse graphs.

Secondly, there is freedom in inference; in principle, there are no limits to what features of the target distribution may be estimated by MCMC, although one needs always to be aware of the Monte Carlo errors unavoidable in such estimates: some features of the target can be computed much more reliably than others. MCMC addresses questions only posed after simulation completed (e.g. ranking and selection) and offers opportunities for simultaneous inference. It allows and even facilitates sensitivity analysis, and addresses questions of model comparison, criticism and choice.

1.12.2 Some weaknesses and dangers

MCMC is not a panacea. In the end it is only a numerical method, and does not displace the need for careful thought about modelling, and about the probable reliability of numerical results obtained in the given context. When other methods are available, MCMC can be relatively extremely expensive; hence the common preference in fields with large data-sets such as signal and image processing for approximations to the full Bayesian paradigm that are amenable to fast numerical calculations for particular outputs of interest.

In qualitative terms, a problem that is insurmountable (at least in estimating expectations and probabilities) is the order $\sqrt{N}$ precision of any simulation method, and for MCMC, the possibility of slow convergence, especially when it is not diagnosable.

Use of MCMC imposes serious responsibilities on the careful researcher, for there is the risk that fitting technology runs ahead of statistical science, so that models are fitted that are not understood, and the risk of overusing the flexibility allowed in inference, leading to undisciplined, selective presentation of posterior information.

1.12.3 Some important lines of continuing research

MCMC remains an important, exciting and challenging field for further research. It is impossible to predict how the field will develop over the next few years, but I believe that the most interesting questions for exploration at present include:

a. Adaptive methods, and other possibilities for automation;

b. Perfect simulation: will this become useful in statistical practice?

c. Getting quantitative results from theoretical analysis;

d. Learning even more from physics.

Acknowledgments

I have presented tutorial lectures on MCMC on several occasions over the last 5 years, to different kinds of audience. Most of these have been in partnership with David Spiegelhalter, who in addition to his wonderful expository powers has enlivened the lecture series with biomedical applications and with material on the BUGS software for Gibbs sampling. I am indebted to him.

I am also grateful to all my collaborators on MCMC research problems: Robert Aykroyd, Julian Besag, Steve Brooks, Carmen Fernández, Arnoldo Frigessi, Paolo Giudici, Xiao-liang Han, Miles Harkness, David Higdon, Matthew Hodgson, Kerrie Mengersen, Antonietta Mira, Duncan Murdoch, Agostino Nobile, Marco Pievatolo, Sylvia Richardson, Claudia Tarantola, and Iain Weir.

Some of this work was supported by the EPSRC.

1.13 References and further reading

Aykroyd, R. G. and Green, P. J. (1991). Global and local priors, and the location of lesions using gamma-camera imagery. *Philosophical Transactions of the Royal Sociecty of London*, A, **337**, 323–342.

Barone, P. and Frigessi, A. (1989). Improving stochastic relaxation for Gaussian random fields. *Probability in the Engineering and Informational Sciences*, **4**, 369–389.

Barone, P., Sebastiani, G. and Stander, J. (1998) Over-relaxation methods and Metropolis-Hastings coupled Markov chains for Monte Carlo simulation. Technical Report, University of Plymouth.

Bernardo, J. M. and Smith, A. F. M. (1994). *Bayesian Theory*, John Wiley, Chichester, United Kingdom.

Besag, J. (1974). Spatial interaction and the statistical analysis of lattice systems (with discussion). *Journal of the Royal Statistical Society*, B, **36**, 192–236.

Besag, J. (1994). Contribution to the discussion of paper by Grenander and Miller (1994). *Journal of the Royal Statistical Society*, B, **56**, 591–592.

Besag, J. and Green, P. J. (1993). Spatial statistics and Bayesian computation (with discussion). *Journal of the Royal Statistical Society*, B, **55**, 25–37.

Besag, J., Green, P. J., Higdon, D. and Mengersen, K. (1995). Bayesian computation and stochastic systems (with discussion), *Statistical Science*, **10**, 3–66.

Besag, J. and York, J. C. (1989). Bayesian restoration of images. In *Analysis of Statistical Information*, (T. Matsunawa, ed.), pp. 491–507. Inst. Statist. Math., Tokyo.

Brooks, S. P. (1998). Markov chain Monte Carlo method and its application. *The Statistician*, **47**, 69–100.

Brooks, S. P. and Gelman, A. (1998). General methods for monitoring convergence of iterative simulations. *Journal of Computational and Graphical Statistics*, **7**, 434–455.

Carlin, B. P. and Chib, S. (1995). Bayesian model choice via Markov chain Monte Carlo methods. *Journal of the Royal Statistical Society*, B, **57**, 473–484.

Clifford, P. (1990). Markov random fields in statistics. In *Disorder in Physical Systems* (G. R. Grimmett and D. J. A. Welsh, eds.), pp. 19–32. Clarendon Press, Oxford.

Creutz, M. (1979). Confinement and the critical dimensionality of space-time. *Physical Review Letters*, **43**, 553–556.

Damien, P., Wakefield, J. and Walker, S. (1999). Gibbs sampling for Bayesian non-conjugate and hierarchical models by using auxiliary variables. *Journal of the Royal Statistical Society*, B, **61**, 331–344.

Denison, D. G. T., Mallick, B. K. and Smith, A. F. M. (1998). Automatic Bayesian curve fitting. *Journal of the Royal Statistical Society*, B, **60**, 333–350.

Edwards, R. G. and Sokal, A. D. (1988). Generalization of the Fortuin-Kastelyn-Swendsen-Wang representation and Monte Carlo algorithm. *Physical Review*, D, **38**, 2009–2012.

Fill, J. A. (1998). An interruptible algorithm for exact sampling via Markov chains. *Annals of Applied Probability*, **8**, 131–162.

Frigessi, A., Martinelli, F. and Stander, J. (1997). Computational complexity of Markov chain Monte Carlo methods for finite Markov random fields. *Biometrika*, **84**, 1–18.

Gelfand, A. E., Sahu, S. K. and Carlin, B. P. (1995). Efficient parametrisations for normal linear mixed models. *Biometrika* **82**, 479–488.

Gelfand, A. E. and Smith, A. F. M. (1990). Sampling-based approaches to calculating marginal densities. *Journal of the American Statistical Association*, **85**, 398–409.

Gelman, A., Roberts, G. O. and Gilks, W. R. (1996). Efficient Metropolis jumping rules. In *Bayesian Statistics 5*, (J. M. Bernardo, J. O. Berger, A. P. Dawid and A. F. M. Smith, eds.), pp. 599–607. Clarendon Press, Oxford.

Geman, S. and Geman, D. (1984). Stochastic relaxation, Gibbs distributions and the Bayesian restoration of images. *IEEE Transactions on Pattern Analysis and Machine Intelligence*, **6**, 721–741.

George, E. I. and McCulloch, R. E. (1997). Approaches for Bayesian variable selection. *Statistica Sinica*, **7**, 339–373.

Geyer, C. J. (1991). Markov chain Monte Carlo maximum likelihood. In *Computing Science and Statistics: Proceedings of the 23rd Symposium on the Interface* (E. M. Keramidas, ed.), pp. 156–163. Interface Foundation, Fairfax Station, Virginia.

Geyer, C. J. (1992). Practical Markov chain Monte Carlo (with discussion). *Statistical Science*, **8**, 473–483.

Geyer, C. J. and Møller, J. (1994). Simulation procedures and likelihood inference for spatial point processes. *Scandinavian Journal of Statistics*, **21**, 359–373.

Geyer, C. J. and Thompson, E. A. (1995). Annealing Markov chain Monte Carlo with applications to ancestral inference. *Journal of the American Statistical Association*, **90**, 909–920.

Gilks, W. R. and Wild, P. (1992). Adaptive rejection sampling for Gibbs sampling. *Applied Statistics*, **41**, 337–348.

Gilks, W. R., Clayton, D. G., Spiegelhalter, D. J., Best, N. G., McNeil, A. J., Sharples, L. D. and Kirby, A. J. (1993). Modelling complexity: applications of Gibbs sampling in medicine (with discussion). *Journal of the Royal Statistical Society*, B, **55**, 39–52.

Gilks, W. R., Richardson, S. and Spiegelhalter, D. J. (eds.) (1996). *Markov Chain Monte Carlo in Practice*, Chapman and Hall, London.

Gilks, W. R. and Roberts, G. O. (1996). Strategies for improving MCMC. In *Practical Markov Chain Monte Carlo* (W. R. Gilks, S. Richardson and D. J. Spiegelhalter, eds.), pp. 89–114. Chapman and Hall, London.

Gilks, W. R., Thomas, A. and Spiegelhalter, D. J. (1994). A language and program for complex Bayesian modelling. *The Statistician*, **43**, 169–178.

Giudici, P. and Green, P. J. (1999). Decomposable graphical Gaussian model determination. *Biometrika*, **86**, 785–801.

Gordon, M. I. and Lifšic, B. A. (1978) The central limit theorem for stationary Markov processes. *Soviet Math. Doklady*, **19**, 392–394.

Green, P. J. (1994). Contribution to the discussion of paper by Grenander and Miller (1994). *Journal of the Royal Statistical Society*, B, **56**, 589–590.

Green, P. J. (1995). Reversible jump Markov chain Monte Carlo computation and Bayesian model determination. *Biometrika*, **82**, 711–732.

Green, P. J. and Han, X.-L. (1992). Metropolis methods, Gaussian proposals and antithetic variables. In *Stochastic Models, Statistical Methods and Algorithms in Image Analysis, Lecture Notes in Statistics*, **74**, 142–164, Springer-Verlag, Berlin.

Green, P. J. and Murdoch, D. J. (1998). Exact sampling for Bayesian inference: towards general purpose algorithms. In *Bayesian Statistics 6*, (J. M. Bernardo, J. O. Berger, A. P. Dawid and A. F. M. Smith, eds.), pp. 301–321. Clarendon Press, Oxford.

Green, P. J. and Richardson, S. (1999). *Modelling heterogeneity with and without the Dirichlet process*. Submitted to the Scand. J. Statist.

Grenander, U. and Miller, M. (1994). Representations of knowledge in complex systems (with discussion). *Journal of the Royal Statistical Society*, B, **56**, 549–603.

Hastings, W. K. (1970). Monte Carlo sampling methods using Markov chains and their applications. *Biometrika*, **57**, 97–109.

Heikkinen, J. and Arjas, E. (1998). Nonparametric Bayesian estimation of a spatial Poisson intensity. *Scandinavian Journal of Statistics*, **25**, 435–450.

Hodgson, M. E. A. (1999). A Bayesian restoration of an ion channel signal. *Journal of the Royal Statistical Society*, B, **61**, 95–114.

Hodgson, M. E. A. and Green, P. J. (1999). Bayesian choice among Markov models of ion channels using Markov chain Monte Carlo. *Proceedings of the Royal Society of London*, A, **455**, 3425–3448.

Holmes, C. C. and Mallick, B. K. (1998). Bayesian radial basis functions of variable dimension. *Neural Computation*, **10**, 1217–1233.

Kass, R. E., Tierney, L. and Kadane, J. B. (1988). Asymptotics in Bayesian computation (with discussion). In *Bayesian Statistics 3* (J. M. Bernardo, M. H. DeGroot, D. V. Lindley and A. F. M. Smith, eds.), pp. 261-278. Clarendon Press, Oxford.

Kipnis, C. and Varadhan, S. R. S. (1986). Central limit theorem for additive functionals of reversible Markov processes and applications to simple exclusions. *Communications in Mathematical Physics*, **104**, 1–19.

Lauritzen, S. L. (1996). *Graphical Models*, Clarendon Press, Oxford.

Lauritzen, S. L. and Spiegelhalter, D. J. (1988). Local computations with probabilities on graphical structures and their application to expert systems (with discussion). *Journal of the Royal Statistical Society*, B, **50**, 157–224.

Madigan, D. and Raftery, A.E. (1994). Model selection and accounting for model uncertainty in graphical models using Occam's window. *Journal of the American Statistical Association*, **89**, 1535–1546.

Marinari, E. and Parisi, G. (1992). Simulated tempering: a new Monte Carlo scheme. *Europhysics Letters*, **19**, 451–458.

Mengersen, K. L. and Tweedie, R. L. (1996). Rates of convergence of the Hastings and Metropolis algorithms. *Annals of Statistics*, **24**, 101–121.

Metropolis, N., Rosenbluth, A. W., Rosenbluth, M. N., Teller, A. H. and Teller, E. (1953). Equations of state calculations by fast computing machines. *Journal of Chemical Physics*, **21**, 1087–1091.

Meyn, S. P. and Tweedie, R. L. (1993). *Markov Chains and Stochastic Stability*. Springer-Verlag, New York.

Mira, A. and Geyer, C. J. (1999). Ordering Monte Carlo Markov chains. Technical Report 632, School of Statistics, University of Minnesota.

Mira, A., Møller, J. and Roberts, G. O. (1999). Perfect slice samplers. Technical Report R-99-2020, Department of Mathematical Sciences, Aalborg University, Denmark.

Møller, J. (1999). Perfect simulation of some conditionally-specified models. *Journal of the Royal Statistical Society*, B, **61**, 251–264.

Møller, J. and Nicholls, G. K. (1999). Perfect simulation for sample-based inference. Technical Report R-99-2011, Department of Mathematical Sciences, Aalborg University, Denmark.

Mooley, D.A. (1981). Applicability of the Poisson probability model to the severe cyclonic storms striking the coast around the Bay of Bengal. *Sankhya*, **43** B, 187–197.

Murdoch, D. J. and Green, P. J. (1998). Exact sampling from a continuous state space. *Scandinavian Journal of Statistics*, **25**, 483–502.

Mykland, P., Tierney, L. and Yu, B. (1995). Regeneration in Markov chain samplers. *Journal of the American Statistical Association*, **90**, 233–241.

Neal, R. M. (1997). Markov chain Monte Carlo methods based on 'slicing' the density function. Technical Report 9722, University of Toronto.

Neal, R. M. (1998). Suppressing random walks in Markov chain Monte Carlo using ordered overrelaxation. In *Learning in Graphical Models* (M. I. Jordan, ed.) pp. 205–228, Kluwer Academic Publishers, Dordrecht.

Newton, M. A. and Raftery, A. E. (1994). Approximate Bayesian inference with the weighted likelihood bootstrap (with discussion). *Journal of the Royal Statistical Society*, B, **56**, 3–48.

Nobile, A. and Green, P. J. (2000). Bayesian analysis of factorial experiments by mixture modelling. *Biometrika*, **87**, 15–35.

O'Hagan, A. (1994). *Bayesian Inference (Kendall's Advanced Theory of Statistics, 2 B)*, Edward Arnold, London.

Peskun, P. H. (1973). Optimum Monte-Carlo sampling using Markov chains. *Biometrika*, **60**, 607–612.

Phillips, D. B. and Smith, A. F. M. (1996). Bayesian model comparison via jump diffusions. In *Practical Markov Chain Monte Carlo* (W. R. Gilks, S. Richardson and D. J. Spiegelhalter, eds.), pp. 215–239, Chapman and Hall, London.

Pievatolo, A. and Green, P. J. (1998). Boundary detection through dynamic polygons. *Journal of the Royal Statistical Society*, B, **60**, 609–626.

Priestley, M. (1981). *Spectral Analysis and Time Series*, Academic Press, London.

Propp, J. G. and Wilson, D. B. (1996). Exact sampling with coupled Markov chains and applications to statistical mechanics. *Random Structures and Algorithms*, **9**, 223–252.

Richardson, S. and Green, P. J. (1997). On the Bayesian analysis of mixtures with an unknown number of components (with discussion). *Journal of the Royal Statistical Society*, B, **59**, 731–792.

Roberts, G. O. and Rosenthal, J. S. (1999). Convergence of slice sampler Markov chains. *Journal of the Royal Statistical Society*, B, **61**, 643–660.

Roberts, G. O. and Sahu, S. K. (1997). Updating schemes, correlation structure, blocking and parameterization for the Gibbs sampler. *Journal of the Royal Statistical Society*, B, **59**, 291–317.

Roberts, G. O. and Tweedie, R. L. (1996). Geometric convergence and central limit theorems for multidimensional Hastings and Metropolis algorithms. *Biometrika*, **83**, 95–100.

Rue, H. and Hurn, M. A. (1999). Bayesian object identification. *Biometrika*, **86**, 649–660.

Rue, H. and Syversveen, A. R. (1998). Bayesian object recognition using Baddeley's delta loss. *Advances in Applied Probability*, **30**, 64–84.

Smith, A. F. M. and Roberts, G. O. (1993). Bayesian computation via the Gibbs sampler and related Markov chain Monte Carlo methods (with discussion). *Journal of the Royal Statistical Society*, B, **55**, 3–23.

Sokal, A. D. (1989). Monte Carlo methods in statistical mechanics: foundations and new algorithms. *Cours de Troisiéme Cycle de la Physique en Suisse Romande*. Lausanne.

Stramer, O. and Tweedie, R. L. (1998). Langevin-type models II: self-targetting candidates for MCMC algorithms. *Methodology and Computing in Applied Probability*, **1**, 307–328.

Suomela, P. (1976). Unpublished Ph.D. thesis. University of Jyväskylä, Finland.

Swendsen, R. H. and Wang, J.-S. (1987). Non-universal critical dynamics in Monte Carlo simulations. *Physical Review Letters*, **58**, 86–88.

Tierney, L. (1994). Markov chains for exploring posterior distributions, *Annals of Statistics*, **22**, 1701–1762.

Tierney, L. (1996). Introduction to general state-space Markov chain theory. In *Practical Markov Chain Monte Carlo* (W. R. Gilks, S. Richardson and D. J. Spiegelhalter, eds.), pp. 59–74, Chapman and Hall, London.

Tierney, L. (1998). A note on Metropolis-Hastings kernels for general state spaces. *Annals of Applied Probability*, **8**, 1–9.

Uimari, P. and Hoeschele, I. (1997). Mapping linked quantitative trait loci using Bayesian analysis and Markov chain Monte Carlo. *Genetics*, **146**, 734–743.

CHAPTER 2

Causal Inference from Graphical Models

Steffen L. Lauritzen
Aalborg University

2.1 Introduction

The introduction of Bayesian networks (Pearl 1986b) and associated local computation algorithms (Lauritzen and Spiegelhalter 1988, Shenoy and Shafer 1990, Jensen, Lauritzen and Olesen 1990) has initiated a renewed interest for understanding causal concepts in connection with modelling complex stochastic systems.

It has become clear that graphical models, in particular those based upon directed acyclic graphs, have natural causal interpretations and thus form a base for a language in which causal concepts can be discussed and analysed in precise terms.

As a consequence there has been an explosion of writings, not primarily within mainstream statistical literature, concerned with the exploitation of this language to clarify and extend causal concepts. Among these we mention in particular books by Spirtes, Glymour and Scheines (1993), Shafer (1996), and Pearl (2000) as well as the collection of papers in Glymour and Cooper (1999).

Very briefly, but fundamentally, the important distinction to be made is the distinction between two types of conditional probability. We refer to these as conditioning by *intervention* and conditioning by *observation* and suggest the notation

$$p(x \,\|\, y) = P(X = x \,|\, Y \leftarrow y), \quad p(x \,|\, y) = P(X = x \,|\, Y = y)$$

for these two notions. Other authors have used expressions such as $do(y)$, $Y = \hat{y}$, and $set(Y = y)$ to denote intervention conditioning.

Much existing controversy and lack of clarity is due to the misconception that these two are identical or even related in a simple fashion although the distinction has also been made both properly, clearly, and explicitly in better expositions of regression, see for example Box (1966) or Section 3.3 of Cox (1984).

In the following, we try to develop the basic ideas needed to make this distinction precise and discuss a number of classical statistical problems where the distinction is important.

There are many important aspects and views of causality and causal inference which are not even touched upon here, as we are only concerned with one particular such aspect: the prediction of the effect of interventions in a given system.

The material is organized as follows. Section 2.2 introduces the necessary graph-terminology. The next three sections are concerned with the very basic elements of graphical models, conditional independence and Markov properties for undirected and directed graphs.

Section 2.6 introduces the notion of a causal Markov field and associated intervention probabilities. The next sections are concerned with the exploitation of this idea in a number of important cases.

We conclude by discussing structural equation models and methods based upon using counterfactual variables or potential responses, and finally give a brief discussion of other issues which are not treated *per se* here.

While writing, I have in particular exploited Pearl (1995a) and Robins (1997).

2.2 Graph terminology

This section introduces some necessary graph terminology. We are basically following the terminology used in Cowell, Dawid, Lauritzen and Spiegel-halter (1999) which is almost identical to that in Lauritzen (1996).

We define a *graph* $\mathcal{G}$ to be a pair $\mathcal{G} = (V, E)$, where V is a finite set of *vertices*, also called *nodes*, of $\mathcal{G}$, and E is a subset of the set $V \times V$ of ordered pairs of vertices, called the *edges* or *links* of $\mathcal{G}$. Thus, as E is a set, the graph $\mathcal{G}$ has no multiple edges. We further require that E consist of pairs of distinct vertices, so that there are no loops.

If both ordered pairs (α, β) and (β, α) belong to E, we say that we have an *undirected* edge between α and β, and write $\alpha \sim \beta$; we also say that α and β are *neighbours*, α is a neighbour of β, or β is a neighbour of α. The set of neighbours of a vertex β is denoted by ne(β).

If $(\alpha, \beta) \in E$ but $(\beta, \alpha) \notin E$ we call the edge *directed*, and write $\alpha \to \beta$. We also say that α is a *parent* of β, and that β is a *child* of α. The set of parents of a vertex β is denoted by pa(β), and the set of children of a vertex α by ch(α). The *family* of β, denoted fa(β), is fa(β) = $\{\beta\} \cup$ pa(β).

If $(\alpha, \beta) \in E$ or $(\beta, \alpha) \in E$ we say that α and β are *joined*. Then $\alpha \nsim \beta$ indicates that α and β are not joined, i.e. both $(\alpha, \beta) \notin E$ and $(\beta, \alpha) \notin E$. We also write $\alpha \nrightarrow \beta$ if $(\alpha, \beta) \notin E$.

If $A \subset V$, the expressions pa(A), ne(A) and ch(A) will denote the collection of parents, children and neighbours, respectively, of the elements of A, but excluding any element in A:

$$\begin{aligned}
\text{pa}(A) &= \bigcup_{\alpha \in A} \text{pa}(\alpha) \setminus A, \\
\text{ne}(A) &= \bigcup_{\alpha \in A} \text{ne}(\alpha) \setminus A, \\
\text{ch}(A) &= \bigcup_{\alpha \in A} \text{ch}(\alpha) \setminus A.
\end{aligned}$$

If all the edges of a graph are directed, we say that it is a *directed graph*. Conversely, if all the edges of a graph are undirected, we say that it is an *undirected graph*.

The *boundary* bd(α) of a vertex α is the set of parents and neighbours of α; the boundary bd(A) of a subset $A \subset V$ is the set of vertices in $V \setminus A$ that are parents or neighbours to vertices in A, i.e. bd(A) = pa $A \cup$ ne A. The *closure* of A is cl(A) = $A \cup$ bd(A).

The *undirected version* $\mathcal{G}^{\sim}$ of a graph $\mathcal{G}$ is the undirected graph obtained by replacing the directed edges of $\mathcal{G}$ by undirected edges.

We call $\mathcal{G}_A = (A, E_A)$ a *subgraph* of $\mathcal{G} = (V, E)$ if $A \subseteq V$ and $E_A \subseteq E \cap (A \times A)$. Thus it may contain the same vertex set but possibly fewer edges. If in addition $E_A = E \cap (A \times A)$, we say that $\mathcal{G}_A$ is the subgraph of $\mathcal{G}$ *induced* by the vertex set A.

A graph is called *complete* if every pair of vertices are joined. We say that a subset of vertices of $\mathcal{G}$ is *complete* if it induces a complete subgraph. A complete subgraph which is maximal (with respect to $\subseteq$) is called a *clique*.

A *path* of length n from α to β is a sequence $\alpha = \alpha_0, \ldots, \alpha_n = \beta$ of distinct vertices such that $(\alpha_{i-1}, \alpha_i) \in E$ for all $i = 1, \ldots, n$. Thus a path can never cross itself and moving along a path never goes against the directions of arrows.

A *cycle* of length n is a path with the modification that the first and last vertex are identical $\alpha_0 = \alpha_n$. The cycle is *directed* if it contains at least one arrow.

A directed graph which contains no cycles is called a *directed acyclic graph*, or DAG.

A *trail* of length n from α to β is a sequence $\alpha = \alpha_0, \ldots, \alpha_n = \beta$ of distinct vertices such that $\alpha_{i-1} \to \alpha_i$, or $\alpha_i \to \alpha_{i-1}$, or $\alpha_{i-1} \sim \alpha_i$ for all $i = 1, \ldots, n$. Thus, moving along a trail could go against the direction of the arrows, in contrast to the case of a path. In other words, a trail in $\mathcal{G}$ is a sequence of vertices that form a path in the undirected version $\mathcal{G}^{\sim}$ of $\mathcal{G}$.

It is always possible to *well order* the nodes of a DAG, by a linear ordering or numbering, such that if two nodes are connected the edge points from the lower to the higher of the two nodes with respect to the ordering.

Given a directed acyclic graph, the set of its vertices α such that $\alpha \mapsto \beta$ but not $\beta \mapsto \alpha$ are the *ancestors* an(β) of β and the *descendants* de(α) of α are the vertices β such that $\alpha \mapsto \beta$ but not $\beta \mapsto \alpha$. The *nondescendants*

$\mathrm{nd}(\alpha)$ of α is the set $V \setminus (\mathrm{de}(\alpha) \cup \alpha)$. If $\mathrm{pa}(\alpha) \subseteq A$ for all $\alpha \in A$ we say that A is an *ancestral* set. The symbol $\mathrm{An}(A)$ denotes the smallest ancestral set containing A.

A subset $C \subseteq V$ is said to be an (α, β)-*separator* if all trails from α to β intersect C. The subset C is said to *separate* A from B if it is an (α, β)-separator for every $\alpha \in A$ and $\beta \in B$. An (α, β)-separator C is said to be *minimal* if no proper subset of C is itself an (α, β)-separator.

For a directed acyclic graph $\mathcal{D}$, we define the *moral graph* of $\mathcal{D}$ to be the undirected graph $\mathcal{D}^m$ obtained from $\mathcal{D}$ by first adding undirected edges between all pairs of vertices which have common children and are not already joined, and then forming the undirected version of the resulting graph.

2.3 Conditional independence

Throughout this text a central notion is that of conditional independence of random variables and groups of these.

We are concerned with the situation where we have a collection of random variables $(X_\alpha)_{\alpha \in V}$ taking values in probability spaces $(\mathcal{X}_\alpha)_{\alpha \in V}$. The probability spaces are either real finite-dimensional vector spaces or finite and discrete sets but could be quite general, just sufficiently well-behaved to ensure the existence of conditional probabilities. For simplicity we mostly consider the discrete case.

For A being a subset of V we let $\mathcal{X}_A = \times_{\alpha \in A} \mathcal{X}_\alpha$ and further $\mathcal{X} = \mathcal{X}_V$. Typical elements of $\mathcal{X}_A$ are denoted as $x_A = (x_\alpha)_{\alpha \in A}$. Similarly $X_A = (X_\alpha)_{\alpha \in A}$.

Formally, if X, Y, Z are random variables with a joint distribution P, we say that X *is conditionally independent of Y given Z under P*, and write $X \perp\!\!\!\perp Y \mid Z$ $[P]$, if, for any measurable set A in the sample space of X, there exists a version of the conditional probability $P(A \mid Y, Z)$ which is a function of Z alone. Usually P will be fixed and omitted from the notation. If Z is trivial we say that X *is independent of Y*, and write $X \perp\!\!\!\perp Y$.

When X, Y, and Z are discrete random variables the condition for $X \perp\!\!\!\perp Y \mid Z$ simplifies as

$$P(X = x, Y = y \mid Z = z) = P(X = x \mid Z = z)P(Y = y \mid Z = z),$$

where the equation holds for all z with $P(Z = z) > 0$. When the three variables admit a joint density with respect to a product measure μ, we have

$$X \perp\!\!\!\perp Y \mid Z \iff f_{XY \mid Z}(x, y \mid z) = f_{X \mid Z}(x \mid z)f_{Y \mid Z}(y \mid z), \qquad (2.1)$$

where this equation is to hold almost surely with respect to P. The condition (2.1) can be rewritten as

$$X \perp\!\!\!\perp Y \mid Z \iff f_{XYZ}(x, y, z)f_Z(z) = f_{XZ}(x, z)f_{YZ}(y, z) \qquad (2.2)$$

and this equality must hold *for all values of z* when the densities are continuous.

The ternary relation $X \perp\!\!\!\perp Y \mid Z$ has the following properties, where h denotes an arbitrary measurable function on the sample space of X:

(C1) if $X \perp\!\!\!\perp Y \mid Z$ then $Y \perp\!\!\!\perp X \mid Z$;

(C2) if $X \perp\!\!\!\perp Y \mid Z$ and $U = h(X)$, then $U \perp\!\!\!\perp Y \mid Z$;

(C3) if $X \perp\!\!\!\perp Y \mid Z$ and $U = h(X)$, then $X \perp\!\!\!\perp Y \mid (Z, U)$;

(C4) if $X \perp\!\!\!\perp Y \mid Z$ and $X \perp\!\!\!\perp W \mid (Y, Z)$, then $X \perp\!\!\!\perp (W, Y) \mid Z$.

Note that the converse to (C4) follows from (C2) and (C3).

If we use f as generic symbol for the probability density of the random variables corresponding to its arguments, the following statements are true:

$$X \perp\!\!\!\perp Y \mid Z \iff f(x, y, z) = f(x, z)f(y, z)/f(z) \tag{2.3}$$

$$X \perp\!\!\!\perp Y \mid Z \iff f(x \mid y, z) = f(x \mid z) \tag{2.4}$$

$$X \perp\!\!\!\perp Y \mid Z \iff f(x, z \mid y) = f(x \mid z)f(z \mid y) \tag{2.5}$$

$$X \perp\!\!\!\perp Y \mid Z \iff f(x, y, z) = h(x, z)k(y, z) \text{ for some } h, k \tag{2.6}$$

$$X \perp\!\!\!\perp Y \mid Z \iff f(x, y, z) = f(x \mid z)f(y, z). \tag{2.7}$$

The equalities above hold apart from a set of triples (x, y, z) with probability zero.

Another property of the conditional independence relation is often used:

(C5) if $X \perp\!\!\!\perp Y \mid Z$ and $X \perp\!\!\!\perp Z \mid Y$ then $X \perp\!\!\!\perp (Y, Z)$.

However (C5) does not hold universally, but only under additional conditions — essentially that there be no non-trivial logical relationship between Y and Z. A trivial counterexample appears when $X = Y = Z$ with $P\{X = 1\} = P\{X = 0\} = 1/2$. We have however

Proposition 2.1 *If the joint density of all variables with respect to a product measure is strictly positive, then the statement* (C5) *will hold true.*

Proof: We assume for simplicity that the variables are discrete with density $f(x, y, z) > 0$ and that $X \perp\!\!\!\perp Y \mid Z$ as well as $X \perp\!\!\!\perp Z \mid Y$. Then (2.6) gives for all values of (x, y, z) that

$$f(x, y, z) = k(x, z)l(y, z) = g(x, y)h(y, z)$$

for suitable strictly positive functions g, h, k, l. Thus we have for all z that

$$g(x, y) = \frac{k(x, z)l(y, z)}{h(y, z)}.$$

Choosing a fixed $z = z_0$ we get $g(x, y) = \pi(x)\rho(y)$ where $\pi(x) = k(x, z_0)$ and $\rho(y) = l(y, z_0)/h(y, z_0)$. Thus $f(x, y, z) = \pi(x)\rho(y)h(y, z)$ and hence $X \perp\!\!\!\perp (Y, Z)$ as desired. $\qquad\square$

In most cases we are specifically interested in conditional independence among groups of random variables such as for example $X_A = (X_\alpha, \alpha \in A)$, where A is a subset of V. We then use the short notation

$$A \perp\!\!\!\perp B \mid C$$

for

$$X_A \perp\!\!\!\perp X_B \mid X_C$$

and so on. We then get the following properties as a consequence of (C1)–(C4):

(C1') if $A \perp\!\!\!\perp B \mid C$ then $B \perp\!\!\!\perp A \mid C$;

(C2') if $A \perp\!\!\!\perp B \mid C$ and $D \subseteq B$, then $A \perp\!\!\!\perp D \mid C$;

(C3') if $A \perp\!\!\!\perp B \mid C$ and $D \subseteq B$, then $A \perp\!\!\!\perp B \mid C \cup D$;

(C4') if $A \perp\!\!\!\perp B \mid C$ and $A \perp\!\!\!\perp D \mid B \cup C$, then $A \perp\!\!\!\perp B \cup D \mid C$.

And similarly the analogue of (C5) is that for disjoint subsets A, B, C, and D, we have

(C5') if $A \perp\!\!\!\perp B \mid C \cup D$ and $A \perp\!\!\!\perp C \mid B \cup D$ then $A \perp\!\!\!\perp B \cup C \mid D$

although (C5') does not hold universally, but only under specific extra assumptions. It holds for example under the assumption that the joint density of the random variables involved is strictly positive.

It is illuminating to think of the properties (C1)–(C5) or in particular their analogues (C1')–(C5') as purely formal expressions, with a meaning that is not necessarily tied to probability. If we interpret the symbols used for random variables as abstract symbols for pieces of knowledge obtained from, say, reading books, and further interpret the symbolic expression $X \perp\!\!\!\perp Y \mid Z$ as:

Knowing Z, reading Y is irrelevant for reading X,

the properties (C1)–(C4) translate to the following:

(I1) if, knowing Z, reading Y is irrelevant for reading X, then so is reading X for reading Y;

(I2) if, knowing Z, reading Y is irrelevant for reading the book X, then reading Y is irrelevant for reading any chapter U of X;

(I3) if, knowing Z, reading Y is irrelevant for reading the book X, it remains irrelevant after having read any chapter U of X;

(I4) if, knowing Z, reading the book Y is irrelevant for reading X and even after having also read Y, reading W is irrelevant for reading X, then reading of both Y and W is irrelevant for reading X.

Thus one can view the relations (C1)–(C4) as pure formal properties of the notion of irrelevance. The property (C5) is slightly more subtle. In a certain sense, also the symmetry (C1) is a somewhat special property of probabilistic conditional independence, rather than general irrelevance.

It is thus tempting to use the relations (C1)–(C4) as formal axioms for conditional independence or irrelevance. A *semi-graphoid* is an algebraic structure which satisfies (C1')–(C4'). If also (C5') holds for disjoint subsets, it is called a *graphoid* (Pearl 1988). Similarly we refer to (C1')–(C4') as the *semi-graphoid axioms* and (C1')–(C5') as the *graphoid axioms*.

2.4 Markov properties for undirected graphs

Conditional independence properties of joint distributions of collections of random variables can be compactly described and expressed as so-called Markov properties for various graphs. In this section we consider the case when the graph is undirected. We refer to Lauritzen (1996) or Cowell et al. (1999) for proofs of all assertions that are not proved here.

Associated with an undirected graph $\mathcal{G} = (V, E)$ and a collection of random variables $(X_\alpha)_{\alpha \in V}$ as above there is a range of different Markov properties. A probability distribution P on $\mathcal{X}$ is said to obey

(P) *the pairwise Markov property*, relative to $\mathcal{G}$, if for any pair (α, β) of non-adjacent vertices

$$\alpha \perp\!\!\!\perp \beta \,|\, V \setminus \{\alpha, \beta\};$$

(L) *the local Markov property*, relative to $\mathcal{G}$, if for any vertex $\alpha \in V$

$$\alpha \perp\!\!\!\perp V \setminus \operatorname{cl}(\alpha) \,|\, \operatorname{bd}(\alpha);$$

(G) *the global Markov property*, relative to $\mathcal{G}$, if for any triple (A, B, S) of disjoint subsets of V such that S separates A from B in $\mathcal{G}$

$$A \perp\!\!\!\perp B \,|\, S.$$

As conditional independence is intimately related to factorization, so are the Markov properties. A probability measure P on $\mathcal{X}$ is said to *factorize* according to $\mathcal{G}$ if for all complete subsets $a \subseteq V$ there exist non-negative functions ψ_a that depend on x through x_a only, and there exists a product measure $\mu = \otimes_{\alpha \in V} \mu_\alpha$ on $\mathcal{X}$, such that P has density f with respect to μ where f has the form

$$f(x) = \prod_{a \text{ complete}} \psi_a(x). \tag{2.8}$$

The functions ψ_a are not uniquely determined. There is arbitrariness in the choice of μ, but also groups of functions ψ_a can be multiplied together or split up in different ways. In fact one can without loss of generality assume — although this is not always practical — that only cliques appear as the sets a, i.e. that

$$f(x) = \prod_{c \in \mathcal{C}} \psi_c(x), \tag{2.9}$$

where $\mathcal{C}$ is the set of cliques of $\mathcal{G}$. If P factorizes, we say that P has property (F). The different Markov properties are related as follows:

Proposition 2.2 *For any undirected graph $\mathcal{G}$ and any probability distribution on $\mathcal{X}$ it holds that*

$$\text{(F)} \implies \text{(G)} \implies \text{(L)} \implies \text{(P)}.$$

Proof: See Lauritzen (1996).   □

For a given graph $\mathcal{G}$ and state space $\mathcal{X} = \times_{\alpha \in V} \mathcal{X}_\alpha$ we denote the set of distributions that satisfy the different Markov properties as $M_F(\mathcal{G})$, $M_G(\mathcal{G})$, $M_L(\mathcal{G})$, and $M_P(\mathcal{G})$. Proposition 2.2 can now be equivalently formulated as

$$M_F(\mathcal{G}) \subseteq M_G(\mathcal{G}) \subseteq M_L(\mathcal{G}) \subseteq M_P(\mathcal{G}).$$

The Markov properties are genuinely different in general, but in the case where P has a positive density it is possible to show that (P) implies (F), and thus that all Markov properties are equivalent. This result has been discovered in various forms by a number of authors (Speed 1979) but is usually attributed to Hammersley and Clifford (1971). More precisely, we have

Theorem 2.3 (Hammersley and Clifford) *A probability distribution P with positive density f with respect to a product measure μ satisfies the pairwise Markov property with respect to an undirected graph $\mathcal{G}$ if and only if it factorizes according to $\mathcal{G}$.*

Proof: See Lauritzen (1996).   □

In fact, if (C5') holds, the global, local, and pairwise Markov properties coincide. This fact is stated in the theorem below, due to Pearl and Paz (1987); see also Pearl (1988).

Theorem 2.4 (Pearl and Paz) *If a probability distribution on $\mathcal{X}$ is such that (C5') holds for disjoint subsets A, B, C, D then*

$$\text{(G)} \iff \text{(L)} \iff \text{(P)}.$$

Proof: See Lauritzen (1996).   □

The global Markov property (G) is important because it gives a general criterion for deciding when two groups of variables A and B are conditionally independent given a third group of variables S. Moreover, it cannot be further strengthened. For example it holds (Frydenberg 1990b) that if all state spaces are binary, i.e. $\mathcal{X}_\alpha = \{1, -1\}$, then

$$A \perp\!\!\!\perp B \mid S \text{ for all } P \in M_F(\mathcal{G}) \iff S \text{ separates } A \text{ from } B.$$

In other words, if A and B are not separated by S then there is a factorizing distribution that makes them conditionally dependent.

2.5 The directed Markov property

We consider the same setup as in the previous section, except that now the graph $\mathcal{D}$ is assumed to be directed and acyclic.

We say that a probability distribution P admits a *recursive factorization* according to $\mathcal{D}$, if there exist (σ-finite) measures μ_α over $\mathcal{X}$ and non-negative functions $k^\alpha(\cdot, \cdot), \alpha \in V$, henceforth referred to as *kernels*, defined on $\mathcal{X}_\alpha \times \mathcal{X}_{\mathrm{pa}(\alpha)}$ such that

$$\int k^\alpha(y_\alpha, x_{\mathrm{pa}(\alpha)})\mu_\alpha(dy_\alpha) = 1$$

and P has density f with respect to the product measure $\mu = \otimes_{\alpha \in V}\mu_\alpha$ given by

$$f(x) = \prod_{\alpha \in V} k^\alpha(x_\alpha, x_{\mathrm{pa}(\alpha)}).$$

We then also say that P *has property* (DF). It is easy to show that, if P admits a recursive factorization as above, then the kernels $k^\alpha(\cdot, x_{\mathrm{pa}(\alpha)})$ are in fact densities for the conditional distribution of X_α, given $X_{\mathrm{pa}(\alpha)} = x_{\mathrm{pa}(\alpha)}$ and thus

$$f(x) = \prod_{\alpha \in V} f(x_\alpha \mid x_{\mathrm{pa}(\alpha)}). \qquad (2.10)$$

We refer to these kernels as the *conditional specifications* for P. It is immediate that if we form the (undirected) moral graph $\mathcal{D}^m$ (see Section 2.2) we have the following:

Lemma 2.5 *If P admits a recursive factorization according to the directed acyclic graph $\mathcal{D}$, it factorizes according to the moral graph $\mathcal{D}^m$ and therefore obeys the global Markov property relative to $\mathcal{D}^m$.*

Proof: The factorization follows from the fact that, by construction, the sets $\{\alpha\} \cup \mathrm{pa}(\alpha)$ are complete in $\mathcal{D}^m$ and we can therefore let $\psi_{\{\alpha\} \cup \mathrm{pa}(\alpha)} = k^\alpha$. $\qquad \square$

This simple lemma has very useful consequences and we shall see several examples of this in the sequel. Also, using the local Markov property on the moral graph $\mathcal{D}^m$ we find that

$$\alpha \perp\!\!\!\perp V \setminus \alpha \mid \mathrm{bl}(\alpha),$$

where $\mathrm{bl}(\alpha)$ is the so-called *Markov blanket* of α. The Markov blanket is the set of neighbours of α in the moral graph $\mathcal{D}^m$. It can be found directly from the original DAG $\mathcal{D}$ as the set of α's parents, children, and children's parents:

$$\mathrm{bl}(\alpha) = \mathrm{pa}(\alpha) \cup \mathrm{ch}(\alpha) \cup \{\beta : \mathrm{ch}(\beta) \cap \mathrm{ch}(\alpha) \neq \emptyset\}. \qquad (2.11)$$

In particular it follows that the so-called full conditionals satisfy

$$\mathcal{L}(X_\alpha \mid X_{V \setminus \alpha}) = \mathcal{L}(X_\alpha \mid X_{\mathrm{bl}(\alpha)})$$

with density given as

$$\mathcal{L}(X_\alpha \,|\, X_{V\setminus\alpha}) = f(x_\alpha \,|\, x_{\mathrm{pa}(\alpha)}) \prod_{\beta \in \mathrm{ch}(\alpha)} f(x_\beta \,|\, x_{\mathrm{pa}(\beta)}).$$

The following result is easily shown:

Proposition 2.6 *If P admits a recursive factorization according to the directed acyclic graph $\mathcal{D}$ and A is an ancestral set, then the marginal distribution P_A admits a recursive factorization according to $\mathcal{D}_A$.*

In combination with Lemma 2.5 this yields:

Corollary 2.7 *Let P factorize recursively according to $\mathcal{D}$. Then*

$$A \perp\!\!\!\perp B \,|\, S$$

whenever A and B are separated by S in $(\mathcal{D}_{\mathrm{An}(A \cup B \cup S)})^m$, the moral graph of the smallest ancestral set containing $A \cup B \cup S$.

Following Lauritzen, Dawid, Larsen and Leimer (1990), the property in Corollary 2.7 will be referred to as the *directed global Markov property* (DG) and a distribution satisfying it is a *directed Markov field* over $\mathcal{D}$.

One can show that the global directed Markov property has the same rôle as the global Markov property does in the case of an undirected graph, in the sense that it gives the sharpest possible rule for reading conditional independence relations off the directed graph. The procedure is illustrated in the following example:

Example 2.8 Consider a directed Markov field on the first graph in Fig. 2.1 and the problem of deciding whether $a \perp\!\!\!\perp b \,|\, S$. The moral graph of the smallest ancestral set containing all the variables involved is shown in the second graph of Fig. 2.1. It is immediate that S separates a from b in this moral graph, implying $a \perp\!\!\!\perp b \,|\, S$.   □

An alternative formulation of the global, directed Markov property was given by Pearl (1986a) with a formal treatment in Verma and Pearl (1990). Recall that a trail in $\mathcal{D}$ is a sequence of vertices that forms a path in the undirected version $\mathcal{D}^\sim$ of $\mathcal{D}$, i.e. when the directions of arrows are ignored. A trail π from a to b in a directed, acyclic graph $\mathcal{D}$ is said to be *blocked* by S if it contains a vertex $\gamma \in \pi$ such that either

$\gamma \in S$ and arrows of π do not meet head-to-head at γ, or

γ and all its descendants are not in S, and arrows of π meet head-to-head at γ.

A trail that is not blocked by S is said to be *active*. Two subsets A and B are said to be *d-separated* by S if all trails from A to B are blocked by S. We then have the following result:

Proposition 2.9 *Let A, B and S be disjoint subsets of a directed, acyclic graph $\mathcal{D}$. Then S d-separates A from B if and only if S separates A from B in $(\mathcal{D}_{\mathrm{An}(A \cup B \cup S)})^m$.*

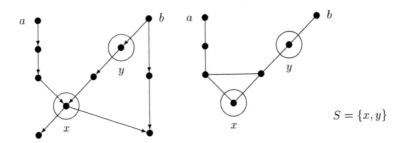

Figure 2.1 *The directed, global Markov property. Is $a \perp\!\!\!\perp b \mid S$? In the moral graph of the smallest ancestral set in the graph containing $\{a\} \cup \{b\} \cup S$, clearly S separates a from b, implying $a \perp\!\!\!\perp b \mid S$.*

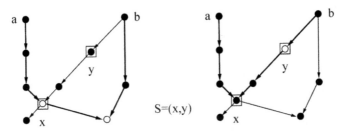

Figure 2.2 *Illustration of Pearl's d-separation criterion. There are two trails from a to b, drawn with thick lines. Both are blocked, but different vertices γ, indicated with open circles, play the rôle of blocking vertices.*

Proof: See Lauritzen (1996). $\qquad\qquad\square$

The global directed Markov property can thus be formulated by requiring that $A \perp\!\!\!\perp B \mid S$ whenever S d-separates A from B thereby making the analogy with the undirected case clearer. It depends on the specific context and purpose whether the pathwise criterion, or the criterion used in the definition of the global directed Markov property is easiest to use.

We illustrate the concept of d-separation by applying it to the query of Example 2.8. As Fig. 2.2 indicates, all trails between a and b are blocked by S, whereby the global Markov property gives that $a \perp\!\!\!\perp b \mid S$.

For further use, we shall use the symbolic expression $A \perp_{\mathcal{D}} B \mid S$ to denote that A and B are d-separated by S or, equivalently, A and B are separated by S in $(\mathcal{D}_{\mathrm{An}(A \cup B \cup S)})^m$. It was shown in Verma and Pearl (1990) that

Lemma 2.10 *For any fixed directed acyclic graph $\mathcal{D}$, the relation $\perp_{\mathcal{D}}$ satisfies the graphoid axioms.*

Geiger and Pearl (1990) show that the criterion of d-separation cannot be improved, in the sense that, for any given directed acyclic graph $\mathcal{D}$, one can find state spaces $\mathcal{X}_\alpha, \alpha \in V$ and a probability distribution P such that

$$A \perp\!\!\!\perp B \mid S \iff A \perp_{\mathcal{D}} B \mid S. \tag{2.12}$$

This result was strengthened by Meek (1995), who showed that if the state spaces were finite and had cardinality at least two, the set of probability distributions P not satisfying (2.12) had Lebesgue measure zero in the set of all directed Markov probability measures.

To complete this section we say that P obeys the *local directed Markov property* (DL) if any variable is conditionally independent of its non-descendants, given its parents:

$$\alpha \perp\!\!\!\perp \mathrm{nd}(\alpha) \mid \mathrm{pa}(\alpha).$$

A seemingly weaker requirement, the *ordered directed Markov property* (DO), replaces all non-descendants of α in the above condition by the predecessors $\mathrm{pr}(\alpha)$ of α in some given well-ordering of the nodes:

$$\alpha \perp\!\!\!\perp \mathrm{pr}(\alpha) \mid \mathrm{pa}(\alpha).$$

In contrast with the undirected case we have that all the four properties (DF), (DL), (DG) and (DO) are equivalent just assuming existence of the density f. This is stated formally as:

Theorem 2.11 *Let $\mathcal{D}$ be a directed acyclic graph. For a probability distribution P on $\mathcal{X}$ which has density with respect to a product measure μ, the following conditions are equivalent:*

(DF)   *P admits a recursive factorization according to $\mathcal{D}$;*
(DG)   *P obeys the global directed Markov property, relative to $\mathcal{D}$;*
(DL)   *P obeys the local directed Markov property, relative to $\mathcal{D}$;*
(DO)   *P obeys the ordered directed Markov property, relative to $\mathcal{D}$.*

Proof: That (DF) implies (DG) is Corollary 2.7. That (DG) implies (DL) follows by observing that $\{\alpha\} \cup \mathrm{nd}(\alpha)$ is an ancestral set and that $\mathrm{pa}(\alpha)$ obviously separates $\{\alpha\}$ from $\mathrm{nd}(\alpha) \setminus \mathrm{pa}(\alpha)$ in $(\mathcal{D}_{\{\alpha\} \cup \mathrm{nd}(\alpha)})^m$. It is trivial that (DL) implies (DO), since $\mathrm{pr}(\alpha) \subseteq \mathrm{nd}(\alpha)$. The final implication is shown by induction on the number of vertices $|V|$ of $\mathcal{D}$. Let α_0 be the last vertex of $\mathcal{D}$. Then we can let k^{α_0} be the conditional density of X_{α_0}, given $X_{V \setminus \{\alpha_0\}}$, which by (DO) can be chosen to depend on $x_{\mathrm{pa}(\alpha_0)}$ only. The marginal distribution of $X_{V \setminus \{\alpha_0\}}$ trivially obeys the ordered directed Markov property and admits a factorization by the inductive assumption.

Combining this factorization with k^{a_o} yields the factorization for P. This completes the proof.   □

Since the four conditions in Theorem 2.11 are equivalent, it makes sense to speak of a *directed Markov field* as one where any of the conditions is satisfied. The set of such distributions for a directed acyclic graph $\mathcal{D}$ is denoted by $M(\mathcal{D})$.

In the particular case when the directed acyclic graph $\mathcal{D}$ is perfect, i.e. all parents are married, the directed Markov property on $\mathcal{D}$ and the factorization property on its undirected version $\mathcal{D}^{\sim}$ coincide.

Proposition 2.12 *Let $\mathcal{D}$ be a perfect directed acyclic graph and $\mathcal{D}^{\sim}$ its undirected version. Then P is directed Markov with respect to $\mathcal{D}$ if and only if it factorizes according to $\mathcal{D}^{\sim}$.*

Proof: See Lauritzen (1996).   □

2.6 Causal Markov models

For simplicity we assume here and in the following that all random variables are discrete and have finite state spaces unless we specifically indicate otherwise. To emphasize the discreteness we use little p as a generic symbol for a probability mass function rather than f for a general density.

2.6.1 Conditioning by observation or intervention

The first important issue is to distinguish between different types of conditioning operations, each of which modify a given probability distribution in response to information obtained. Conditional probabilities are sometimes defined and calculated as

$$p(y \mid x) = P(Y = y \mid X = x) = \frac{P(Y = y, X = x)}{P(X = x)}.$$

We refer to this type of conditioning as *conditioning by observation* or *conventional conditioning*. In many cases this represents the way in which a probability distribution, $P(Y = y)$, should be modified when the information $X = x$ is revealed. Paradoxes appear when it is unclear how the information about X is revealed (Shafer 1985, 1996), but that is a different discussion.

When discussing causal issues it is important to realize that this is typically not the way the distribution of Y should be modified if we intervene externally and force the value of X to be equal to x. We refer to this type of modification as *conditioning by intervention* or *conditioning by action*. To make the distinction clear we use different symbols for this conditioning,

as indicated below

$$p(y \,||\, x) = P(Y = y \,|\, X \leftarrow x).$$

Generally, the two quantities will be different

$$p(y \,||\, x) \neq p(y \,|\, x)$$

and the quantity on the left-hand side cannot be calculated from the probability measure P alone, without additional assumptions. To judge whether these assumptions are reasonable in any given context one needs a specification of the precise way in which the intervention is made, just as conventional conditioning needs a specification about how the information is revealed.

In a moment we will give a precise meaning to a directed acyclic graph being causal. This will imply that in the graph below to the left

$$X \qquad Y \qquad\qquad X \qquad Y$$

we will have that $p(y \,||\, x) = p(y \,|\, x)$ and $p(x \,||\, y) = p(x)$, whereas these relations are reversed in the graph to the right, i.e. there it holds that $p(y \,||\, x) = p(y)$ and $p(x \,||\, y) = p(x \,|\, y)$.

2.6.2 Causal graphs

A directed acyclic graph $\mathcal{D}$ is said to be *causal* for a probability distribution P with respect to a subset $B \subseteq V$, if P is Markov with respect to $\mathcal{D}$, i.e.

$$p(x) = \prod_{\alpha \in V} p(x_\alpha \,|\, x_{\mathrm{pa}(\alpha)})$$

and it further holds for any $A \subseteq B$ that

$$
\begin{aligned}
p(x \,||\, x_A^*) &= \left. \prod_{\alpha \in V \setminus A} p(x_\alpha \,|\, x_{\mathrm{pa}(\alpha)}) \right|_{x_A = x_A^*} \\
&= \left. \frac{p(x)}{\prod_{\alpha \in A} p(x_\alpha^* \,|\, x_{\mathrm{pa}(\alpha)})} \right|_{x_A = x_A^*}.
\end{aligned}
\tag{2.13}
$$

Alternatively, one can think of the right-hand side of (2.13) as the mathematical definition of the intervention probability on the left-hand side.

If $B = V$ we simply say that $\mathcal{D}$ is *causal* or *fully causal* for P. We also use the expression that P is a *causal directed Markov field* with respect to $\mathcal{D}$ or say that P is *causally Markov* with respect to $\mathcal{D}$. Note that the causal Markov property thus gives a relation between different probability measures, each representing the probability law associated with a specific intervention.

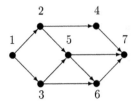

Figure 2.3 *Illustration of causal directed acyclic graph.*

We will refer to (2.13) as the *intervention formula*. It appeared in various forms in Pearl (1993) and Spirtes et al. (1993). It is implicit in Robins (1986) and in other literature.

There are many ways in which this causal interpretation of a directed Markov model can be justified. But it is also important to realize that there are many other ways in which one can associate causal relationships with directed acyclic graphs. This is in particular apparent in the highly interesting book of Shafer (1996) who develops a language for causal interpretation of probabilities through event trees. This leads to events being more natural as direct causes than variables. A variety of causal relationships between variables can then be derived as consequences of the formalism.

In a more general setting one would be interested in allowing other types of intervention than those described. For example, one could wish to control the value of a variable in a way that depends on previously observed variables. But for simplicity we only consider the case of simple interventions.

One should contrast the intervention formula (2.13) with conventional conditioning using Bayes' formula:

$$p(x \mid x_A^*) = \left.\frac{p(x)}{p(x_A^*)}\right|_{x_A = x_A^*} = \left.\frac{p(x)}{\sum_{y : y_A = x_A^*} p(y)}\right|_{x_A = x_A^*}, \tag{2.14}$$

which differs from the intervention formula in the denominator, where the product of conditional specifications is replaced by the marginal probability $p(x_A^*)$. This implies in particular that if intervention takes place on a single variable without parents, observation and intervention have identical effects:

Corollary 2.13 *If $\alpha \in V$ has no parents, i.e. $\mathrm{pa}(\alpha) = \emptyset$, then it holds that $p(x \,||\, x_\alpha^*) = p(x \mid x_\alpha^*)$.*

We illustrate the similarities and differences by intervening on variable 5 in Figure 2.3. If this graph is causal, we have that the intervention $X_5 \leftarrow x_5^*$ produces the distribution

$$p(x \,||\, x_5^*) = p(x_1)p(x_2 \mid x_1)p(x_3 \mid x_1)p(x_4 \mid x_2)p(x_6 \mid x_3, x_5^*)p(x_7 \mid x_4, x_5^*, x_6)$$

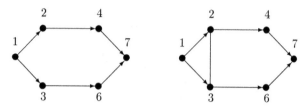

Figure 2.4 *The intervention $X_5 \leftarrow x_5^*$ in Fig. 2.3 produces a causal directed Markov field with respect to the graph on the left. The observation $X_5 = x_5^*$ produces a distribution which satisfies the chain graph Markov property with respect to the graph to the right.*

whereas the observation $X_5 = x_5^*$ leads to

$$p(x \mid x_5^*) \quad \propto \quad p(x_1)p(x_2 \mid x_1)p(x_3 \mid x_1)p(x_4 \mid x_2)$$
$$p(x_5^* \mid x_2, x_3)p(x_6 \mid x_3, x_5^*)p(x_7 \mid x_4, x_5^*, x_6). \qquad (2.15)$$

The modified distribution $P(\cdot \mid X_A \leftarrow x_A^*)$ is again a directed causal Markov field over the subgraph $\mathcal{D}_{V \setminus A}$ induced by the remaining variables. The corresponding conditional specifications are just modified such that

$$p(x_\alpha \mid x_{\text{pa}_{V \setminus A}(\alpha)} \| x_A^*) = p(x_\alpha \mid x_{\text{pa}(\alpha) \setminus A}, x_{\text{pa}(\alpha) \cap A}^*).$$

Expressed in words, the causal assumption is that the conditional specifications are unchanged for variables which are not used for intervention. In the example above, where we have intervened on variable 5, the only modifications of the specifications involve children of the intervention variables, i.e. variables 6 and 7, where we get

$$p(x_6 \mid x_3 \| x_5^*) = p(x_6 \mid x_3, x_5^*), \quad p(x_7 \mid x_4, x_6 \| x_5^*) = p(x_7 \mid x_4, x_5^*, x_6).$$

The corresponding subgraph is displayed as the graph to the left in Fig. 2.4. This is again to be contrasted with the effect of observation of variable 5, which creates a dependence structure determined by the chain graph (Lauritzen 1996) to the right in the same figure. This is due to the factor $p(x_5^* \mid x_2, x_3)$ creating a function depending on (x_2, x_3) in the factorization (2.15). Note that, in general, the conditional independence structure induced by conditioning by observation will not be in perfect correspondence with a chain graph.

It is important to realize that successive conditioning operations of the same type commute whereas intervention and observation in general cannot be interchanged. We therefore adopt the convention that all operations are performed from the right to the left. Thus

$$p(x \| y \mid z) = P(X = x \mid Y \leftarrow y, Z = z)$$

denotes the modified probability obtained by first observing $Z = z$ and subsequently intervening as $Y \leftarrow y$, whereas

$$p(x \mid z \mid\mid y) = P(X = x \mid Z = z, Y \leftarrow y)$$

reflects that the intervention is performed before the observation. As an example, consider the graph

$$\begin{array}{ccc} X & Y & Z \end{array}$$

Intervening with $X \leftarrow x^*$ and then subsequently observing $Y = y$ leads to

$$p(z \mid y \mid\mid x^*) = p(z \mid y),$$

whereas additional assumptions are to be made to predict the effect of the intervention $X \leftarrow x^*$ after observation of $Y = y$. Such assumptions could for example be that X, Y, and Z are functionally related in a structural equation model, see Section 2.9. This assumption would lead to the equality

$$p(z \mid\mid x^* \mid y) = p(z \mid y \mid\mid x^*) = p(z \mid y)$$

as then X and Z are functionally unrelated, once the value of y is known or has been fixed.

Generally, to ensure unambiguous meaning of intervention conditioning without introducing assumptions beyond those already made, intervention at a node α must always be made before any variables corresponding to its descendant nodes have been observed.

2.7 Assessment of treatment effects in sequential trials

The following example is adapted from Robins (1997) and is the simplest example where traditional approaches to assessment of treatment effects give incorrect results, whereas the methods described here coincide with those developed by Robins (1986), known as G-computation, and give the correct answer.

Consider a study made in a population of AIDS patients. Let us imagine the population being so large that sampling error can be ignored for practical purposes. The study involves 4 binary variables. In our notation, a is the label for an initial, randomized treatment, where $X_a = 1$ denotes that the patient has been treated with AZT, and $X_a = 0$ indicates placebo. After a given period it is for each patient observed whether the patient develops pneumonia, corresponding to the variable l, where $X_l = 1$ indicates that this is the case. We assume that all patients survive up to this point. Subsequently a secondary treatment with antibiotics is contemplated, corresponding to the variable b. For ethical reasons, all patients who have developed pneumonia are treated with antibiotics, i.e. $P(X_b = 1 \mid X_l = 1) = 1$, whereas the treatment is randomized for the patients with $X_l = 0$. Finally, after a given period it is registered whether a

Figure 2.5 *Graph displaying causal relationships between variables in a particular sequential trial. The graph is only assumed causal with respect to interventions at $\{a, b\}$. The missing arrow from a to b reflects that b is assigned by randomization.*

given patient has survived up to that time, corresponding to the variable s, where $X_s = 1$ denotes that the patient has survived.

The question is now to assess the effect on survival of a combined treatment with AZT and antibiotics of a new patient. In other words, we wish to calculate

$$P(X_s = 1 \mid X_a \leftarrow 1, X_b \leftarrow 1) = p(1_s \,\|\, 1_a, 1_b).$$

This is done in the following way. The relevant graph is displayed in Figure 2.5, where missing arrows reflect the randomized allocation of treatments. This graph is not fully causal as there may be unobserved variables (confounders) that simultaneously affect l and r. It is only assumed causal with respect to interventions at $\{a, b\}$.

Note that not all effects are estimable as there are no observations with $X_l = 1$ and $X_b = 0$. For example, the effect of treating with AZT only cannot be assessed. We find

$$
\begin{aligned}
p(1_s \,\|\, 1_a, 1_b) &= \sum_{x_l} p(1_s, x_l \,\|\, 1_a, 1_b) \\
&= \sum_{x_l} p(1_s \mid x_l \,\|\, 1_a, 1_b) p(x_l \,\|\, 1_a, 1_b) \\
&= \sum_{x_l} p(1_s \mid x_l, 1_a, 1_b) p(x_l \mid 1_a).
\end{aligned}
$$

As pointed out by Robins, conventional wisdom gives ambiguous or incorrect answers: The variable l is affected by the treatment and one should therefore not adjust for it but simply use the estimate of the conditional probability

$$p(1_s \,\|\, 1_a, 1_b) \sim \hat{p}(1_s \mid 1_a, 1_b).$$

On the other hand, the covariate l is also a confounder for the effect of the treatment on survival. Thus adjustment for l is needed and one should rather use

$$p(1_s \,\|\, 1_a, 1_b) \sim \sum_{x_l} \hat{p}(1_s \mid x_l, 1_a, 1_b) \hat{p}(x_l).$$

Both answers disagree with the correct calculation as given above.

The calculation is a special case of more general situations described by Robins, where randomized treatment allocations and intermediate responses alternate as $t_1, r_1, t_2, r_2, \ldots, t_k, r_k$ and where, for example, the effect of a combined treatment, fixing $t_1, \ldots, t_k$, on the final response r_k is desired. This is then found by G-computation, involving two steps:

1. modifying the joint distribution of all variables (corresponding to a complete DAG in many cases) by the intervention formula (2.13);

2. calculating the marginal distribution of X_{r_k} by a recursive forward computation, possibly using Monte Carlo methods.

We refrain here from describing the more general situation where the suggested treatment regime is allowed to depend on previous treatments and recordings, but emphasize that this does not create essentially new problems.

2.8 Identifiability of causal effects

This section will be concerned with the problem of identifying the effects of interventions from partial observation of a causal system, expressed in the form of a causal directed Markov field. It is largely based on ideas in Pearl (1993) and Pearl (1995a).

2.8.1 The general problem

Consider as usual a finite set of variables V, one of which is labelled t and designated the treatment variable, and another group of variables are considered to be the response, labelled R. We also assume that there is a directed acyclic graph $\mathcal{D}$ such that the joint distribution of all the variables V is causally Markov with respect to $\mathcal{D}$.

The object of interest is the *causal effect* of t on the group of response variables R, represented by the intervention distribution

$$P(X_R = x_R \mid X_t \leftarrow x_t^*) = p(x_R \,\|\, x_t^*).$$

The remaining variables are partitioned into $C \subseteq V$ and $U \subseteq V$, where C is a set of observed *covariates* whereas the variables in U are to remain unobserved. In principle one could discuss multiple treatments and responses, but this will not be done here.

Thus from an experimental or observational study we obtain information about the joint distribution of the observed variables, t, R and C. Ignoring sampling error, can the causal effect of t on R be determined from this information? Or, phrased in another way, which variables C are needed in order to determine this effect? If the causal effect can be determined from the observed distribution, then how can it be calculated, i.e. is there an analogue of the G-computation that gives the correct answer? If the causal effects cannot be precisely determined, can we at least give inequalities that these numbers must satisfy?

To make the discussion precise, we say that C *identifies* the causal effect of t on R if for any pair P_1, P_2 of distributions that are causally Markov with respect to $\mathcal{D}$ it holds that

$$p_1(x_t, x_C, x_R) \equiv p_2(x_t, x_C, x_R) \implies p_1(x_R \,\|\, x_t) \equiv p_2(x_R \,\|\, x_t).$$

The most basic question above can now be phrased as determining whether a given set covariates C identifies the causal effect of t on r. Clearly, if $C' \supseteq C$ and C identifies the effect of t on r, so does C', so we are interested in minimal sets of identifying covariates.

Generally C will not identify causal effects unless the conditional distributions are identified by the joint distribution. Thus, *throughout this section we will assume that*

$$p(x_t, x_C) > 0 \quad \text{for all combinations of } x_t \text{ and } x_C, \qquad (2.16)$$

unless we explicitly state otherwise.

2.8.2 Intervention graphs

When the effect of potential intervention are to be discussed, it is convenient to represent these explicitly in the associated graph of the model considered. As also done in Pearl (1993) and Spirtes et al. (1993), this is done through an *intervention graph* $\mathcal{D}'$, which is formed by augmenting each node representing a variable where intervention is contemplated, with an additional parent.

We denote this additional parent of a vertex α by α'. The corresponding random variable $X_{\alpha'}$ is, when no ambiguity results, just denoted by F_α. The variable F_α has state space $\mathcal{X}_\alpha \cup \{\phi\}$ and the conditional distributions of X_α given its parents in the intervention graph are given by

$$p'(x_\alpha \,|\, x_{\mathrm{pa}(\alpha)}, f_\alpha) = \begin{cases} p(x_\alpha \,|\, x_{\mathrm{pa}(\alpha)}) & \text{if } f_\alpha = \phi \\ \delta_{x_\alpha, x_\alpha^*} & \text{if } f_\alpha = x_\alpha^*, \end{cases} \qquad (2.17)$$

where δ_{xy} is Kronecker's symbol

$$\delta_{xy} = \begin{cases} 1 & \text{if } x = y \\ 0 & \text{otherwise.} \end{cases}$$

A more general setup would let f_α vary in the set of all (randomized) decision policies, but here we only consider the simpler case.

This approach to the representation of causal effects is related to so-called *influence diagrams* (Howard and Matheson 1984, Shachter 1986, Smith 1989, Oliver and Smith 1990) and taking this connection to its consequence gives yet an alternative basis for causal interpretation of graphical models (Heckerman and Shachter 1995).

Each of the variables $F_\alpha, \alpha \in A$, where A is the set of variables for which intervention is contemplated, can be given an arbitrary distribution with positive probability of all states. We then clearly have

$$p(x) = p'(x \mid F_\alpha = \phi, \alpha \in A),$$

but it also holds for any subset $B \subseteq A$ that

$$
\begin{aligned}
p(x \,\|\, x_B^*) &= P(X = x \mid X_B \leftarrow x_B^*) \\
&= P'(x \mid F_\alpha = x_\alpha^*, \alpha \in B, F_\alpha = \phi, \alpha \in B \setminus A), \quad (2.18)
\end{aligned}
$$

since it follows from Corollary 2.13 that

$$
\begin{aligned}
P'(X = x \mid F_\alpha \leftarrow x_\alpha^*, \alpha \in B, F_\alpha \leftarrow \phi, \alpha \in B \setminus A) = \\
P'(X = x \mid F_\alpha = x_\alpha^*, \alpha \in B, F_\alpha = \phi, \alpha \in B \setminus A)
\end{aligned}
$$

because the variables α' do not have parents. The importance of the relation (2.18) is that it gives a simple connection between intervention conditioning in the original graph and conventional conditioning in the intervention graph.

2.8.3 Three inference rules

The operations needed to find groups of identifying covariates typically involve a sequence of operations that gradually transform expressions involving intervention probabilities to expressions involving ordinary conditional probabilities, the latter being in principle accessible by empirical observation.

We are considering the simple case, where intervention at a node t is contemplated, its effect on a group of variables R is studied, in a context where X_A is observed to be x_A. We let $\perp_{\mathcal{D}'}$ denote d-separation in the intervention graph $\mathcal{D}'$ obtained by augmenting $\mathcal{D}$ with an intervention variable t' as an additional parent of t, and possibly other intervention variables, if also other interventions are contemplated, as described above. We then have the following three inference rules:

Neutral observation of X_t:

$$R \perp_{\mathcal{D}'} t \mid A \implies p(x_r \mid x_A, x_t) = p(x_R \mid x_A) \qquad (2.19)$$

Neutral intervention at t:

$$R \perp_{\mathcal{D}'} t' \mid A \implies p(x_r \mid x_A \,\|\, x_t) = p(x_R \mid x_A) \qquad (2.20)$$

Equivalence of observation and intervention at t:

$$R \perp_{\mathcal{D}'} t' \mid \{A, t\} \implies p(x_R \mid x_A \,\|\, x_t) = p(x_R \mid x_A, x_t). \qquad (2.21)$$

Each of these can be derived from the directed Markov property of P and P' combined with the fact that intervention probabilites can be obtained by appropriate observation conditioning in the intervention graph.

For example, to derive (2.19) we observe that $R \perp_{D'} t \mid A$ implies $R \perp_D t \mid A$. This holds because all trails from t to R in D are also trails in D' and if one is blocked by A in D', it is also blocked by A in D. Therefore the global directed Markov property for D entails that

$$R \perp_{D'} t \mid A \implies R \perp\!\!\!\perp t \mid A,$$

whereby (2.19) follows.

The relations (2.20) and (2.21) follow directly from the fact that intervention conditioning at t in D is equivalent to observation conditioning at t' in D'. These rules are also direct consequences of Theorem 7.1 of Spirtes et al. (1993)

Although Pearl (1995a) formulates these inference rules somewhat differently, he conjectures that the three inference rules are complete, in the sense that a set of covariates is identifying for the effect of t on R if and only if all terms involving intervention conditioning in the expression for the intervention distribution can be changed to terms involving observational conditioning by successive application of these three rules. We shall see examples of this in the next subsection, where a number of classical concepts from epidemiology will be illustrated.

2.8.4 The back-door formulae

One of the classic conditions for a set of covariates to be identifying is captured in the theorem below, known as the back-door theorem and formula.

As earlier we contemplate the effect of t on a group of variables R and plan to observe these together with a set of covariates C, whereas the remaining variables in the system are unobserved. Also, as above, D' denotes the intervention graph obtained from D by augmenting with an intervention variable t' as an additional parent of t and $\perp_{D'}$ denotes d-separation in D'.

We then have the following theorem, which can also be derived directly from Theorem 7.1 of Spirtes et al. (1993):

Theorem 2.14 (Back-door) *Assume $C \supseteq C_0$, where C_0 satisfies*

(BD1) *The covariates in C_0 are unaffected by an intervention: $C_0 \perp_{D'} t'$;*

(BD2) *An intervention only affects the response through the treatment itself, as modified by the observed covariates: $R \perp_{D'} t' \mid C_0 \cup \{t\}$.*

Then C identifies the effect of the treatment t on R as

$$p(x_R \,||\, x_t^*) = \sum_{x_{C_0}} p(x_R \mid x_{C_0}, x_t^*) p(x_{C_0}). \qquad (2.22)$$

Proof: The proof is a simple application of the inference rules. If we partition according to X_{C_0} and then apply first (2.21) for $A = C_0$ and then (2.20) for $R = C_0$ and $A = \emptyset$, we get

$$p(x_R \,||\, x_t^*) = \sum_{x_{C_0}} p(x_R \,|\, x_{C_0} \,||\, x_t^*) p(x_{C_0} \,||\, x_t^*)$$

$$= \sum_{x_{C_0}} p(x_R \,|\, x_{C_0}, x_t^*) p(x_{C_0} \,||\, x_t^*)$$

$$= \sum_{x_{C_0}} p(x_R \,|\, x_{C_0}, x_t^*) p(x_{C_0}).$$

$\square$

Condition (BD1) might as well have been formulated by demanding that none of the covariates in C_0 are descendants of t. This is a condition which can then be checked in $\mathcal{D}$ rather than $\mathcal{D}'$.

Note that the positivity assumption (2.16) is important for the joint distribution to identify $p(x_R \,|\, x_{C_0}, x_t^*)$ for all combinations of its arguments.

In the formulation given, the name 'back-door theorem' is not obvious. The lemma below clarifies the reason for the name. A *back-door trail* from t to R in $\mathcal{D}$ is a trail from t to R that does not involve an arrow emanating from t, i.e. leaves t through the 'back door'. Similarly, we let a *front-door trail* from t to R be a trail that begins with an arrow emanating from t.

Lemma 2.15 *If no covariates in C_0 are descendants of t, $r \perp_{\mathcal{D}'} t' \,|\, C_0 \cup \{t\}$ if and only if all back-door trails from t to R are blocked by C_0 in $\mathcal{D}$.*

Proof: Assume $r \perp_{\mathcal{D}'} t' \,|\, C_0 \cup \{t\}$. Each trail from t' to R in $\mathcal{D}'$ corresponds uniquely to a trail from t to R in $\mathcal{D}$. Since the descendants of t are identical in $\mathcal{D}$ and $\mathcal{D}^*$ and none of these are in C_0, t is blocking all trails from t' to R in $\mathcal{D}$ that correspond to front-door trails and it is not blocking any trails corresponding to back-door trails. Consequently, these are be blocked by $C_0 \cup t$ in $\mathcal{D}'$ if and only if they are blocked by C_0 in $\mathcal{D}$. $\square$

The condition in Lemma 2.15 is also phrased in terms of the original graph $\mathcal{D}$ rather than the intervention graph.

James Robins (personal communication) gives the following heuristic argument for the criterion: The treatment effect can be identified if, conditionally on C_0, there is no association beyond causation. Removing arrows out of t eliminates causation. One must thus demand that no conditional association remains after these arrows have been removed.

The formula (2.22) is the classical formula which adjusts for covariates that are not affected by the treatment.

Theorem 2.14 has a slightly more general version, extended by a recursive argument.

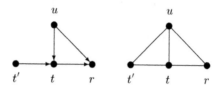

Figure 2.6 *Intervention graph and associated moral graph for experiment with an unobserved confounder. There is a path in the moral graph from t' to r circumventing t so the back-door criterion is violated and the effect of t on r cannot be identified from observations of t and r.*

Theorem 2.16 (Extended back-door) *Assume $C \supseteq C_0$, where C_0 satisfies*

(EBD1) *The effect of the treatment t on the covariates in C_0 is identified by C;*

(BD2) *An intervention only affects the response through the treatment itself, modified by the observed covariates: $R \perp_{\mathcal{D}'} t' \,|\, C_0 \cup \{t\}$.*

Then C identifies the effect of the treatment t on R as

$$p(x_R \,\|\, x_t^*) = \sum_{x_{C_0}} p(x_R \,|\, x_{C_0}, x_t^*) p(x_{C_0} \,\|\, x_t^*). \qquad (2.23)$$

Proof: This is shown exactly as for Theorem 2.14, just omitting the last step in the calculation. □

Confounding

The first situation to be considered in the light of the back-door theorem is the classical case of a *confounder*, which in the current context is defined to be an unobserved quantity that simultaneously affects the treatment and the response. Thus, in a causal graph, a confounder is a common ancestor to the treatment and response. The literature in epidemiology contains a wealth of more or less precise definitions of the term.

This situation is illustrated in Figure 2.6, displaying the corresponding intervention graph and its associated moral graph. The conditional distribution of X_r after intervention at t cannot be determined from the joint distribution of (X_t, X_r).

Randomization

The next example illustrates how randomization overcomes the identification problem caused by the confounder. Instead of just observing (X_t, X_r), the treatment X_t is now allocated by a known random mechanism, possibly

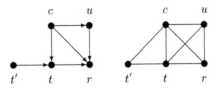

Figure 2.7 *Intervention graph and associated moral graph for experiment with randomized treatment allocation. The moral graph reveals that $r \perp_{\mathcal{D}'} t' \mid \{c, t\}$ so the back-door criterion is satisfied and the treatment effect can be assessed.*

depending on an observed covariate X_c, leading to the diagram described in Figure 2.7. The randomization ensures that there is no arrow pointing from u to t, i.e. the treatment X_t is conditionally independent of X_u given the covariates X_c. To see that (BD1) of Theorem 2.14 is satisfied, we form the ancestral set in $\mathcal{D}'$ generated by $\{c, t'\}$. This is equal to $\{c, t', u\}$ and the associated moral graph has only one edge between c and u. Thus $c \perp_{\mathcal{D}'} t'$. The ancestral set generated by $\{c, t, t', r\}$ is equal to the full set of variables, and the associated moral graph is also displayed in Figure 2.7. Clearly r is separated from t' by $\{c, t\}$ in this graph, so (BD2) is satisfied. Note that if u had been allowed to have an influence on the treatment allocation, the corresponding arrow from u to t in the graph to the left would have induced an edge in the moral graph between t' and u, who were common parents of t, thus violating (BD2) and confounding the relation between t and r.

Sufficient covariate

The next situation to be considered is an observational study where we have no control over the treatment allocation mechanism, but we are able to find a *sufficient* set of covariates, i.e. a set of covariates which is so informative about the response mechanism that the response is conditionally independent of the unobserved variable given the treatment and the covariates. The corresponding intervention graph and associated moral graph is displayed in Figure 2.8.

The ancestral set generated by c and t' is equal to $\{c, t', u\}$ and the associated moral graph has only one edge between c and u and thus $c \perp_{\mathcal{D}'} t'$. The ancestral set generated by $\{c, t, t', r\}$ is equal to the full set of variables, and the associated moral graph is displayed in Figure 2.8. Clearly r is separated from t' by $\{c, t\}$ in this graph, so (BD2) is satisfied.

Partial compliance

The next example describes a study in which treatments are assigned completely at random to individuals, but not all individuals are complying with

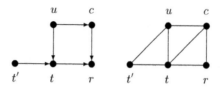

Figure 2.8 *Intervention graph and associated moral graph for an observational study with a sufficient covariate. The moral graph reveals that* $r \perp_{\mathcal{D}'} t' \mid \{c, t\}$ *so the back-door criterion is satisfied and the treatment effect can be assessed.*

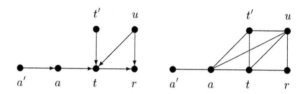

Figure 2.9 *Intervention graph and associated moral graph for a study with partial compliance. The moral graph reveals that* $r \perp_{\mathcal{D}'} a' \mid \{a, t\}$ *so the effect of the treatment assignment is identified. However, r is not separated from t' by t so the effect of the treatment itself cannot be assessed.*

the assignments so that some receive a treatment different from the one assigned. The situation is displayed in Figure 2.9, where a is labelling the assignment and t the actual treatment received. The response, treatment assigned, and treatment received are all observed. From inspection of the moral graph it clearly follows that $r \perp_{\mathcal{D}'} a' \mid \{a, t\}$ so the effect of the treatment assigned is identified via the back-door formula. The corresponding assessment is commonly referred to as "analysis by intention-to-treat".

However, r is not separated from t' by t so the effect of the treatment itself is not identifiable from these observations. We shall later see how to derive bounds for the effects in this particular case.

2.8.5 The front-door formula

Theorem 2.17 below describes yet another situation where the causal effect of a treatment can be identified. Here the observed covariates are to be considered as the active agent determining the response. A basic example to have in mind could be the effect of smoking on lung cancer, the active agent being the tar content in the lungs. There could be an unobserved, say genetic, feature that influenced both the response and the tendency to smoke. The corresponding diagram is displayed in Figure 2.10 and the

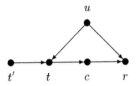

Figure 2.10 *Intervention graph associated with a situation where Theorem 2.17 applies. The covariate c is capturing the way in which t is affecting r, possibly modified by the unobserved variable u.*

conditions in the Theorem 2.17 reflect conditional independence relations following from the directed Markov property of this diagram.

Theorem 2.17 (Front-door formula) *Assume that $C \supseteq C_0$, and there is a subset $D \subseteq V \setminus (C_0 \cup R \cup \{t\})$ such that*

(FD1) *The variables in D are unaffected by an intervention: $D \perp_{\mathcal{D}'} t'$;*

(FD2) *An intervention only affects the covariate through the treatment itself, independently of the variables in D: $C_0 \perp_{\mathcal{D}'} (D \cup \{t'\}) \mid t$;*

(FD3) *An intervention only affects the response through the covariate, as modified by the variables in D: $R \perp_{\mathcal{D}'} t' \mid C_0 \cup D$.*

Then C identifies the effect of the treatment t on R as

$$p(x_R \,\|\, x_t^*) = \sum_{x_{C_0}} p(x_{C_0} \mid x_t^*) \sum_{x_t} p(x_R \mid x_{C_0}, x_t) p(x_t). \qquad (2.24)$$

Proof: We assume without loss of generality that $C = C_0$ and $D = U$. Then we have

$$
\begin{aligned}
p(x_R \,\|\, x_t^*) &= \sum_{x_C, x_U} p(x_R \mid x_C, x_U \,\|\, x_t^*) p(x_C, x_U \,\|\, x_t^*) \\
&= \sum_{x_C, x_U} p(x_R \mid x_C, x_U) p(x_C \mid x_U \,\|\, x_t^*) p(x_U \,\|\, x_t^*),
\end{aligned}
$$

where we have used (FD3) together with (2.20) to deduce that the intervention in the first term of the product is neutral.

Next, (FD2) combined with the semi-graphoid properties (C3') and (C2') (which are satisfied by the relation $\perp_{\mathcal{D}'}$ by Lemma 2.10) yields

$$C \perp_{\mathcal{D}'} t' \mid (U \cup \{t\}) \text{ and } C \perp_{\mathcal{D}'} U \mid t.$$

Thus (2.21) yields that intervention in the second term may be substituted with ordinary conditioning, and (2.19) that conditioning with x_U then can be ignored.

Further, (FD1) with (2.20) gives that the third term in the product is independent of x_t^* so that we have

$$p(x_R \,\|\, x_t^*) = \sum_{x_U, x_C} p(x_R \,|\, x_U, x_C) p(x_C \,|\, x_t^*) p(x_U).$$

If we now rewrite $p(x_U)$ by partitioning according to x_t and note that (FD3) with (2.19) allows further conditioning on x_t in the first term, we get

$$
\begin{aligned}
p(x_R \,\|\, x_t^*) &= \sum_{x_U, x_C, x_t} p(x_R \,|\, x_U, x_C, x_t) p(x_C \,|\, x_t^*) p(x_U \,|\, x_t) p(x_t) \\
&= \sum_{x_C} p(x_C \,|\, x_t^*) \sum_{x_t} p(x_R \,|\, x_C, x_t) p(x_t)
\end{aligned}
$$

and the proof is complete.   □

Both the formula and its name are due to Pearl (1995a), where it is given as part of Theorem 2. Pearl's front-door conditions are formulated rather differently and it is not obvious that they are equivalent to those given here, but we believe them to be. In Pearl's formulation (and our terminology), the conditions are:

(FD1') All directed paths from t to R intersect C_0;

(FD2') All trails from t to C_0 in $\mathcal{D}$ contain an arrow out of t;

(FD3') All back-door trails from C_0 to R are blocked by t in $\mathcal{D}$.

The justification for the name is not so obvious, neither in Pearl's formulation nor in the formulation given here. Although it is true that (FD1) and (FD3) together reflect that C_0 blocks front-door paths from t to R, (FD2) rather reflects that back-door paths from t to C_0 are blocked.

Again we note the importance of the positivity assumption (2.16). Without this, we could always take $c = t$ and satisfy all conditions in Theorem 2.17. However, then the necessary conditional distributions would not be identified by the marginal distributions.

2.8.6 Additional examples

We conclude the section on identifiability of treatment effects by discussing a number of additional examples, illustrating the potential use of the front- and back-door formulae.

Example 2.18 Consider the example with intervention graph displayed in Figure 2.11. In the case where c and l are observed together with the treatment t and the response r, the back-door formula applies with c as the covariate and the front-door formula with l as the covariate. The extended back-door theorem applies with both covariates observed. Thus it holds

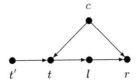

Figure 2.11 *Intervention graph associated with an example where both of the front-door and back-door formulae apply when l and c are observed.*

that

$$p(x_r \,\|\, x_t^*) = \sum_{x_c} p(x_r \mid x_c, x_t^*) p(x_c)$$

$$= \sum_{x_l} p(x_l \mid x_t^*) \sum_{x_t} p(x_r \mid x_l, x_t) p(x_t)$$

$$= \sum_{x_c, x_l} p(x_r \mid x_l, x_c) p(x_c) p(x_l \mid x_t^*).$$

However, although the treatment effect can be identified in all three observational situations, it is not true that the corresponding maximum likelihood estimates are equally efficient in the case where these are estimated from data. There is clearly loss of information associated with not observing all four variables.

To illustrate this, assume that all variables are discrete and a potential sample of n independent and identically distributed cases with counts $n(x_t, x_r, x_l, x_c)$ are observed, and contrast this with the corresponding incomplete samples only giving $n(x_t, x_r, x_c)$ or $n(x_t, x_r, x_l)$.

In the first situation, the maximum likelihood estimate in the model which is only restricted by satisfying the directed Markov property on the graph is equal to

$$\hat{p}(x_t, x_r, x_l, x_c) = \frac{n(x_c)n(x_t, x_c)n(x_t, x_l)n(x_r, x_l, x_c)}{n\, n(x_c)n(x_t)n(x_l, x_c)}$$

$$= \frac{n(x_t, x_c)n(x_t, x_l)n(x_r, x_l, x_c)}{n\, n(x_t)n(x_l, x_c)},$$

as each of the conditional probabilities of a variable given its parents is estimated by the corresponding observed relative frequencies (Lauritzen 1996, Theorem 4.36). Using the extended back-door formula, i.e. the last relation above, we therefore get

$$\hat{p}(x_r \,\|\, x_t^*) = \sum_{x_c, x_l} \hat{p}(x_r \mid x_l, x_c) \hat{p}(x_c) \hat{p}(x_l \mid x_t^*)$$

$$= \sum_{x_c, x_l} \frac{n(x_r, x_l, x_c)}{n(x_l, x_c)} \frac{n(x_c)}{n} \frac{n(x_l, x_t^*)}{n(x_t^*)}.$$

The similar expression in the back-door case, i.e. when only $n(x_r, x_c, x_t)$ is observed, becomes

$$\hat{p}(x_r \,\|\, x_t^*) = \sum_{x_c} \hat{p}(x_r \,|\, x_c, x_t^*) \hat{p}(x_c)$$

$$= \sum_{x_c} \hat{p}(x_r \,|\, x_c, x_t^*) \frac{n(x_c)}{n}.$$

Note that in this case it is not generally true that we have

$$\hat{p}(x_r \,|\, x_c, x_t^*) = \frac{n(x_r, x_c, x_t^*)}{n(x_c, x_t^*)}$$

because the model induces restrictions on this conditional probability. However, it is obvious that

$$\tilde{p}(x_r \,\|\, x_t^*) = \sum_{x_c} \frac{n(x_r, x_c, x_t^*)}{n(x_c, x_t^*)} \frac{n(x_c)}{n}$$

is still a reasonable estimate of the intervention probability. The latter estimate is also the traditional estimate used, and it also applies in the more general case, where the conditional independence $c \perp\!\!\!\perp l \,|\, t$ is violated. Presumably this estimate will be less efficient if indeed the condition $c \perp\!\!\!\perp l \,|\, t$ were known to hold.

In the front-door case, when $n(x_r, x_l, x_t)$ is observed, we similarly have

$$\hat{p}(x_r \,\|\, x_t^*) = \sum_{x_l} \hat{p}(x_l \,|\, x_t^*) \sum_{x_t} \hat{p}(x_r \,|\, x_l, x_t) \hat{p}(x_t)$$

$$= \sum_{x_l} \frac{n(x_l, x_t^*)}{n(x_t^*)} \sum_{x_t} \hat{p}(x_r \,|\, x_l, x_t) \frac{n(x_t)}{n},$$

where again the maximum likelihood estimate of the second conditional probability may not be equal to the corresponding relative frequency. However, it is obvious that a reasonable estimate of the treatment effect is equal to

$$\tilde{p}(x_r \,\|\, x_t^*) = \sum_{x_l} \frac{n(x_l, x_t^*)}{n(x_t^*)} \sum_{x_t} \frac{n(x_r, x_l, x_t)}{n(x_l, x_t)} \frac{n(x_t)}{n}.$$

It would be interesting to compare the loss of efficiency by not observing c vs. not observing l. $\qquad\qquad\qquad\qquad\qquad\qquad\qquad\qquad\qquad\square$

Example 2.19 The next example is taken from Pearl (1995a) and is somewhat more complex. It is illustrated in Figure 2.12.

As the figure shows, l suffices as a covariate for the front-door formula in Theorem 2.17 to apply.

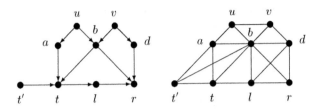

Figure 2.12 *Intervention graph associated with Example 2.19 with its correspond-ing moral graph.*

Figure 2.13 *Graphical model expressing that i is an instrumental variable. Note the similarity with the situation of partial compliance described in Figure 2.9, where the assignment variable a is an instrument.*

From the moral graph it is seen by direct inspection that observing b is necessary but not sufficient to satisfy the back-door Theorem 2.14. It needs to be supplemented with any non-empty subset of the variables a, u, v, and d for its union with t to separate t' from r in this graph.

If, for example b, d, and l are observed together with t and r, the extended back-door formula (2.23) yields that the treatment effect is to be estimated from complete data counts as

$$\hat{p}(x_r \,\|\, x_t^*) = \sum_{x_b, x_l, x_d} \frac{n(x_r, x_b, x_d, x_l) n(x_l, x_t^*)}{n(x_b, x_d, x_l) n(x_t^*)} \hat{p}(x_b, x_d),$$

where the latter probability ideally should be estimated by taking into ac-count the relevant restrictions induced by the model, rather than using the empirical relative frequencies directly. We leave it to the reader to consider estimation of treatment effects under different observational schemes.   □

Figure 2.13 displays the situation in which the variable i is an *instru-mental variable* or *instrument* for assessing the effect of t on r. This notion is important in econometrics (Bowden and Turkington 1984, Angrist, Im-bens and Rubin 1996). An instrumental variable is one which affects the treatment, but is uncorrelated with unobserved factors. An instrumental variable can be used to derive bounds for treatment effects as we shall show in Section 2.10.1 below.

But here we show an inequality which provides a good example of the restrictions that conditional independence constraints imply for marginal distributions. More precisely, it holds for any discrete treatment variable t that if the independence assumptions associated with the diagram in Figure 2.13 hold, then

$$\sup_{x_t} \int_{x_r} \sup_{x_i} f(x_r, x_t \mid x_i) \, \mu_r(dx_r) \leq 1, \tag{2.25}$$

where f is a generic symbol for the appropriate (conditional) density. This *instrumental inequality* was apparently first derived by Pearl (1995b) and we give the (quite simple) proof below.

The conditional independence restrictions imply that

$$f(x_r, x_t \mid x_i) = \int_{x_u} p(x_t \mid x_i, x_u) f(x_r \mid x_t, x_u) \, P_u(dx_u), \tag{2.26}$$

where P_u denotes the marginal distribution of X_u and the remaining entities are appropriate densities. Since the treatment variable is discrete, we have

$$p(x_t \mid x_i, x_u) \leq 1$$

and this must also hold for its supremum

$$h(x_t, x_u) = \sup_{x_i} p(x_t \mid x_i, x_u) \leq 1.$$

Now we get from (2.26) that

$$\sup_{x_i} f(x_r, x_t \mid x_i) = \int_{x_u} h(x_t, x_u) f(x_r \mid x_t, x_u) \, P_u(dx_u)$$

$$\leq \int_{x_u} f(x_r \mid x_t, x_u) \, P_u(dx_u),$$

whereby

$$\int_{x_r} \sup_{x_i} f(x_r, x_t \mid x_i) \, \mu_r(dx_r) \leq \int_{x_r} \int_{x_u} f(x_r \mid x_t, x_u) \, P_u(dx_u) \mu_r(dx_r)$$

$$= \int_{x_u} \int_{x_r} f(x_r \mid x_t, x_u) \, \mu_r(dx_r) P_u(dx_u)$$

$$= 1,$$

and (2.25) follows.

The importance of the inequality (2.25) is that it makes the assumption that i is an instrument *falsifiable* from observations of (X_i, X_t, X_r).

Note that the discreteness of the variable t is used at a very critical point in the proof of (2.25). At the time of writing it is not known whether the assumption of i being an instrument is falsifiable in the general case. In other words, given an arbitrary joint distribution Q of variables (X_i, X_t, X_r), does there exist a random variable X_u and a distribution P of (X_u, X_i, X_t, X_r) which is Markov with respect to the graph in Figure 2.13 and has Q as its marginal to (X_i, X_t, X_r)? In the case where Q is multivariate Gaussian, the answer to the last question is known to be positive, i.e. such a distribution always exists and 'instrumentality' is therefore not falsifiable in the Gaussian case.

2.9 Structural equation models

As mentioned in Section 2.6, the assumption that the intervention formula (2.13) applies is an additional model assumption that does not follow from the basic axioms of probability. There are different ways of justifying this assumption and in any given context, subject matter knowledge must play an essential rôle in this justification process.

A particular modelling formulation, leading to causal Markov models, has documented its relevance in several areas of application. Structural equation models (Bollen 1989) were invented in the context of genetics (Wright 1921, 1923, 1934) , and exploited in economics (Haavelmo 1943, Wold 1954) and social sciences (Goldberger 1972), see for example Pearl (1998) and Spirtes, Richardson, Meek, Scheines and Glymour (1998) for further discussion.

They were used as the main justification and motivation for studying causal Markov models in Kiiveri, Speed and Carlin (1984) and Kiiveri and Speed (1982), as well as in Pearl (1995a) and Pearl (2000).

Most commonly, structural equation models have been assumed linear although there are important exceptions (Goldfeld and Quandt 1972). Here we consider a general structural equation system associated with a directed acyclic graph $\mathcal{D}$. More precisely we consider a system of 'equations'

$$X_v \leftarrow g_v(X_{\mathrm{pa}(v)}, U_v), v \in V, \tag{2.27}$$

where the assignments have to be carried out sequentially, in a well-ordering of the directed acyclic graph $\mathcal{D}$, so that at all times, when X_v is about to be assigned a value, all variables in pa(v) have already been assigned a value.

The variables $U_v, v \in V$ are assumed to be independent. In the literature, correlation is generally allowed among the 'disturbances' U_v. Also non-recursive systems are often studied. Such systems do not correspond to directed Markov models and they are not studied here. Conditional independence properties for cyclic linear structural equation systems have been studied, for example, by Spirtes (1995), Richardson (1996), Spirtes et al. (1998), and Koster (1996, 1999a, 1999b).

The term 'structural equation system' is really misplaced, and 'structural assignment system' would have been much more appropriate. Much controversy in the literature, in particular concerning calculation of intervention effects, is due to treating the assignment systems as equation systems, 'solving' them and uncritically moving variables between the right-hand side and the left-hand side of (2.27). In particular, this matters when interventions are considered.

It is an important aspect of structural equation models that they also specify the way in which intervention is to be carried out. As is implicit in much literature and, for example, quite explicit in Strotz and Wold (1960), the effect of the intervention $X_a \leftarrow x_a^*$ on a variable with label a is simply that the corresponding line in (2.27) is replaced with the assignment described by the intervention. We refer to this process as *intervention by replacement*. Clearly, the justification that this is a reasonable assumption in any given context is no less difficult than the direct justification of the causal Markov assumption, since the latter follows from (2.27), as stated formally below.

Theorem 2.20 *Let* $X = (X_v)_{v \in V}$ *be determined by a structural equation system corresponding to a given directed acyclic graph* $\mathcal{D}$ *and let* P *denote its distribution. If intervention is carried out by replacement, then* P *is causally Markov with respect to* $\mathcal{D}$.

Proof: Let the vertices of $\mathcal{D}$ be well-ordered as $v_1, \ldots, v_n$ so that the assignments in (2.27) are carried out in the corresponding order. As the variables U_{v_i} are assumed independent, we clearly have

$$U_{v_i} \perp\!\!\!\perp (X_{v_1}, \ldots, X_{v_{i-1}})$$

and thus, from (C2) and (C3)

$$U_v \perp\!\!\!\perp X_{\mathrm{pr}(v)} \mid X_{\mathrm{pa}(v)}.$$

Using (2.27) with (C2) gives

$$X_v \perp\!\!\!\perp X_{\mathrm{pr}(v)} \mid X_{\mathrm{pa}(v)},$$

i.e. the distribution of X satisfies the ordered directed Markov property (DO). Theorem 2.11 now yields that P is directed Markov on $\mathcal{D}$.

As intervention in a structural equation system is made by replacement, it is clear that all conditional distributions except those involving interventions are preserved. Hence the intervention formula (2.13) applies.   □

Note that neither the functions g_v nor the random disturbances U_v are uniquely determined from the distribution P, and not even if P is known to be causally Markov. Thus assuming a specific structural equation model (2.27) is generally stronger — in a way which is typically not empirically testable — than just assuming the causal Markov property, as captured in (2.13).

Some authors seem to prefer to use a structural equation model as justification for the causal Markov property, rather than taking this property as a primitive assumption that must stand usual scientific testing. In view of the above, this may not be reasonable unless specific subject matter knowledge naturally leads to such equations.

2.10 Potential responses and counterfactuals

As mentioned, any causal Markov model for a given DAG $\mathcal{D}$ can be represented by a structural equation system, although this can be done in many different ways.

One type of representation deserves particular attention. Observe first that in each of the equations in (2.27), the values of U_v do not matter beyond what they prescribe as values for g_v, for each fixed value of possible parent configurations $x_{\mathrm{pa}(v)}$. Taking this to its consequence, we can introduce maps ω_v

$$\omega_v : \mathcal{X}_{\mathrm{pa}(v)} \to \mathcal{X}_v.$$

Then each pair (g_v, u_v) in (2.27) determines a map ω_v as

$$\omega_v(x_{\mathrm{pa}(v)}) = g_v(x_{\mathrm{pa}(v)}, u_v)$$

and, conversely, for each set of maps ω_v, we can define g_v as

$$g_v(x_{\mathrm{pa}(v)}, \omega_v) = \omega_v(x_{\mathrm{pa}(v)}).$$

Denoting a random map by Ω_v, we can thus define a structural equation system by

$$X_v \leftarrow g_v(X_{\mathrm{pa}(v)}, \Omega_v), v \in V,$$

and such a system is said to have *canonical form*. The random variables $\Omega_v(x_{\mathrm{pa}(v)})$ describes the *potential response*, i.e. the value of X_v that would have been observed, had the parent configuration been equal to $x_{\mathrm{pa}(v)}$. In this sense, the sets of random variables

$$\{\Omega_v(x_{\mathrm{pa}(v)}) : x_{\mathrm{pa}(v)} \in \mathcal{X}_{\mathrm{pa}(v)}\}$$

are *counterfactual*. The variables Ω_v were called *mapping variables* by Heckerman and Shachter (1995).

This approach to causal inference was for example used by Neyman (1923), Rubin (1974, 1978), and Holland (1986), and it plays a fundamental rôle in the methods developed by Robins (1996, 1997), although it is usually introduced in a slightly different context. Counterfactual objects have at all times been at the basis for causal reasoning (Lewis 1973).

Note that in the formulation given above, the variables Ω_v are no more and no less counterfactual than the ω used when a random variable X is considered to be a deterministic function $X(\omega)$ of a random element ω. This has proved useful in many contexts, although it has also lead to paradoxes, when consequences have been taken too far.

Dawid (2000) argues strongly against the use of counterfactual random variables as for any given individual it is impossible to observe more than one of the variables $\Omega_v(x_{\mathrm{pa}(v)})$; the counterfactual variables are *comple-mentary*. Thus it is dangerous to make assumptions concerning the joint distribution of $\{\Omega_v(x_{\mathrm{pa}(v)}) : x_{\mathrm{pa}(v)} \in \mathcal{X}_{x_{\mathrm{pa}(v)}}\}$, as such distributions are purely metaphysical. And, as it seems that all interesting results concerning causal inference can be derived without counterfactuals, the pitfalls associated with their use can be avoided.

2.10.1 Partial compliance revisited

In this section we show how to use counterfactual variables to get bounds for treatment effects in the case of partial compliance, corresponding to the situation displayed in Figure 2.9. Although as mentioned, the bound can be derived without using conterfactual random variables, they seem to yield a simple method for deriving these bounds in the present example.

With the same notation as earlier, we are interested in the intervention probabilities

$$p(x_r \,\|\, x_t^*) = \int_{x_u} p(x_r \,|\, x_t, x_u)\, P_u(dx_u). \qquad (2.28)$$

However, only joint observations of a, t, and r are possible. Assuming that we have an infinite sample, we can observe all combinations of

$$p(x_r, x_t \,|\, x_a) = \int_{x_u} p(x_r \,|\, x_t, x_u) p(x_t \,|\, x_a, x_u)\, P_u(dx_u). \qquad (2.29)$$

As neither of the back-door or front-door criterions apply, the treatment effect appears not to be identifiable, but it is possible to derive bounds for the intervention probabilities in (2.28) subject to the 'constraints' given in (2.29).

For simplicity we assume that all observed values are binary taking the values 0 or 1. In this case there are a total of six independent constraints, three for each group of treatment assignment.

Bounds for the probabilities involved can be derived in many ways. For example, the bounds (2.25) derived for instrumental variables apply to the observed frequencies here since a is indeed an instrument. Thus this part of the assumptions can and should be checked with observed data. Bounds for treatment effects were also derived by Robins (1989) and Manski (1990). However, it is not always easy to check that the bounds derived are sharp and indeed Balke and Pearl (1994, 1997) derive sharper bounds and show that the bounds cannot be improved. Their argument is based upon the use of counterfactual variables and we shall sketch their argument below.

It may be illuminating to phrase the arguments in terms of the example also considered by Imbens and Rubin (1997) and Balke and Pearl (1997). The example considered is thus the study of the effects on child mortality of vitamin A supplementation in Sumatra, as described by Sommer, Tarwotjo,

Djunaedi, West, Leodin, Tilden and Mele (1986) and Sommer and Zeger (1991).

Also here the first part of the argument is that it is not the value or nature of X_u that matters, but only the way in which it affects the two responses t and r. Thus — as was also done by Imbens and Rubin (1997) — we can without loss of generality assume that the unobserved variable is the pair of *potential responses* $\omega = (\omega_t, \omega_r)$, where $\omega_t(x_a)$ denotes the treatment taken by an individual with assigned treatment x_a, and $\omega_r(x_t)$ indicates the response of an individual with treatment x_t.

Each of the potential response variables varies in a space of four elements, so the unobserved variable ω has a total of 16 possible values. The four values of the first variable ω_t may well be called

$$\{always\ taker, never\ taker, complier, defier\},$$

so that we have *always taker* $(x_a) = 1$, where 1 denotes that vitamin A is taken, *complier* $(x_a) = x_a$ etc. Similarly the four values of ω_t may be called

$$\{always\ cured, never\ cured, beneficial, damaging\}.$$

In these terms we can rewrite the equations (2.28) and (2.29) as

$$p(x_r \,\|\, x_t^*) = \sum_\omega p(x_r \,|\, x_t, \omega)p(\omega). \tag{2.30}$$

and

$$p(x_r, x_t \,|\, x_a) = \sum_\omega p(x_r \,|\, x_t, \omega)p(x_t \,|\, x_a, \omega)p(\omega). \tag{2.31}$$

The difference between these and those above are that the conditional probabilities in (2.30) and (2.31) are known and equal to one or zero. Thus the problem of finding bounds can be solved by linear programming methods that also identify the best possible bounds. If we let $p_{ij.k} = p(i_r, j_t \,|\, k_a)$ and $q_{ij} = p(i_r \,\|\, j_a)$, the bounds were found to be

$$\left.\begin{array}{c} p_{10.1} \\ p_{01.0} \\ p_{10.0} + p_{11.0} - p_{00.1} - p_{11.1} \\ p_{01.0} + p_{10.0} - p_{00.1} - p_{01.1} \end{array}\right\} \le q_{10} \le \left\{\begin{array}{c} 1 - p_{00.1} \\ 1 - p_{00.0} \\ p_{01.0} + p_{10.0} + p_{10.1} + p_{11.1} \\ p_{10.0} + p_{11.0} + p_{01.1} + p_{10.1} \end{array}\right.$$

and the remaining bounds are obtained by suitable index substitution.

The bounds turn out to be quite wide in the example mentioned and thus the analysis is inconclusive in this case. Imbens and Rubin (1997), make a full Bayesian analysis of the model, by imposing prior assumptions on the distribution of the potential responses, and thereby obtains the conclusion that the effect of vitamin A is beneficial on average. However, such prior assumptions are untestable and may therefore be questionable. See also Chickering and Pearl (1999) for a further discussion of this example.

As demonstrated in Balke and Pearl (1997), the bounds are sometimes tight and sharp conclusions therefore available. This holds for example for data concerning lipids and coronary heart disease analysed by Efron and Feldman (1991).

2.11 Other issues

2.11.1 Extension to chain graphs

The intervention calculus can be extended to more general graphical models than those given by directed acyclic graphs. Chain graph models are given by graphs that have both directed and undirected links, but no cycles that can be traversed only in one direction without going against the arrows.

The *chain components* $\mathcal{T}$ of such graphs are undirected graphs that are obtained by removing all directed arrows from a chain graph. They naturally unify directed acyclic graphs and undirected graphs in that undirected graphs are chain graphs with only one chain component, and directed acyclic graphs are chain graphs with all chain components being singletons. There is a corresponding set of Markov properties associated with chain graphs (Frydenberg 1990a, Lauritzen 1996). In terms of factorization, the chain graph Markov property manifests itself through an outer factorization

$$f(x) = \prod_{\tau \in \mathcal{T}} f\left(x_\tau \mid x_{\mathrm{pa}(\tau)}\right), \tag{2.32}$$

where each factor further factorizes according to the graph $\mathcal{G}^*(\tau)$ as

$$f\left(x_\tau \mid x_{\mathrm{pa}(\tau)}\right) = Z^{-1}\left(x_{\mathrm{pa}(\tau)}\right) \prod_{A \in \mathcal{A}(\tau)} \phi_A(x_A), \tag{2.33}$$

where $\mathcal{A}(\tau)$ are the complete sets in $\mathcal{G}^*(\tau)$ and

$$Z\left(x_{\mathrm{pa}(\tau)}\right) = \sum_{x_\tau} \prod_{A \in \mathcal{A}(\tau)} \phi_A(x_A).$$

The graph $\mathcal{G}^*(\tau)$ is obtained from $\mathcal{G}_{\tau \cup \mathrm{pa}(\tau)}$ by dropping directions on edges and adding edges between any pair of members of $\mathrm{pa}(\tau)$.

If the intervention $X_\alpha \leftarrow x_\alpha^*$ is made, the corresponding intervention formula can be argued to be

$$p(x \,\|\, x_\alpha^*) = \left. \frac{p(x)}{\sum_{y_{\tau_\alpha} : y_\alpha = x_\alpha^*} p(y_{\tau_\alpha} \mid x_{\mathrm{pa}(\tau_\alpha)})} \right|_{x_\alpha = x_\alpha^*} \tag{2.34}$$

where τ_α is the chain component including α. This formula specializes to (2.13) in the fully directed case and (2.14) in the undirected case. This intervention formula corresponds to the analogy with decision networks based on chain graphs as discussed in Cowell et al. (1999). Lauritzen and Richardson (2000) are investigating dynamic regimes that lead to such an intervention calculus and their potential use as an alternative interpretation of simultaneous equation systems.

2.11.2 Causal discovery

Another and more controversial aspect of causal inference from graphical models is associated with identifying causal relationships from data. Ever since the appearance of Glymour, Scheines, Spirtes and Kelly (1987) and the first version of the corresponding program TETRAD, this has been the subject of sometimes quite heated discussions (Freedman 1997, Humphreys and Freedman 1996, Robins and Wasserman 1999, Glymour, Spirtes and Richardson 1999, Humphreys and Freedman 2000).

Basically there have been two different types of approach. The constraint-based approach (Spirtes et al. 1993) is generally conceived to take place in an ideal environment where the joint distribution P of a system X of random variables is known completely without error, whereas the causal graph $\mathcal{D}$ which has generated the distribution is unknown.

Apart from the assumption that such a causal directed acyclic graph $\mathcal{D}$ exists, it is also assumed that P is *faithful* to $\mathcal{D}$, in other words there are no conditional independence relationships between the variables that do not follow from the directed Markov property:

$$A \perp\!\!\!\perp B \,|\, S \implies A \perp_{\mathcal{D}} B \,|\, S.$$

As previously mentioned, results of Meek (1995) indicate that most distributions are indeed faithful.

On the assumption above, Spirtes et al. (1993) provide several algorithms that from a relatively modest number of tests identifies the causal graph up to Markov equivalence, i.e. produce a graph $\mathcal{D}'$ with the property that for all disjoint subsets A, B, and S of V

$$A \perp_{\mathcal{D}'} B \,|\, S \iff A \perp_{\mathcal{D}} B \,|\, S \iff A \perp\!\!\!\perp B \,|\, S.$$

They also give variants of these algorithms that do not assume the entire system of variables to be observed. These results are supplemented with conditions for identifiability of causal effects and give methods for identifying causal effects that remain invariant over such an equivalence class.

Richardson and Spirtes (1999) extend the approach to situations involving feedback.

Little has been done to explore the statistical properties of these and similar methods applied to cases where knowledge about the distribution of X is only obtained through finite samples. Although Spirtes et al. (1993) contains a small simulation study, this area deserves to be better explored.

Another line of this research is based on a pure Bayesian approach to learning the structure of a Bayesian network, as initiated by Cooper and

Herskovits (1992) and Heckerman, Geiger and Chickering (1995). This approach has been further pursued by Heckerman, Meek and Cooper (1999).

See Cooper (1999) for an overview of the current state of the art within this area.

Acknowledgments

The necessary background for the material presented here was acquired while the author was a Fellow at the Center for Advanced Study in Behavioral Sciences, Stanford, California, with financial support partly provided by National Science Foundation, Grant number SBR-9022192. The work was completed while the author was a Fellow at the Fields Institute for Research in Mathematical Sciences, Toronto, Canada. I am grateful to these institutions for the excellent working conditions they provide.

The research was also supported by the Danish Research Councils through their PIFT programme.

I am indebted to Thomas Richardson for providing detailed and constructive critical comments to earlier versions of this manuscript.

2.12 References

Angrist, J. D., Imbens, G. W. and Rubin, D. B.: 1996, Identification of causal effects using instrumental variables (with discussion), *Journal of the American Statistical Association* **91**, 444–472.

Balke, A. and Pearl, J.: 1994, Nonparametric bounds on causal effects from partial compliance data, *in* R. L. de Mantaras and D. Poole (eds), *Proceedings of the 10th Conference on Uncertainty in Artificial Intelligence*, Morgan Kaufmann Publishers, San Francisco, CA, pp. 46–54.

Balke, A. and Pearl, J.: 1997, Bounds on treatment effects from studies with imperfect compliance, *Journal of the American Statistical Association* **92**, 1171–1176.

Bollen, K. A.: 1989, *Structural Equations with Latent Variables*, John Wiley and Sons, New York.

Bowden, R. J. and Turkington, D. A.: 1984, *Instrumental Variables*, Cambridge University Press, Cambridge, UK.

Box, G. E. P.: 1966, Use and abuse of regression, *Technometrics* **8**, 625–629.

Chickering, D. M. and Pearl, J.: 1999, A clinician's tool for analyzing noncompliance, *in* C. Glymour and G. F. Cooper (eds), *Computation, Causation, and Discovery*, MIT Press, Cambridge, MA, pp. 407–424.

Cooper, G. F.: 1999, An overview of the representation and discovery of causal relationships using Bayesian networks, *in* C. Glymour and G. F. Cooper (eds), *Computation, Causation, and Discovery*, MIT Press, Cambridge, MA, pp. 3–62.

Cooper, G. F. and Herskovits, E.: 1992, A Bayesian method for the induction of probabilistic networks from data, *Machine Learning* **9**, 309–347.

Cowell, R. G., Dawid, A. P., Lauritzen, S. L. and Spiegelhalter, D. J.: 1999, *Probabilistic Networks and Expert Systems*, Springer-Verlag, New York.

Cox, D. R.: 1984, Design of experiments and regression, *Journal of the Royal Statistical Society, Series A* **147**, 306–315.

Dawid, A. P.: 2000, Causal inference without counterfactuals, *Journal of the American Statistical Association* **95**, to appear.

Efron, B. and Feldman, D.: 1991, Compliance as an explanatory variable in clinical trials, *Journal of the American Statistical Association* **86**, 9–26.

Freedman, D.: 1997, From association to causation via regression, *in* V. McKim and S. Turner (eds), *Causality in Crisis?*, University of Notre Dame Press, South Bend, IN, pp. 113–182.

Frydenberg, M.: 1990a, The chain graph Markov property, *Scandinavian Journal of Statistics* **17**, 333–353.

Frydenberg, M.: 1990b, Marginalization and collapsibility in graphical interaction models, *Annals of Statistics* **18**, 790–805.

Geiger, D. and Pearl, J.: 1990, On the logic of causal models, *in* R. D. Shachter, T. S. Levitt, L. N. Kanal and J. F. Lemmer (eds), *Uncertainty in Artificial Intelligence IV*, North-Holland, Amsterdam, The Netherlands, pp. 136–147.

Glymour, C. and Cooper, G. F.: 1999, *Computation, Causation, and Discovery*, MIT Press, Cambridge, MA.

Glymour, C., Scheines, R., Spirtes, P. and Kelly, K.: 1987, *Discovering Causal Structure*, Academic Press, New York.

Glymour, C., Spirtes, P. and Richardson, T.: 1999, On the possibility of inferring causation from association without background knowledge, *in* C. Glymour and G. F. Cooper (eds), *Computation, Causation, and Discovery*, MIT Press, Cambridge, MA, pp. 323–331.

Goldberger, A. S.: 1972, Structural equation models in the social sciences, *Econometrica* **40**, 979–2001.

Goldfeld, S. M. and Quandt, R. E.: 1972, *Non-linear Methods in Econometrics*, North-Holland, Amsterdam, The Netherlands.

Haavelmo, T.: 1943, The statistical implications of a system of simultaneous equations, *Econometrica* **11**, 1–12.

Hammersley, J. M. and Clifford, P. E.: 1971, Markov fields on finite graphs and lattices, unpublished manuscript.

Heckerman, D. and Shachter, R.: 1995, Decision-theoretic foundations for causal reasoning, *Journal of Artificial Intelligence Research* **3**, 405–430.

Heckerman, D., Geiger, D. and Chickering, D. M.: 1995, Learning Bayesian networks: the combination of knowledge and statistical data, *Machine Learning* **20**, 197–243.

Heckerman, D., Meek, C. and Cooper, G.: 1999, A Bayesian approach to causal discovery, *in* C. Glymour and G. F. Cooper (eds), *Computation, Causation, and Discovery*, MIT Press, Cambridge, MA, pp. 141–165.

Holland, P.: 1986, Statistics and causal inference, *Journal of the American Statistical Association* **81**, 945–960.

Howard, R. A. and Matheson, J. E.: 1984, Influence diagrams, *in* R. A. Howard and J. E. Matheson (eds), *Readings in the Principles and Applications of Decision Analysis*, Strategic Decisions Group, Menlo Park, CA.

Humphreys, P. and Freedman, D.: 1996, The grand leap, *British Journal for the Philosophy of Science* **47**, 113–123.

Humphreys, P. and Freedman, D.: 2000, Are there algorithms that discover causal structure?, *Synthese*. In press.

Imbens, G. W. and Rubin, D. B.: 1997, Bayesian inference for causal effects in randomized experiments with noncompliance, *Annals of Statistics* **25**, 305–327.

Jensen, F. V., Lauritzen, S. L. and Olesen, K. G.: 1990, Bayesian updating in causal probabilistic networks by local computation, *Computational Statistics Quarterly* **4**, 269–282.

Kiiveri, H. and Speed, T. P.: 1982, Structural analysis of multivariate data: A review, *in* S. Leinhardt (ed), *Sociological Methodology*, Jossey-Bass, San Francisco.

Kiiveri, H., Speed, T. P. and Carlin, J. B.: 1984, Recursive causal models, *Journal of the Australian Mathematical Society, Series A* **36**, 30–52.

Koster, J. T. A.: 1996, Markov properties of non-recursive causal models, *Annals of Statistics* **24**, 2148–2177.

Koster, J. T. A.: 1999a, Linear structural equations and graphical models, Lecture Notes. The Fields Institute, Toronto.

Koster, J. T. A.: 1999b, On the validity of the Markov interpretation of path diagrams of Gaussian structural equation systems with correlated errors, *Scandinavian Journal of Statistics* **26**, 413–431.

Lauritzen, S. L.: 1996, *Graphical Models*, Clarendon Press, Oxford.

Lauritzen, S. L. and Richardson, T. S.: 2000, Chain graph models for intervention, Manuscript in preparation.

Lauritzen, S. L. and Spiegelhalter, D. J.: 1988, Local computations with probabilities on graphical structures and their application to expert systems (with discussion), *Journal of the Royal Statistical Society, Series B* **50**, 157–224.

Lauritzen, S. L., Dawid, A. P., Larsen, B. N. and Leimer, H.-G.: 1990, Independence properties of directed Markov fields, *Networks* **20**, 491–505.

Lewis, D.: 1973, *Counterfactuals*, Harvard University Press, Cambridge, MA.

Manski, C. F.: 1990, Nonparametric bounds on treatment effects, *American Economic Review, Papers and Proceedings* **80**, 319–323.

Meek, C.: 1995, Strong completeness and faithfulness in Bayesian networks, *Proceedings of the Eleventh Annual Conference on Uncertainty in Artificial Intelligence (UAI-95)*, Morgan Kaufmann Publishers, San Francisco, CA, pp. 411–418.

Neyman, J.: 1923, On the application of probability theory to agricultural experiments. Essay on principles, in Polish. English translation of Section 9 by D. Dabrowska and T. P. Speed in *Statistical Science* **5** (1990), 465–480.

Oliver, R. M. and Smith, J. Q.: 1990, *Influence Diagrams, Belief Nets and Decision Analysis*, John Wiley and Sons, Chichester, UK.

Pearl, J.: 1986a, A constraint–propagation approach to probabilistic reasoning, *in* L. N. Kanal and J. F. Lemmer (eds), *Uncertainty in Artificial Intelligence*, North-Holland, Amsterdam, The Netherlands, pp. 357–370.

Pearl, J.: 1986b, Fusion, propagation and structuring in belief networks, *Artificial Intelligence* **29**, 241–288.

Pearl, J.: 1988, *Probabilistic Inference in Intelligent Systems*, Morgan Kaufmann Publishers, San Mateo, CA.

Pearl, J.: 1993, Graphical models, causality and intervention, *Statistical Science* **8**, 266–269. Comment to Spiegelhalter *et al.* (1993).

Pearl, J.: 1995a, Causal diagrams for empirical research, *Biometrika* **82**, 669–710.

Pearl, J.: 1995b, Causal inference from indirect experiments, *Artificial Intelligence in Medicine* **7**, 561–582.

Pearl, J.: 1998, Graphs, causality, and structural equation models, *Sociological Methods and Research* **27**, 226–284.

Pearl, J.: 2000, *Causality: Models, Reasoning, and Inference*, Cambridge University Press, Cambridge, UK.

Pearl, J. and Paz, A.: 1987, Graphoids: A graph based logic for reasoning about relevancy relations, *in* B. D. Boulay, D. Hogg and L. Steel (eds), *Advances in Artificial Intelligence—II*, North-Holland, Amsterdam, The Netherlands, pp. 357–363.

Richardson, T. and Spirtes, P.: 1999, Automated discovery of linear feedback models, *in* C. Glymour and G. F. Cooper (eds), *Computation, Causation, and Discovery*, MIT Press, Cambridge, MA, pp. 253–304.

Richardson, T. S.: 1996, Models of Feedback: Interpretation and Discovery, PhD thesis, Carnegie-Mellon University, Pittsburgh, PA.

Robins, J. M.: 1986, A new approach to causal inference in mortality studies with sustained exposure periods — application to control of the healthy worker survivor effect, *Mathematical Modelling* **7**, 1393–1512.

Robins, J. M.: 1989, The analysis of randomized and non-randomized AIDS treatment trials using a new approach to causal inference in longitudinal studies, *in* L. Sechrest, H. Freeman, and A. Mulley (eds), *Health Service Research Methodology: A Focus on AIDS*, U.S. Public Health Service, Washington D.C., pp. 113–159.

Robins, J. M.: 1997, Causal inference from complex longitudinal data, *in* M. Berkane (ed.), *Latent Variable Modelling and Applications to Causality*, Vol. 120 of *Lecture Notes in Statistics*, Springer-Verlag, New York, pp. 69–117.

Robins, J. M. and Wasserman, L.: 1999, On the impossibility of inferring causation from association without background knowledge, *in* C. Glymour and G. F. Cooper (eds), *Computation, Causation, and Discovery*, MIT Press, Cambridge, MA, pp. 305–321.

Rubin, D. B.: 1974, Estimating causal effects of treatments in randomized and non-randomized studies, *Journal of Educational Psychology* **66**, 688–701.

Rubin, D. B.: 1978, Bayesian inference for causal effects. The role of randomization, *Annals of Statistics* **6**, 34–58.

Shachter, R. D.: 1986, Evaluating influence diagrams, *Operations Research* **34**, 871–882.

Shafer, G.: 1985, Conditional probability, *International Statistical Review* **53**, 261–277.

Shafer, G.: 1996, *The Art of Causal Conjecture*, MIT Press, Cambridge, MA.

Shenoy, P. P. and Shafer, G.: 1990, Axioms for probability and belief-function propagation, *in* R. D. Shachter, T. S. Levitt, L. N. Kanal and J. F. Lemmer (eds), *Uncertainty in Artificial Intelligence IV*, North-Holland, Amsterdam, The Netherlands, pp. 169–198.

Smith, J. Q.: 1989, Influence diagrams for Bayesian decision analysis, *European Journal of Operational Research* **40**, 363–376.

Sommer, A. and Zeger, S. L.: 1991, On estimating efficacy from clinical trials, *Statistics in Medicine* **10**, 45–52.

Sommer, A., Tarwotjo, I., Djunaedi, E., West, K. P., Leodin, A. A., Tilden, R. and Mele, L.: 1986, Impact of vitamin A supplementation on childhood mortality: a randomized controlled community trial, *The Lancet*, pp. 1169–1173.

Speed, T. P.: 1979, A note on nearest-neighbour Gibbs and Markov probabilities, *Sankhyā, Series A* **41**, 184–197.

Spiegelhalter, D. J., Dawid, A. P., Lauritzen, S. L. and Cowell, R. G.: 1993, Bayesian analysis in expert systems (with discussion), *Statistical Science* **8**, 219–283.

Spirtes, P.: 1995, Directed cyclic graphical representations of feedback models, *in* P. Besnard and S. Hanks (eds), *Proceedings of the Eleventh Annual Conference on Uncertainty in Artificial Intelligence (UAI–95)*, Morgan Kaufmann Publishers, San Francisco, CA, pp. 491–498.

Spirtes, P., Glymour, C. and Scheines, R.: 1993, *Causality, Prediction and Search*, Springer-Verlag, New York.

Spirtes, P., Richardson, T., Meek, C., Scheines, R. and Glymour, C.: 1998, Using path diagrams as a structural modelling tool, *Sociological Methods and Research* **27**, 182–225.

Strotz, R. H. and Wold, H. O. A.: 1960, Recursive versus nonrecursive systems: an attempt at synthesis, *Econometrica* **28**, 417–427.

Verma, T. and Pearl, J.: 1990, Causal networks: Semantics and expressiveness, *in* R. D. Shachter, T. S. Levitt, L. N. Kanal and J. F. Lemmer (eds), *Uncertainty in Artificial Intelligence IV*, North-Holland, Amsterdam, The Netherlands, pp. 69–76.

Wold, H. O. A.: 1954, Causality and econometrics, *Econometrica* **22**, 162–177.

Wright, S.: 1921, Correlation and causation, *Journal of Agricultural Research* **20**, 557–585.

Wright, S.: 1923, The theory of path coefficients: a reply to Niles' criticism, *Genetics* **8**, 239–255.

Wright, S.: 1934, The method of path coefficients, *Annals of Mathematical Statistics* **5**, 161–215.

CHAPTER 3

State Space
and
Hidden Markov Models

Hans R. Künsch
ETH-Zürich

3.1 Introduction

State space and hidden Markov models have the common feature that they assume the observations to be incomplete and noisy functions of some underlying unobservable process, called the state process, which is assumed to have a simple Markovian dynamic. They originated in engineering in the early sixties, the most famous names being Kalman and Bucy with their filter (Kalman (1960), Kalman and Bucy (1961)) and Baum with what is nowadays called the EM-algorithm (Baum and Petrie (1966), Baum et al., (1970)). Whereas these methods were immediately widely used by control engineers and people working in speech recognition, their importance was recognized by statisticians only later. This is surprising because ideas of latent variables and recursive estimation have been around in statistics for a long time, and special cases of the Kalman filter can be found in the statistical literature before 1960, see Plackett (1950) and Lauritzen (1981). State space methods appeared in the time series literature in the seventies (Akaike (1974), Harrison and Stevens (1976)) and became established during the eighties. Finally during the last decade they have become a focus of interest. This is due on the one hand to an even wider range of applicability, including for instance molecular biology and genetics, and on the other hand to the possibility of dealing with many nonlinear and non-Gaussian situations with modern Monte Carlo methods.

These lectures have the following three purposes:

- To show the wide variety of applications of state space and hidden Markov models.

- To give a unified presentation of some basic recursions for these models.

- To present some of the recently developed Monte Carlo methods which allow to fit these models without restriction to special cases and distributions.

A large part of these lectures is about modeling and computing issues since they are central for all applications. In order to achieve some balance, we will look also at a few more mathematical issues like the effect of changes in the initial distribution (Section 3.3.6), the error in recursive Monte Carlo methods (Section 3.5.4) and asymptotics for maximum likelihood estimators in state space models (Section 3.6.5). However these parts are not essential for understanding the rest. In a final Section we present extensions of the basic models: We discuss the corresponding models in a spatial setting in some detail and give a brief introduction to stochastic grammars.

I have tried to give adequate references wherever possible. But the field is very wide, the literature is immense and there are many connections to other areas and additional topics. It is therefore impossible to be complete and I apologize for any omissions.

3.2 The general state space model

The general state-space assumes that the observed time series (Y_t) is derived from an unobservable state process (X_t) having simple, i.e. Markovian dependence through independent and instantaneous transitions. This means

(M1) $X_0, X_1, X_2, \ldots$ is a Markov chain.

(M2) Conditionally on (X_t), the Y_t's are independent and Y_t depends on X_t only.

In addition, we assume that the initial distribution of X_0 and the relevant conditional distributions are absolutely continuous with respect to reference measures μ and ν on the spaces $(\mathcal{X}, \mathcal{F})$ and $(\mathcal{Y}, \mathcal{G})$ respectively where X_t and Y_t take their values. In all our examples, these spaces are either discrete or open or compact subsets of an Euclidean space. The reference measure is then either the counting or Lebesgue measure. In general we will not worry about issues of measurability.

Thus we assume

$$X_0 \quad \sim \quad a_0(x)d\mu(x), \tag{3.1}$$

$$X_t \mid X_{t-1} = x_{t-1} \quad \sim \quad a_t(x_{t-1}, x)d\mu(x), \tag{3.2}$$

$$Y_t \mid X_t = x_t \quad \sim \quad b_t(x_t, y)d\nu(y). \tag{3.3}$$

When the state variables are discrete, one usually calls this model a *hidden Markov model*. The term *state space model* is mainly used for continuous state variables. Their development has been largely separated although the basic recursions and algorithms are closely related. We will reserve here the term hidden Markov for discrete state variables, but use the term state space model for all cases.

The main tasks are to make inferences about the unobserved states or future observations based on a part of the observation sequence, or to make inferences about the model itself (estimating unknown parameters in the transitions a_t and b_t, comparing different models, assessing how well a model fits). In Sections 3.3 till 3.5 we deal with inference about the states and future observations, assuming the transition densities a_t and b_t to be given. In Section 3.6 we will let these densities depend on parameters θ and discuss their estimation.

The joint density of $(X_0, X_1, \ldots X_T, Y_1, \ldots Y_T)$ can easily be written down

$$a_0(x_0) \prod_{t=1}^{T} a_t(x_{t-1}, x_t) b_t(x_t, y_t). \tag{3.4}$$

This is a special case of a graphical model on a directed acyclic graph (see Lauritzen (1996)) as shown in the following scheme

$$
\begin{array}{ccccccccc}
\text{state process} & \ldots & \to & X_{t-1} & \to & X_t & \to & X_{t+1} & \to & \ldots \\
& & & \downarrow & & \downarrow & & \downarrow & & \\
\text{observation} & \ldots & & Y_{t-1} & & Y_t & & Y_{t+1} & & \ldots
\end{array}
$$

Because the joint density factorizes into a product of terms containing only pairs of variables, it is also a Markov random field on the undirected graph obtained by dropping all arrows. This means that two groups of variables are conditionally independent given a third group when the set of vertices corresponding to the third group separates the two other groups of vertices, see Lauritzen (1996), Section 3.2.

In particular, the joint process (Z_t) where $Z_t = (X_t, Y_t)$ is also Markovian. Note that the density of $(Y_1, \ldots Y_T)$ is obtained by integrating out all x-variables from the joint density (3.4). It is thus not available in closed form, and the observation process (Y_t) is in general not Markovian. Although the dependence structure of (Y_t) is more complicated than for a Markov process, it still inherits various properties like ergodicity or strong mixing from the corresponding properties of (Z_t). In particular, state space models are not suitable for modeling long-range dependence, see also Chan and Palma (1998).

Often, the basic model is extended by introducing additional arrows in the dependence graph. Usually it is not difficult to extend the results we will present also to these slightly more complex situations. In many instances, one can reduce the more complex models to the basic model by enlarging the state space, but often the condition of absolute continuity of the transitions is then no longer satisfied.

3.2.1 Examples

A prominent role is played by the linear state space model

$$X_t = G_t X_{t-1} + V_t \tag{3.5}$$

$$Y_t = H_t X_t + W_t, \tag{3.6}$$

where the states X_t are k-dimensional, the observations Y_t are l-dimensional, G_t is a $k \times k$-matrix, H_t is a $l \times k$-matrix and $(V_t), (W_t)$ are two independent white noise sequences with mean zero and covariance matrices Σ_t and Ω_t respectively. In contrast to this, the general state space model can be written in the form

$$X_t = g_t(X_{t-1}, V_t) \tag{3.7}$$
$$Y_t = h_t(X_t, W_t) \tag{3.8}$$

with arbitrary functions g_t and h_t. It is thus much more flexible.

ARMA models as state space models

It can be shown that Gaussian ARMA and linear Gaussian state space models are equivalent in the time-invariant case, see e.g. Wei (1990), Chapter 15, for a simple proof and Akaike (1974) or Hannan and Deistler (1988) for more details. It is instructive to see the basic idea in the univariate case. Let (Y_t) be a Gaussian causal and invertible ARMA(p,q)-process (see Brockwell and Davis (1991), Chapter 3, for the definitions)

$$Y_t = \sum_{j=1}^{p} \phi_j Y_{t-j} + \sum_{j=0}^{q} \theta_j \varepsilon_{t-j}.$$

Here, (ε_t) is a Gaussian white noise. By causality ε_t is independent of $(Y_s, s < t)$ and by invertibility ε_t can be expressed with $(Y_s, s \le t)$. Denoting by $Y_{t|s}$ the conditional expectation of Y_t given Y_u for all $u \le s$, it follows from this and linearity of the conditional expectation

$$Y_{t+j|t} = \sum_{i=1}^{\min(j-1,p)} \phi_i Y_{t+j-i|t} + \sum_{i=j}^{p} \phi_i Y_{t+j-i} + \sum_{i=j}^{q} \theta_i \varepsilon_{t+j-i}, \tag{3.9}$$

empty sums being taken as zero. Thus for $j \ge k = \max(p, q+1)$, $Y_{t+j|t}$ is linearly dependent with $(Y_t, Y_{t+1|t}, \dots Y_{t+k-1|t})$. We therefore define the state vector

$$X_t = (Y_t, Y_{t+1|t}, \dots Y_{t+k-1|t})'.$$

Then (3.6) obviously holds with $H_t = (1, 0, \dots 0)$ and $W_t \equiv 0$. Moreover, again by standard properties of the ARMA model and of the conditional expectation

$$Y_{t+i|t+1} = Y_{t+i|t} + b_i \varepsilon_{t+1} \tag{3.10}$$

for some coefficients b_i. Combining (3.9) with (3.10), we obtain

$$Y_{t+k|t+1} = \sum_{j=1}^{p} \phi_j Y_{t+k-j|t} + b_k \varepsilon_{t+1},$$

so (3.5) also holds.

In this example, the state variables consist of everything that is needed to predict the future from the present and the past. (Because of the Gaussian assumption, it is sufficient to know the conditional expectations of future observations). This intuitive interpretation of the state variables holds quite generally. Note the close connection with second order differential equations in deterministic modeling where one only needs the position and velocity to obtain the whole future.

An important advantage of the state space representation of ARMA models is the ease with which missing data can be handled. If the observation at time t is missing, we simply put $H_t = (0, \ldots 0)'$. The methods we are going to discuss allow then to compute the exact likelihood function for ARMA models with missing observations which is difficult to do otherwise, see e.g. Jones (1980).

Replacing the Gaussian errors with heavy tailed errors can be used to model different types of outliers. An unusually large value in V_t corresponds to an innovative outlier which affects also subsequent observations Y_s for $s > t$. On the other hand, an unusually large value in W_t corresponds to an additive outlier which affects only Y_t. In general, one can construct good robust estimators by using maximum likelihood with a heavy tailed instead of a Gaussian error distribution. This principle can be used here to obtain robust estimators for Gaussian ARMA models. At the same time, we will be able to identify outliers. However, the whole idea is not so easy to implement, because it is difficult to deal with non-Gaussian linear state space models. Approximate methods are due to Masreliez (1975) and Martin and Thompson (1982). But these methods do not allow one to estimate the parameters simultaneously. One has to use an iterative scheme which alternates between removing outliers and estimating the parameters from the cleaned data. Therefore the methods discussed later in this chapter provide a useful alternative for robust time series analysis.

Introducing heavy tails in the noise distribution generates only isolated outliers since the noises are independent for different times. For many applications, this is not sufficient. One can however also model groups of outliers, by adding a discrete component $Z_t \in \{0, 1, \ldots, m\}$ to the state variable. We let the noises have light tails if $Z_t = 0$ and heavy tails if $Z_t > 0$ and take (Z_t) to be a Markov chain with transitions according to the following scheme:

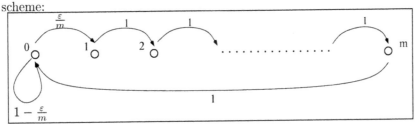

Obviously, this creates groups of outliers of size m. By changing the transition structure of (Z_t), we can also obtain groups of outliers of random size.

Structural time series models

This class of models is described in detail in Harvey (1989), West and Harrison (1997) and Kitagawa and Gersch (1996). The observations are modeled as a superposition of unobservable components, namely a trend T_t, a seasonal component S_t and noise W_t. The state variable X_t consists then of the trend and the seasonal component at time t and some previous time points. An additive superposition gives the observation equation

$$Y_t = T_t + S_t + W_t,$$

but one can also consider GLIM type models where

$$b_t(x_t, y_t) = \exp(y_t(t_t + s_t) - d(t_t + s_t))c(y_t).$$

In this way one can in particular model count data by using a conditional Poisson distribution for Y_t. For the use of state space models for count data, see also MacDonald and Zucchini (1997).

The state transitions are obtained by writing down a difference equation for a deterministic trend and seasonal component and adding some noise to it. A simple deterministic trend is constant or linear, i.e. $T_t = T_{t-1}$ or $T_{t+1} - T_t = T_t - T_{t-1}$. This leads then to the following models for a stochastic trend

$$T_t = T_{t-1} + V_t \tag{3.11}$$

and

$$T_t = T_{t-1} + A_{t-1}, \quad A_t = A_{t-1} + V_t \tag{3.12}$$

respectively. The transitions in model (3.12) are not absolutely continuous unless we add also a small noise to the first equation. We will discuss in Section 3.3.5 how to do the computations for the model (3.12).

A deterministic seasonal component (with monthly data) satisfies

$$S_t = S_{t-12}$$

or – if we want to add the requirement that a constant belongs to the trend and not to the seasonal component –

$$S_t + S_{t-1} + \ldots + S_{t-11} = 0.$$

Note that the former condition follows from the latter because

$$S_t - S_{t-12} = (S_t + \ldots + S_{t-11}) - (S_{t-1} + \ldots + S_{t-12}).$$

By adding noises to these equations, we obtain stochastic seasonal components which are Markovian of higher order. They can be converted to first order by stacking consecutive values into a vector.

An alternative way begins with the Fourier representation

$$S_t = \sum_{j=1}^{5}(A_j \cos(\frac{\pi j}{6}t) + B_j \sin(\frac{\pi j}{6}t)) + A_6(-1)^t. \qquad (3.13)$$

A stochastic seasonal is then obtained by making the coefficients A_j and B_j time dependent, but approximately constant as in (3.11).

Again, noises need not be Gaussian. With heavy tailed errors we can represent outliers in the data as well as jumps (level shifts) in the trend or the seasonal component. Moreover, the irregular component W_t need not be white, but could also be modeled as an ARMA process by using the results from Section 3.2.1.

The trend plus noise model without the seasonal component can be used to describe equidistant noisy observations of a smooth curve $(f(u); 0 \le u \le 1)$ in a Bayesian way, by letting $T_t = f(t/T)$. As we shall see in Section 3.3.3, there is a connection with discrete splines.

Autoregressive models with time varying coefficients have also been considered, see e.g. Kitagawa and Gersch (1996), Chapter 11. The autoregressive coefficients $(\phi_{1,t}, \dots \phi_{p,t})'$ form the state variables and the model is

$$\begin{aligned}
\phi_{i,t} &= \phi_{i,t-1} + V_{i,t} \quad (i = 1, \dots p) \\
Y_t &= (\phi_{1,t}, \dots \phi_{p,t})(Y_{t-1}, \dots Y_{t-p})' + W_t.
\end{aligned}$$

Note that here (Y_t) is p-th order Markov given (x_t), i.e. there are additional arrows in the basic dependence graph.

Engineering examples

Because state space and hidden Markov models mainly came from engineering, examples are abundant, see e.g. Anderson and Moore (1979). For instance in the *tracking problem*, the state consists of the position and velocity of a moving object in two or three dimensions. In the simplest case, the state equation is taken as a multidimensional version of the trend model (3.12), but more complicated physics can of course be incorporated. The observation consists of the position or some part of it with added noise. The task is then to infer the position and the velocity of the object based on the data. In the bearings only tracking, the observation is given by the angle under which the object is seen by an observer at a fixed position. See for instance Aidala (1979), Weiss and Moore (1980) or Gordon et al. (1993) for more details and concrete examples.

State space models are also used frequently in *control problems*. There the state transitions depend on an additional control process (U_t). The task is then to choose the control U_t at time t as a function of the observations up to time $t-1$, $(Y_1, \dots Y_{t-1})$, in such a way that the expectation of a cost function $C_t(U_t, X_t, z_t)$ is minimal. Here z_t is the target of the process, and the cost functional involves usually a distance between X_t and z_t. For more details see e.g. Whittle (1996).

Hidden Markov models are used extensively in *speech recognition*. There the recorded acoustic signal is divided into short time intervals (called frames) with some overlap, and a feature extraction process assigns each time frame to one of a finite number of categories. The sequence of these categories is then taken as the observations (Y_t). In an isolated word recognition problem, one has a hidden Markov model for each word in the vocabulary. The states X_t of this model are the different phonemes (or different stages of a phoneme) occurring while pronouncing a given word. The transition probabilities satisfy usually $a(i, j) = 0$ for $j < i$, meaning that the states are passed through sequentially. But the time spent in a given state is random, and certain states can be skipped altogether. In this way variation in the pronunciation is taken into account. For each possible word, one computes the posterior probability of the most likely state sequence given the observations and then selects the word that has highest posterior probability. Often there is also a whole hierarchy of states containing for instance the words or sentences spoken in addition to the phonemes. For more details, see e.g. Rabiner and Juang (1993).

Biological examples

Ion channels are proteins located in the cell membrane of plant and animal cells which allow the transport of selected ions in and out of the cell. This transport of ions induces a small current which governs a number of biological functions. The current can be measured at the level of a protein by the so-called "patch clamp" technique. An example is shown in Figure 1.1. It is seen that open and close periods alternate and that measurements are noisy. A hidden Markov model with binary states describing whether the channel is open or closed is thus a natural model. The observed current Y_t is then modeled as

$$Y_t = m(X_t) + \varepsilon_t.$$

Here $m(0)$ and $m(1)$ are the two levels of the measurements and (ε_t) is Gaussian white noise. However, a closer look at data like in Figure 1.1 reveals that this model is too simple. First, the periods during which the channel is closed are obviously not geometric as they should be for a binary Markov chain. This is usually handled by introducing several types of closed states, e.g. a "fast" and a "slow" closed state or a sequence of closed states which are visited sequentially, similar to the diagram in Section 3.2.1. Moreover, the noise is not white and the variance is state dependent. Finally in the case of very short closures the signal does not reach the level zero. The last phenomenon is explained as the effect of a filter in the measurement process. This makes Y_t depend not only on X_t, but also on a few previous values X_{t-j}, i.e. additional arrows are introduced in the depen-

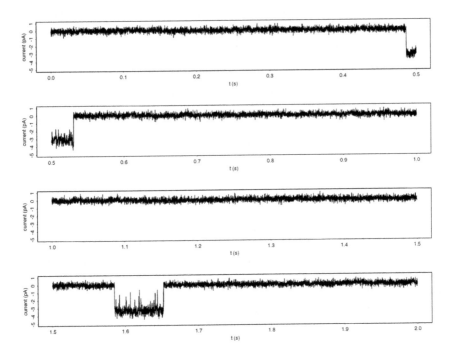

Figure 3.1 *Two seconds of a potassium ion-channel in a leaf cell of barley with sampling frequency 10 kHz. Data kindly provided by Sake Vogelzang at the Biology faculty of the Free University.*

dence graph. The task is then to identify the closed and open periods and to make inference about the process switching between open and closed states.

For an introduction to ion channel models and a bibliography up to 1992 see Ball and Rice (1992). We will discuss a model incorporating some of the features mentioned above in Section 3.5.2, following De Gunst et al. (1998). There one can also find some more recent references.

Hidden Markov models have also found wide applications in genetics and molecular sequence analysis, see e.g. Durbin et al. (1998) and Thompson (2000) in Chapter 4 of this volume. In order to give some flavor of the kind of problems which are solved using hidden Markov models, we give two examples.

CG-islands, see Durbin et al. (1998). One strand of the genome is a sequence of letters from the DNA alphabet $\{A, C, G, T\}$. These letters are not independent, so one has to use at least a first order Markov model. Moreover, the transitions are not constant over long sequences. In particular, in some parts called *CG* islands, the pairs *CG* occur more frequently than in other parts. A simple model for this is a hidden Markov model where the state belongs to $\{A, C, G, T\} \times \{0, 1\}$, 1 and 0 indicating whether we are in a CG-island or not. The observation consists of the letter only, without the indicator variable, i.e. Y_t is a deterministic function of X_t. The task is then to identify the CG-islands based on the sequence alone.

The transition matrix for the states has the form

$$\begin{pmatrix} (1 - \varepsilon)P_0 & \varepsilon Q \\ \eta Q & (1 - \eta)P_1 \end{pmatrix}.$$

Here P_0 contains the transition probabilities outside a CG-island and P_1 the transition probabilities within CG-islands. The parameter ε is the probability for entering a CG-island and η the probability of leaving a CG-island. The matrix Q contains the conditional transition probabilities given that a switch occurs. The matrices P_0 and P_1 can be obtained empirically from sequences where the segmentation in CG-islands has been done by an ad-hoc algorithm. For simplicity one can let Q have identical rows corresponding to the frequencies of the four letters. For more general models of this kind and their use in segmenting a DNA sequence into homogeneous regions, see Muri (1998).

Evolutionary trees from proteins, see Goldman et al. (1996). Protein sequences are formed from the alphabet of the 20 amino acids. Proteins are however not linear, but arranged in complicated three-dimensional structures. One thus considers secondary structure categories like helices, sheets, turns and coils which in addition may be buried or exposed. The secondary structure along a sequence of amino acids are the states X_t in this example. The observations Y_t are aligned sequences of closely related proteins. It is assumed that during evolution only some sequence letters, but no secondary structure category has been changed. The probability of the observed data depends on the mutation rates at the different positions and the phylogenetic tree, and the mutation rates at each position depend on the secondary structure of that position, i.e. on the state. The phylogenetic tree is considered as an unknown parameter. Transitions between secondary structure categories and mutation rates given the secondary structure are determined empirically from proteins whose secondary structure is known. The task is to infer the secondary structure and the phylogenetic tree.

Examples from finance mathematics

If P_t denotes the price of an asset at time t, then at least to a first approximation the log return $Y_t = \log(P_t/P_{t-1})$ has conditional mean zero given the past, but its conditional variance – called volatility – depends on the past. The ARCH and GARCH models describe this conditional variance as a function of the past values Y_s for $s < t$ whereas *stochastic volatility models* consider it as an exogenous random process. This leads to a state space model with the exogenous random process as the state variable. The simplest form is the following

$$
\begin{aligned}
X_t &= m + \phi X_{t-1} + V_t, \\
Y_t &= \exp(X_t)\, W_t.
\end{aligned}
$$

Here (V_t) and (W_t) are two independent white noises. Often W_t has heavier tails than the normal distribution. The task is to estimate the parameters (m, ϕ, σ_V^2) and to make predictions about the occurrence of large negative values of Y_t for assessing risk. Because the likelihood of (Y_t) is not available in closed form, fitting these models is typically more demanding than for an ARCH/GARCH model. However it seems more easy to model the log returns of a large collection of assets with the stochastic volatility idea. In the simplest case one would model the volatility of $Y_{i,t}$ as $\exp(Z_t + X_{i,t})$ where Z_t is the collective volatility and $X_{i,t}$ are individual volatilities which are independent for different assets. One could in addition also consider group volatilities affecting the assets in homogeneous groups, with independence between groups. Multivariate ARCH and GARCH models are more difficult because the number of parameters is too large in general.

For a more general discussion of stochastic volatility models, see Shephard (1996). More complicated models in finance are often formulated in terms of stochastic differential equations, observed at discrete time points with possibly some additional observation noise, see e.g. Gallant and Tauchen (1998). Because the solutions of stochastic differential equations have the Markov property, these models fit in principle in the framework considered here. However, the transition densities a_t for the states are not available in explicit form, which causes additional difficulties.

Geophysical examples

Zucchini and Guttorp (1991), Hughes and Guttorp (1999), and Hughes et al. (1999) have used hidden Markov models for modeling daily rainfall at different stations in an area. The different states correspond to different weather situations, and given the states the rainfall amounts at different stations are independent. The transition probabilities for the weather states depend in addition on meteorological variables.

State space models are also used extensively in meteorology for the so-called data assimilation, see e.g. Ghil and Ide (1997). The state at time t denotes the values of the atmospheric variables on a regular grid which are used as starting values for the numerical integration of the fluid dynamical equations. The basic task is to combine the predictions from 6 or 12 hours before with the actual observations to obtain better starting values for the next integration interval. Difficulties arise from the complexity of the fluid dynamical equations and the high dimension of states and observations.

3.2.2 General remarks on model building

One of the great flexibility of state space models is that states can either be latent variables in a frequentist sense or unknown parameters which change in time and are considered as random variables in a Bayesian sense. In these notes we stay more closely to the frequentist viewpoint and consider the states usually as latent variables. In addition to the states there may also be unknown parameters whose estimation is discussed in Section 3.6.

For any Markov chain with a finite number of states, the length of the visits in a particular state has a geometric distribution. This is often not satisfied in applications. If we split a state into several new states which all have the same conditional distribution for the observation, we look effectively at the length of the visit in a group of states. The class of possible distributions is then much larger. This has been used explicitly in the example of ion channels above. But it is also an important feature of the hidden Markov model in many other examples. For instance Goldman et al. (1996) use 38 and not 8 states for the secondary structure.

In most examples above, the states had an intuitive meaning and often were defined a priori from subject considerations. For this reason, the allowed transitions between states are often restricted. In such situations the state space model is usually successful. One can also use the state space model in a non-parametric way where the states are simply a device to encode more complicated dependencies and thus have no intuitive meaning. The number of states and the transition probabilities have then to be inferred from the observations alone. There is only limited information on this approach reported in the literature, but it seems much harder (see e.g. Stolcke and Omohundro (1993) or Künsch et al. (1995)). The difficulties come from the fact that non-parametric approaches leave too much freedom in large state spaces. For binary observations and M possible states, one can assume that Y_t is a fixed deterministic function of X_t. But even then, the transition matrix for the states has $M(M-1)$ parameters, and adding one state gives $2M$ additional parameters which can be difficult to estimate without a reasonable starting values.

Related to the difficulty with a non-parametric approach is the inherent non-uniqueness of state-space models. States can always be transformed by smooth invertible maps which do not change the distribution of the observations if the densities b_t are transformed accordingly. An additional non-uniqueness occurs if the state space is too big so that for instance two different states $x \neq x'$ give exactly the same distribution of the observation, i.e. $b_t(x,.) = b_t(x',.)$. In such a case different transition probabilities give the same distribution of the observations. Finding necessary and sufficient conditions for uniqueness in a parametric family of state space models is therefore difficult. In case of linear Gaussian models, one prefers to work with the so-called innovation form, see Section 3.4.3. For further discussion of the equivalence problem for linear systems, see e.g. Hannan and Deistler (1988), Chapter 2. For the case of hidden Markov models, see Itô et al. (1992).

3.3 Filtering and smoothing recursions

For $s \leq t$ we define $y_s^t = (y_s, y_{s+1}, \ldots y_t)$ and similarly x_s^t. We denote the conditional density of X_t given $Y_1^s = y_1^s$ by $f_{t|s}(x_t \mid y_1^s)$ and distinguish between prediction ($s < t$), filtering ($s = t$) and smoothing ($s > t$). Otherwise we use the sloppy, but useful notation $p(u \mid v)$ for the conditional density of a stochastic vector U given another stochastic vector V. Here U and V will be indicated by the arguments of p and the reference measure will be clear from the context.

3.3.1 Filtering

Lemma 3.1 i) *Prediction densities for the states can be computed from the filter density according to the following recursion in k*

$$f_{t+k|t}(x_{t+k} \mid y_1^t) = \int a_{t+k}(x, x_{t+k}) f_{t+k-1|t}(x \mid y_1^t) d\mu(x). \qquad (3.14)$$

ii) *The prediction densities for the observations can be computed from the prediction densities for the states according to*

$$p(y_{t+k} \mid y_1^t) = \int b_{t+k}(x, y_{t+k}) f_{t+k|t}(x \mid y_1^t) d\mu(x). \qquad (3.15)$$

iii) *The filtering densities can be computed according to the following forward recursion in t (starting with $f_{0|0} = a_0$)*

$$f_{t+1|t+1}(x_{t+1} \mid y_1^{t+1}) = \frac{b_{t+1}(x_{t+1}, y_{t+1}) f_{t+1|t}(x_{t+1} \mid y_1^t)}{p(y_{t+1} \mid y_1^t)}$$

$$= \frac{b_{t+1}(x_{t+1}, y_{t+1}) \int a_{t+1}(x, x_{t+1}) f_{t|t}(x \mid y_1^t) d\mu(x)}{p(y_{t+1} \mid y_1^t)}. \qquad (3.16)$$

Proof. i) follows by conditioning on x_{t+k-1} and the conditional independence of X_{t+k} and Y_1^t given x_{t+k-1}.

ii) follows by conditioning on x_{t+k} and by the conditional independence of Y_{t+k} and Y_1^t given x_{t+k}.

For iii), we apply Bayes' formula to the conditional distribution given y_1^t

$$f_{t+1|t+1}(x_{t+1} \mid y_1^{t+1}) = p(x_{t+1} \mid y_{t+1}, y_1^t) = \frac{p(y_{t+1} \mid x_{t+1}, y_1^t) p(x_{t+1} \mid y_1^t)}{p(y_{t+1} \mid y_1^t)}.$$

By conditional independence of Y_{t+1} and Y_1^t given x_{t+1}, the first factor in the numerator is $b_{t+1}(x_{t+1}, y_{t+1})$. The second equality is (3.14). □

Note that the denominator in (3.16) is just a normalization. As a by-product of these recursions we obtain also the joint density of the observations since

$$p(y_1^T) = \prod_{t=1}^{T} p(y_t \mid y_1^{t-1}) \qquad (3.17)$$

and $p(y_t \mid y_1^{t-1})$ is given in (3.15). This will be used in Section 3.6.

3.3.2 Smoothing

Lemma 3.2 i) *Conditional on y_1^T, $(X_0, X_1, \ldots, X_T)$ is an inhomogeneous Markov chain with forward transition densities*

$$p(x_t \mid x_{t-1}, y_1^T) = p(x_t \mid x_{t-1}, y_t^T) \qquad (3.18)$$

$$= \frac{a_t(x_{t-1}, x_t) b_t(x_t, y_t) p(y_{t+1}^T \mid x_t)}{p(y_t^T \mid x_{t-1})} \qquad (3.19)$$

where

$$p(y_t^T \mid x_{t-1}) = \int a_t(x_{t-1}, x_t) b_t(x_t, y_t) p(y_{t+1}^T \mid x_t) d\mu(x_t). \qquad (3.20)$$

The backward transition densities of this chain are given by

$$p(x_t \mid x_{t+1}, y_1^T) = p(x_t \mid x_{t+1}, y_1^t) = \frac{a_{t+1}(x_t, x_{t+1}) f_{t|t}(x_t \mid y_1^t)}{f_{t+1|t}(x_{t+1} \mid y_1^t)}. \qquad (3.21)$$

ii) *The smoothing densities $f_{t|T}$ can be computed according to the following backward recursion in t (starting with $f_{T|T}$)*

$$f_{t|T}(x_t \mid y_1^T) = f_{t|t}(x_t \mid y_1^t) \int \frac{a_{t+1}(x_t, x)}{f_{t+1|t}(x \mid y_1^t)} f_{t+1|T}(x \mid y_1^T) d\mu(x). \qquad (3.22)$$

Proof. For i) observe first that X_t and (X_1^{t-2}, Y_1^{t-1}) are conditionally independent given (x_{t-1}, y_t^T). Thus

$$p(x_t \mid x_1^{t-1}, y_1^T) = p(x_t \mid x_{t-1}, y_t^T)$$

and the Markov property follows. By Bayes' formula applied to the conditional distribution given x_{t-1}

$$p(x_t \mid x_{t-1}, y_t^T) = \frac{p(y_t^T \mid x_t, x_{t-1}) p(x_t \mid x_{t-1})}{p(y_t^T \mid x_{t-1})}.$$

Because Y_t^T and X_{t-1} as well as Y_t and Y_{t+1}^T are conditionally independent given x_t, we have

$$p(y_t^T \mid x_t, x_{t-1}) = p(y_t^T \mid x_t) = b_t(x_t, y_t) p(y_{t+1}^T \mid x_t),$$

and the formula (3.19) follows. Integrating (3.19) with respect to x_t gives (3.20).

Similarly as in the forward case, we obtain

$$p(x_t \mid x_{t+1}^T, y_1^T) = p(x_t \mid x_{t+1}, y_1^t).$$

Moreover, by Bayes' formula applied to the conditional distribution given y_1^t we obtain

$$p(x_t \mid x_{t+1}, y_1^t) = \frac{p(x_{t+1} \mid x_t, y_1^t)p(x_t \mid y_1^t)}{p(x_{t+1} \mid y_1^t)}.$$

By conditional independence of X_{t+1} and Y_1^t given x_t, the first factor in the numerator is $a_{t+1}(x_t, x_{t+1})$ and the formula for the backward transitions follows.

ii) is a straightforward consequence of i). $\square$

For actual computations, one prefers (3.21) whenever the filtering densities have been calculated already. We will see applications of (3.19) in Section 3.3.6 on the influence of different initial distributions.

Obviously, Lemma 3.2 provides also the conditional distribution of X_s^t given y_1^T for any $0 \le s < t \le T$ because

$$p(x_s^t \mid y_1^T) = f_{t\mid T}(x_t \mid y_1^T) \prod_{r=s}^{t-1} p(x_r \mid x_{r+1}, y_1^T).$$

The structure of the recursions for filtering and smoothing is summarized in the following diagram. A letter B indicates an application of Bayes' formula, and a letter M indicates a Markov transition.

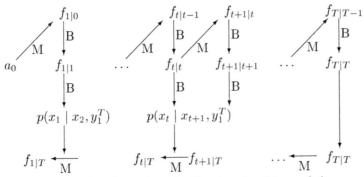

The recursion (3.22) requires the filtering densities and the one-step prediction densities for the states. There is also a recursion which requires the filtering densities and the one-step prediction densities for the observations which are easier to store (because the observations are fixed, these are just numbers). For this purpose, we introduce

$$r_{t\mid T}(x_t, y_1^T) = \frac{p(y_{t+1}^T \mid x_t)}{p(y_{t+1}^T \mid y_1^t)}, \qquad r_{T\mid T} \equiv 1. \tag{3.23}$$

Then we have

Lemma 3.3 i) *The smoothing density satisfies*

$$f_{t|T}(x_t \mid y_1^T) = f_{t|t}(x_t \mid y_1^t) r_{t|T}(x_t, y_1^T). \tag{3.24}$$

ii) *There is the following recursion for* $r_{t|T}$

$$r_{t|T}(x_t, y_1^T) = \frac{\int b_{t+1}(x, y_{t+1}) a_{t+1}(x_t, x) r_{t+1|T}(x, y_1^T) d\mu(x)}{p(y_{t+1} \mid y_1^t)}. \tag{3.25}$$

Proof. By Bayes' formula applied to the conditional distribution given y_1^t we obtain

$$p(x_t \mid y_1^T) = \frac{p(y_{t+1}^T \mid x_t, y_1^t) p(x_t \mid y_1^t)}{p(y_{t+1}^T \mid y_1^t)}.$$

By conditional independence of Y_1^t and Y_{t+1}^T given x_t, the first term in the numerator is equal to $p(y_{t+1}^T \mid x_t)$ and i) follows.

The recursion ii) follows from (3.20) and the relation

$$p(y_{t+1}^T \mid y_1^t) = p(y_{t+1} \mid y_1^t) p(y_{t+2}^T \mid y_1^{t+1}).$$

$\square$

By equation (3.21), the conditional density of X_t given X_{t+1} and Y_1^T does not change if a new observation Y_{T+1} becomes available. Because of this, it is possible to obtain forward recursions for conditional expectations with respect to the smoothing distribution. Let h be an arbitrary (measurable) function on $\mathcal{X}^s$ such that $\mathbb{E}[|h(X_1^s)|] < \infty$ and define for $t \geq s$

$$m_t(x_t, y_1^{t-1}) = \mathbb{E}[h(X_1^s) \mid X_t = x_t, Y_1^{t-1} = y_1^{t-1}].$$

Then we have

Lemma 3.4 *For* $t \geq s$ *it holds*

$$m_{t+1}(x_{t+1}, y_1^t) = \frac{\int m_t(x_t, y_1^{t-1}) f_{t|t}(x_t \mid y_1^t) a_{t+1}(x_t, x_{t+1}) d\mu(x_t)}{\int f_{t|t}(x_t \mid y_1^t) a_{t+1}(x_t, x_{t+1}) d\mu(x_t)} \tag{3.26}$$

$$\mathbb{E}[h(X_1^s) \mid Y_1^t] = \int m_t(x_t, y_1^{t-1}) f_{t|t}(x_t \mid y_1^t) d\mu(x_t) \tag{3.27}$$

Proof. By iterated expectation and conditional independence of (X_1^s, Y_1^{t-1}) and Y_t given x_t we have

$$\mathbb{E}[h(X_1^s) \mid Y_1^t] = \mathbb{E}[\mathbb{E}[h(X_1^s) \mid X_t, Y_1^t] \mid Y_1^t] = \mathbb{E}[\mathbb{E}[h(X_1^s) \mid X_t, Y_1^{t-1}] \mid Y_1^t].$$

From this the second equation (3.27) follows. Similarly, we have for the first equation (3.26)

$$\begin{aligned} \mathbb{E}[h(X_1^s) \mid X_{t+1}, Y_1^t] &= \mathbb{E}[\mathbb{E}[h(X_1^s) \mid X_t, X_{t+1}, Y_1^t] \mid X_{t+1}, Y_1^t] \\ &= \mathbb{E}[\mathbb{E}[h(X_1^s) \mid X_t, Y_1^{t-1}] \mid X_{t+1}, Y_1^t]. \end{aligned}$$

The proof is then completed by applying (3.21). $\square$

In order to compute the starting value $m_s(x_s, y_1^s)$, one still needs the backward recursions, but this is usually simple. In general, Lemma 3.4 is advantageous if one needs $\mathbb{E}[h(X_1^s) \mid Y_1^T]$ only for one or a few functions h.

3.3.3 Posterior mode and dynamic programming

Based on the smoothing densities derived in Section 3.3.2, one can compute several estimates of the state X_t, e.g. the conditional mean or median or the marginal posterior mode $\arg\max_{x_t} f_{t|T}(x_t \mid y_1^T)$. But we cannot obtain the most likely *state sequence*

$$\hat{x}_0^T = \arg\max_{x_0^T} p(x_0^T \mid y_1^T) \tag{3.28}$$

given the observations, the posterior mode or MAP. It can be computed by dynamic programming methods and involves also a forward and a backward recursion. By (3.4) we have

$$\log p(x_0^T \mid y_1^T) \quad \propto \quad \log p(x_0^T, y_1^T)$$
$$= \quad \log a_0(x_0) + \sum_{t=1}^{T} (\log a_t(x_{t-1}, x_t) + \log b_t(x_t, y_t)).$$

The observations y_1^T are considered to be fixed, so we drop them here from the notation. Maximization is done iteratively, by first maximizing over x_0, for fixed x_1^T, then maximizing over x_1 for fixed x_2^T etc. Because of the special structure of $\log p(x_0^T, y_1^T)$ the most likely value of x_0 depends only on x_1, and after maximizing over x_0 the most likely value of x_1 depends only on x_2 etc.. This leads to the following forward recursion

$$\psi_0(x_0) \quad = \quad \log a_0(x_0)$$
$$\psi_t(x_t) \quad = \quad \max_{x_{t-1}}(\psi_{t-1}(x_{t-1}) + \log a_t(x_{t-1}, x_t)) + \log b_t(x_t, y_t)$$
$$\xi_{t-1}(x_t) \quad = \quad \arg\max_{x_{t-1}}(\psi_{t-1}(x_{t-1}) + \log a_t(x_{t-1}, x_t))$$
$$\hat{x}_T \quad = \quad \arg\max_{x_T} \psi_T(x_T).$$

Then the maximal value of $\log p(x_0^T, y_1^T)$ is simply $\psi_T(\hat{x}_T)$ and the posterior mode $\hat{x}_0^T$ is obtained from the backward recursion

$$\hat{x}_{t-1} = \xi_{t-1}(\hat{x}_t).$$

This is most useful in the case where the number of states is finite, say M, because implementation is straightforward. It is then usually called the *Viterbi algorithm*, see Viterbi (1967). It involves TM maximizations over M numbers. For continuous states, one has to discretize the state space first and then typically use some pruning methods in order to keep computations manageable.

The most likely state sequence is of special interest for a local linear trend as in (3.12) with additive observation noise:

$$
\begin{aligned}
X_t &= 2X_{t-1} - X_{t-2} + V_t \\
Y_t &= X_t + W_t.
\end{aligned}
$$

In this case, the posterior mode is nothing else than a discrete smoothing spline. To see this, take the initial distribution of (X_0, X_1) to be flat and denote the log of the densities of V_t and W_t by ρ_V and ρ_W. Then the most likely state sequence (3.28) becomes

$$
\arg\max \sum_{t=1}^{T} (\rho_V(y_t - x_t) + \rho_W(x_t - 2x_{t-1} + x_{t-2})).
$$

In a continuous time formulation for (x_t), the second term which penalizes the lack of smoothness becomes

$$
\int \rho_W(x''(t))dt.
$$

The case $\rho_W(x) = |x|$ has been studied in Mammen and van de Geer (1997).

3.3.4 The reference probability method

In the probability literature, the basic filter recursions of Lemma 3.1 are often derived via the so-called reference probability method, see e.g. Kunita (1971) and Elliott et al. (1995). The idea is to first consider the case where the states and observations are independent so that conditional expectations are easy to compute and then to obtain the case of interest by an absolutely continuous change of measure. More precisely, let $\mathbb{P}$ denote the distribution of the hidden Markov model (3.1) - (3.3) and $\overline{\mathbb{P}}$ denote the distribution of a model where (X_t) and (Y_t) are independent, (X_t) satisfies (3.1) and (3.2) and the Y_t's are i.i.d. $\sim g(y)d\nu(y)$ (the density g is arbitrary, it will not influence the results). Then the likelihood ratio of $\mathbb{P}$ with respect to $\overline{\mathbb{P}}$ restricted to the first t variables is

$$
\Lambda_t = \frac{p(x_0^t, y_1^t)}{\overline{p}(x_0^t, y_1^t)} = \prod_{s=1}^{t} \frac{b_s(x_s, y_s)}{g(y_s)}.
$$

By a standard result on conditional expectations under absolutely continuous measure transformations (see e.g. Loève (1978), Section 27.4, Formula (1')), we have for any bounded measurable function $\phi : \mathcal{X} \to \mathbb{R}$

$$
\mathbb{E}[\phi(X_t) \mid Y_1^t] = \frac{\overline{\mathbb{E}}[\Lambda_t \phi(X_t) \mid Y_1^t]}{\overline{\mathbb{E}}[\Lambda_t \mid Y_1^t]}.
$$

($\mathbb{E}$ and $\overline{\mathbb{E}}$ denote expectation under $\mathbb{P}$ and $\overline{\mathbb{P}}$ respectively). Of course, this is nothing else than a generalization of Bayes' formula in an abstract setting.

The key point is that under $\overline{\mathbb{P}}$, the conditional expectations are trivial: all values Y_s are fixed and the distribution of X_0^t is the same as without conditioning. Therefore

$$
\begin{aligned}
\mathbb{E}[\phi(X_t) \mid Y_1^t = y_1^t] &= \frac{\overline{\mathbb{E}}[\Lambda_t \phi(X_t) \mid Y_1^t = y_1^t]}{\overline{\mathbb{E}}[\Lambda_t \mid Y_1^t = y_1^t]} = \frac{\overline{\mathbb{E}}[\prod_{s=1}^t b_s(X_s, y_s)\phi(X_t)]}{\overline{\mathbb{E}}[\prod_{s=1}^t b_s(X_s, y_s)]} \\
&= \frac{\overline{\mathbb{E}}[\prod_{s=1}^{t-1} b_s(X_s, y_s)\overline{\mathbb{E}}[b_t(X_t, y_t)\phi(X_t) \mid X_{t-1}]]}{\overline{\mathbb{E}}[\prod_{s=1}^{t-1} b_s(X_s, y_s)\overline{\mathbb{E}}[b_t(X_t, y_t) \mid X_{t-1}]]} \\
&= \frac{\overline{\mathbb{E}}[\prod_{s=1}^{t-1} b_s(X_s, y_s) h_{t-1}(X_{t-1}, \phi)]}{\overline{\mathbb{E}}[\prod_{s=1}^{t-1} b_s(X_s, y_s) h_{t-1}(X_{t-1}, 1)]} \\
&= \frac{\overline{\mathbb{E}}[\Lambda_{t-1} h_{t-1}(X_{t-1}, \phi) \mid Y_1^{t-1} = y_1^{t-1}]}{\overline{\mathbb{E}}[\Lambda_{t-1} h_{t-1}(X_{t-1}, 1) \mid Y_1^{t-1} = y_1^{t-1}]} \\
&= \frac{\mathbb{E}[h_{t-1}(X_{t-1}, \phi) \mid Y_1^{t-1} = y_1^{t-1}]}{\mathbb{E}[h_{t-1}(X_{t-1}, 1) \mid Y_1^{t-1} = y_1^{t-1}]}
\end{aligned}
$$

where

$$
\begin{aligned}
h_{t-1}(x, \phi) &= \mathbb{E}[b_t(X_t, y_t)\phi(X_t) \mid X_{t-1} = x] \\
&= \int b_t(x_t, y_t)\phi(x_t)a_t(x, x_t)d\mu(x_t).
\end{aligned}
$$

Assuming that the conditional distribution of X_{t-1} given Y_1^{t-1} has a density $f_{t-1|t-1}$, it follows by exchanging the order of integration that

$$
\begin{aligned}
\mathbb{E}[\phi(X_t) \mid Y_1^t = y_1^t] &= \frac{\int h_{t-1}(x, \phi) f_{t-1|t-1}(x \mid y_1^{t-1})d\mu(x)}{\int h_{t-1}(x, 1) f_{t-1|t-1}(x \mid y_1^{t-1})d\mu(x)} \\
&= \frac{\int \phi(x_t)b_t(x_t, y_t) \int f_{t-1|t-1}(x \mid y_1^{t-1})a_t(x, x_t)d\mu(x) \, d\mu(x_t)}{\int b_t(x_t, y_t) \int f_{t-1|t-1}(x \mid y_1^{t-1})a_t(x, x_t)d\mu(x) \, d\mu(x_t)}.
\end{aligned}
$$

But this implies that also the conditional distribution of X_t given y_1^t has a density which is given by

$$
f_{t|t}(x \mid y_1^t) \propto b_t(x_t, y_t) \int f_{t-1|t-1}(x \mid y_1^{t-1})a_t(x, x_t)d\mu(x),
$$

i.e. we have recovered the filtering recursion (3.16).

The main advantage of this method is that it generalizes to the time continuous case.

3.3.5 Transitions that are not absolutely continuous

It is easy to see that for filtering, we do not need to assume the state transitions to have densities. If (X_t) is a Markov chain with transition kernels $a_t(x_{t-1}, dx_t)$ and if we denote the conditional distributions of X_t

given y_1^t by $\nu_t(dx_t \mid y_1^t)$, then the following recursion holds:

$$\nu_t(dx_t \mid y_1^t) \propto b_t(x_t, y_t) \int \nu_{t-1}(dx \mid y_1^{t-1}) a_t(x, dx_t).$$

But when densities for the conditional distributions of Y_t given x_t do not exist, there is no general filtering formula. Similarly, there is no general formula for smoothing if the state transitions are not absolutely continuous. However, in most practical cases where densities do not exist, we have a simple structure. Namely, conditional on x_t, Y_t or X_{t+1} are restricted to a simple lower-dimensional subspace (often a linear subspace) and the conditional distributions restricted to this subspace are absolutely continuous. In such cases it is usually straightforward how to modify the filtering and smoothing recursions.

A second order Markov model for the state process (X_t) can be converted into a first order model by considering the new state variable $Z_t = (X_{t-1}, X_t)$. But for this state process, transitions are never absolutely continuous. A simple way out is to proceed in steps of size two, i.e. to go in the filtering directly from Z_t to Z_{t+2} and in the smoothing backwards from Z_t to Z_{t-2}. Transitions are then absolutely continuous provided the conditional distribution of X_t given (x_{t-1}, x_{t-2}) is.

3.3.6 Forgetting of the initial distribution

We study here how much the filtering densities with the same observations, but different initial distributions differ. This is of interest in itself because often it is not clear which initial distribution is appropriate. Moreover, because of the recursive nature of the filtering procedure, this will also give us the effect of a change in the filtering density at an arbitrary time t on future filtering densities. It is thus the key for other questions like propagation of the error in approximate filtering recursions (see Section 3.5.6) or the dependence of $p(y_t \mid y_1^{t-1})$ on y_s for $s < t$ (see Section 3.6.5).

We begin by introducing some notation. The two initial densities are denoted by a_0 and $\bar{a}_0$ respectively, and the resulting filtering densities are $f_{t|t}$ and $\bar{f}_{t|t}$. These filtering densities are computed by alternating Markov transitions and applications of Bayes formula as discussed in Section 3.3.1. It is convenient to describe these steps in operator notation. The Markov transitions are given by the operators A_t^* mapping the space of probability densities with respect to μ into itself

$$A_t^* f(x) = \int f(x') a_t(x', x) d\mu(x'). \tag{3.29}$$

(The star indicates that this is an adjoint operator). The Bayes operator

B assigns to any prior f and likelihood b the corresponding posterior:

$$B(f,b)(x) = \frac{f(x)b(x)}{\int f(x)b(x)d\mu(x)} \tag{3.30}$$

With these definitions, we can describe the filtering recursion (3.16) in the compact notation

$$f_{t|t} = B(A_t^* f_{t-1|t-1}, b_t(\cdot, y_t)). \tag{3.31}$$

The same recursion holds also for $\overline{f}_{t|t}$, the only difference being the initial condition.

It should now be clear that the filtering densities forget about the initial distribution if the operators A_t^* and $B(\cdot, b)$ are contracting for some norm on densities. We investigate whether contractivity holds for the L_1-norm. Unfortunately, the answer will be negative.

The L_1-norm for densities is twice the total variation norm between the corresponding distributions, see Devroye (1987), Section 1.1. Its definition and a few equivalent forms are as follows:

$$
\begin{aligned}
\|f - g\|_1 &= \int |f(x) - g(x)|d\mu(x) \\
&= 2\int (f(x) - g(x))^+ d\mu(x) \tag{3.32} \\
&= 2 \sup_A |F(A) - G(A)| \tag{3.33} \\
&= 2 \sup_{\Delta(\psi)\leq 1} \left| \int \psi(x)(f(x) - g(x))d\mu(x) \right|. \tag{3.34}
\end{aligned}
$$

Here $x^+ = \max(x,0)$ is the positive part and $\Delta(\psi) = \sup_{x,x'} |\psi(x) - \psi(x')|$ is the variation of ψ. The equality (3.32) is Scheffé's theorem (see Devroye (1987), p.2). The other equalities are easy consequences of (3.32). For further reference, we note that $f(x) \leq cg(x)$ implies by (3.32)

$$\|f-g\|_1 = 2\int (f(x)-g(x))^+ d\mu(x) \leq 2\int f(x)(1-1/c)d\mu(x) = 2(1-1/c). \tag{3.35}$$

The contractivity of Markov operators is well known, see Dobrushin (1956), Section 3. Namely, if we define the contraction coefficient ρ of an arbitrary Markov operator A^* given by a transition density $a(x,x')$ as in (3.29) by

$$\rho(A^*) = \frac{1}{2} \sup_{x,x'} \|a(x,\cdot) - a(x',\cdot)\|_1, \tag{3.36}$$

then it holds

Lemma 3.5

$$\|A^*f - A^*\overline{f}\|_1 \leq \rho(A^*) \|f - \overline{f}\|_1.$$

Proof. Choose a function ψ with $\Delta(\psi) = 1$. Then by an exchange of the order of integration we obtain

$$\int \psi(x)(A^* f(x) - A^* \overline{f}(x))d\mu(x) = \int A\psi(x)(f(x) - \overline{f}(x))d\mu(x)$$

where

$$A\psi(x) = \int a(x, x')\psi(x')d\mu(x').$$

By the definition of ρ and (3.34) it follows that

$$\Delta(A\psi) = sup_{x,x'}|\int (a(x, x'') - a(x', x''))\psi(x'')d\mu(x'')| \leq \rho(A^*).$$

Thus the Lemma follows from (3.34). $\square$

Note that always $\rho(A^*) \leq 1$, and at least if we assume the state space to be compact, typically $\rho(A^*) < 1$. Thus the Markov operators are contractive. Unfortunately this is not the case for the Bayes operator although naively one might think that the posteriors should be closer than the priors if the same likelihood, i.e. the same data are used. But note that the asymptotic behavior of Bayes estimators is a delicate problem, see e.g. Chapter 5 of Le Cam and Yang (1990), which would be much easier if Bayes formula were contracting.

Even in the simplest binary case, the relative change

$$\frac{||B(f, b) - B(\overline{f}, b)||_1}{||f - \overline{f}||_1}$$

can easily be greater than one. Things become worse for a continuous state space. The following Lemma which is a polished version of results in Hürzeler (1998) shows that the Bayes operator is always expanding in some direction. We have been unable to find this result in the literature.

Lemma 3.6 i) *We always have*

$$\frac{||B(f, b) - B(\overline{f}, b)||_1}{||f - \overline{f}||_1} \leq \frac{||b||_\infty}{\max(\int b(x)f(x)d\mu(x), \int b(x)\overline{f}(x)d\mu(x))}$$

ii) *If the state space is a compact interval, μ is the Lebesgue measure, f and b are continuous, b is not constant and f is strictly positive, then*

$$\lim_{\delta \to 0} \sup_{||f - \overline{f}||_1 \leq \delta} \frac{||B(f, b) - B(\overline{f}, b)||_1}{||f - \overline{f}||_1} = \frac{||b||_\infty}{\int b(x)f(x)dx}.$$

Proof. Let $c(f, b)$ denote $\int b(x)f(x)d\mu(x)$. For i) we can assume without loss of generality that $c(f, b) \geq c(\overline{f}, b) = 1$. Then we have

$$(B(f, b)(x) - B(\overline{f}, b)(x))^+$$
$$= c(f, b)^{-1}(f(x)b(x) - c(f, b)\overline{f}(x)b(x))^+$$

$$\leq \quad c(f,b)^{-1}(f(x)b(x) - \overline{f}(x)b(x))^+$$
$$\leq \quad c(f,b)^{-1}||b||_\infty (f(x) - \overline{f}(x))^+.$$

Thus i) follows from (3.32).

For ii), we choose x_1 such that $b(x_1) = ||b||_\infty$ and x_2 such that $b(x_2) < b(x_1)$. Then we set

$$\overline{f}_\delta(x) = f(x) - \delta K((x - x_1)/\delta) + \delta K((x - x_2)/\delta),$$

where K is a density with support $[-1, 1]$. For δ small enough, $\overline{f}_\delta$ is a density and

$$||f - \overline{f}_\delta||_1 = 2\delta^2.$$

Furthermore

$$c(\overline{f}_\delta, b) = c(f, b) - \delta^2(b(x_1) - b(x_2)) + o(\delta^2) < c(f, b)$$

for δ small enough. In that case, it is easily seen that

$$(f(x)b(x) - \frac{c(f,b)}{c(\overline{f}_\delta, b)}\overline{f}_\delta(x)b(x))^+$$
$$\geq \quad (f(x) - \overline{f}_\delta(x))^+ b(x) - (\frac{c(f,b)}{c(\overline{f}_\delta, b)} - 1)\overline{f}_\delta(x)b(x)1_{[|x-x_1|\leq\delta]}.$$

(If $|x - x_1| > \delta$, both sides are zero, and if $|x - x_1| \leq \delta$, the right-hand side is equal to the left-hand side before taking the positive part.) Integrating both sides of this inequality over x implies

$$c(f, b)||B(f, b) - B(\overline{f}_\delta, b)||_1 \geq ||b||_\infty ||f - \overline{f}_\delta||_1 + o(\delta^2)$$

and ii) follows. □

Note that the bound in this Lemma is always larger than one. Also the sequence $\overline{f}_\delta$ constructed in the proof has uniformly bounded derivatives if K is smooth. Therefore, the "dangerous directions" are not entirely unrealistic. An extension of this Lemma to higher dimension is straightforward.

As a consequence of Lemmas 3.5 and 3.6, the first approach to show that the initial distribution is forgotten does not work. One can try to use different norms, but we have not been successful to find one for which contractivity holds. Another alternative starts from the fact that the sequence $\overline{f}_\delta$ depends on the likelihood b. One might thus try to show that with high probability the observations are such that contractivity holds. But this looks very delicate and complicated. A third approach which actually succeeds consists in looking not at one, but at several iterations. This approach is due to Arapostathis and Marcus (1990), see also Douc and Matias (1999) and LeGland and Mevel (2000).

The key observation is that if we look at several iterations, we can replace alternating applications of Markov and Bayes operators by one application of a Bayes operator followed by several applications of Markov operators.

Then the contractivity of the Markov operators beats the expansion by the Bayes operator.

First we show that for any $s < t$ we can obtain $f_{t|t}$ from $f_{s|s}$ by one application of Bayes formula followed by $t - s$ Markov transitions. Define for $s \leq t$ the Markov operator

$$P^*_{s|t} f(x_s) = \int f(x_{s-1}) p(x_s \mid x_{s-1}, y^t_s) d\mu(x_{s-1})$$

and the likelihood

$$\ell_{s|t}(x_s) = p(y^t_{s+1} \mid x_s).$$

(We suppress the dependence on y^t_s because the observations are assumed to be fixed). Then we have

Lemma 3.7 *For any* $s < t$

$$f_{t|t} = P^*_{t|t}(P^*_{t-1|t} \cdots P^*_{s+1|t}(B(f_{s|s}, \ell_{s|t})) \cdots). \tag{3.37}$$

Proof. By the law of total probability

$$f_{t|t}(x_t \mid y^t_1) = p(x_t \mid y^t_1) = \int p(x_t \mid x_s, y^t_1) p(x_s \mid y^t_1) d\mu(x_s)$$

and by (3.24)

$$p(x_s \mid y^t_1) = f_{s|t}(x_s \mid y^t_1) = \frac{p(y^t_{s+1} \mid x_s) f_{s|s}(x_s \mid y^s_1)}{p(y^t_{s+1} \mid y^s_1)}.$$

Furthermore by Lemma 3.2 i)

$$p(x_t \mid x_s, y^t_1) = \int \cdots \int \prod_{r=s+1}^{t} p(x_r \mid x_{r-1}, y^t_r) \prod_{r=s+1}^{t-1} d\mu(x_r).$$

□

Note that both the operators $P^*_{s|t}$ and the likelihood $\ell_{s|t}$ depend only on y^t_s and on the transition densities a_r and b_r for $s \leq r \leq t$. In particular the expression (3.37) holds also for $\overline{f}_{t|t}$.

We now need conditions which guarantee that the expansion coefficient from Lemma 3.6 i) for the likelihood $\ell_{s|t}$ is bounded and the Markov operators are strict contractions. We do not strive here for the most general result, but make the following assumption that is simple to formulate and leads to simple proofs:

(A1) There exist a constant $C_a < \infty$ such that for all t and all x, x', x''

$$\frac{a_t(x, x'')}{a_t(x', x'')} \leq C_a.$$

Clearly, $1/C_a$ is then a lower bound for these ratios. Also, by (3.35) the assumption **(A1)** implies

$$\rho(A_t^*) \le 1 - \frac{1}{C_a}.$$

Lemma 3.8 *Under the assumption* **(A1)** *we have for all* $s < t$ *and all* y_s^t

$$\frac{\ell_{s|t}(x_s)}{\ell_{s|t}(x_s')} \le C_a, \quad \rho(P_{s|t}^*) \le 1 - C_a^{-2}.$$

Proof. The first claim follows directly from (3.20) in Lemma 3.2. Furthermore, by formula (3.19) in the same Lemma

$$\frac{p(x_s \mid x_{s-1}, y_s^t)}{p(x_s \mid x_{s-1}', y_s^t)} = \frac{a_s(x_{s-1}, x_s)p(y_s^t \mid x_{s-1}')}{a_s(x_{s-1}', x_s)p(y_s^t \mid x_{s-1})} \le C_a^2.$$

Therefore the second claim follows by (3.35). $\square$

Now Lemma 3.7 and the bounds in Lemmas 3.5, 3.6 and 3.8 prove our main result that the filter forgets the initial condition exponentially fast. We formulate it in a slightly more general form that tells how fast changes in the filtering density at a fixed time disappear when the updates are made with the same observations.

Theorem 3.9 *Under assumption* **(A1)**

$$\|f_{t|t} - \overline{f}_{t|t}\|_1 \le C_a(1 - C_a^{-2})^{t-s}\|f_{s|s} - \overline{f}_{s|s}\|_1.$$

3.4 Exact and approximate filtering and smoothing

3.4.1 Hidden Markov models

If the number of states is finite, say M, all the integrals in the filtering and smoothing recursions are simply sums. Calculation is then straightforward. The recursions (3.16) and (3.25) are usually called the forward and backward recursions. We have presented here a form which avoids problems of underflow during the iterations. In the literature, recursions are often given for $p(x_t, y_1^t)$ instead of $f_{t|t}$ and $p(y_{t+1}^T \mid x_t)$ instead of $r_{t|T}$. Then some simplification occurs because there is no denominator in the analogue of (3.16) and (3.25). But there is underflow in the computations which has to be taken care of.

It is easy to see that all these recursions require $O(TM^2)$ operations, the bulk of computations occurring in (3.14). Because of the linear dependence on T, they are feasible also for quite long series. Note that a naive computation of $f_{t|T}$ by summing the joint distribution (3.4) over x_1^{t-1} and x_{t+1}^T would require an exponential number of operations.

3.4.2 Kalman filter

Another case where the general recursions simplify considerably is the linear state space model (3.5)-(3.6) with Gaussian errors and Gaussian initial distribution a_0. Then it is easy to see that all $f_{t|s}$ are again Gaussian densities. It thus suffices to compute the conditional means $m_{t|s}$ and covariances $R_{t|s}$. By completing the square, one obtains from (3.14) and (3.16)

$$
\begin{align}
m_{t|t-1} &= G_t m_{t-1|t-1} \tag{3.38}\\
R_{t|t-1} &= \Sigma_t + G_t R_{t-1|t-1} G_t' \tag{3.39}\\
R_{t|t} &= (H_t'\Omega_t^{-1}H_t + R_{t|t-1}^{-1})^{-1} \tag{3.40}\\
m_{t|t} &= R_{t|t}(H_t'\Omega_t^{-1}y_t + R_{t|t-1}^{-1}m_{t|t-1}) \\
&= m_{t|t-1} + R_{t|t}H_t'\Omega_t^{-1}(y_t - H_t m_{t|t-1}). \tag{3.41}
\end{align}
$$

The filtering equations (3.40) and (3.41) can be simplified by noting that

$$(H_t'\Omega_t^{-1}H_t + R_{t|t-1}^{-1})R_{t|t-1}H_t' = H_t'\Omega_t^{-1}(\Omega_t + H_t R_{t|t-1}H_t').$$

From this we obtain

$$
\begin{align}
m_{t|t} &= m_{t|t-1} + R_{t|t-1}H_t'M_t^{-1}(y_t - H_t m_{t|t-1}) \tag{3.42}\\
R_{t|t} &= R_{t|t-1} - R_{t|t-1}H_t'M_t^{-1}H_t R_{t|t-1} \tag{3.43}\\
\text{where} \quad M_t &= \Omega_t + H_t R_{t|t-1}H_t'. \tag{3.44}
\end{align}
$$

This is the famous Kalman filter. Note that (3.41) and (3.42) have the intuitive form "filter mean equal prediction mean plus a correction depending on how much the new observation differs from its prediction". In most textbooks, e.g. Anderson and Moore (1979), the Kalman filter is derived by properties of orthogonal projections in the Hilbert space of linear combinations of $(X_1, \ldots, X_T)$. The approach based on the general recursions (3.14) and (3.16) has also been used by Meinhold and Singpurwalla (1983).

Similarly we can show using (3.21) that conditional on x_{t+1} and y_1^T, x_t is Gaussian with mean $\overline{m}_t$ and covariance matrix $\overline{R}_t$ where

$$
\begin{align}
\overline{m}_t &= m_{t|t} + \overline{R}_t^{-1}G_{t+1}'\Sigma_{t+1}^{-1}(x_{t+1} - m_{t+1|t}) \\
&= m_{t|t} + R_{t|t}G_{t+1}'R_{t+1|t}^{-1}(x_{t+1} - m_{t+1|t}), \tag{3.45}\\
\overline{R}_t &= (G_{t+1}'\Sigma_{t+1}^{-1}G_{t+1} + R_{t|t}^{-1})^{-1} \\
&= R_{t|t} - R_{t|t}G_{t+1}'R_{t+1|t}^{-1}G_{t+1}R_{t|t}. \tag{3.46}
\end{align}
$$

From this we obtain the Kalman smoother

$$
\begin{align}
m_{t|T} &= m_{t|t} + S_t(m_{t+1|T} - m_{t+1|t}) \tag{3.47}\\
R_{t|T} &= R_{t|t} - S_t(R_{t+1|t} - R_{t+1|T})S_t' \tag{3.48}\\
\text{where} \quad S_t &= R_{t|t}G_{t+1}'R_{t+1|t}^{-1}. \tag{3.49}
\end{align}
$$

3.4.3 The innovation form of state space models

For linear systems, one works usually with the so-called innovation form of the state space model. For this define the innovations of the observation process as

$$\varepsilon_t = Y_t - H_t m_{t|t-1} = Y_t - \mathbb{E}[Y_t \mid Y_1^{t-1}].$$

and choose as new state variables

$$Z_t = m_{t|t-1}.$$

Then by (3.38) and (3.42)

$$Z_{t+1} = G_{t+1} Z_t + G_{t+1} R_{t|t-1} H_t' M_t^{-1} \varepsilon_t.$$

By the fundamental property of conditional expectation, ε_t is uncorrelated with Y_s, $s < t$, and therefore also with ε_s, $s < t$. By the Gaussian assumption, the innovations are therefore an independent sequence. Thus, defining K_t appropriately, (Z_t) and (Y_t) satisfy the model

$$
\begin{aligned}
Z_{t+1} &= G_{t+1} Z_t + K_{t+1} \varepsilon_t, & (3.50) \\
Y_t &= H_t Z_t + \varepsilon_t. & (3.51)
\end{aligned}
$$

This is the innovation form of a linear state space model. Note that in this form, the dynamics of the states is no longer independent from the observations. The main advantage is that in this formulation all components of the state vector on which we cannot obtain any information from the observations are automatically removed. It is thus in some sense a minimal model.

3.4.4 Exact computations in other cases

There have been attempts to find more models where the recursions can be computed in closed form. Unfortunately, there are negative results in this direction (Chaleyat-Maurel and Michel (1983)), showing that in general recursions with a fixed finite dimension do exist only for linear Gaussian state space models. Still one can look for recursions which are finite-dimensional, but the dimension increases with the number of steps in the recursion. For this one needs to find a model and a family $\mathcal{M}$ of densities on the state space such that each member of $\mathcal{M}$ is described by finitely many parameters and such that for all y_1^{t+1} $f_{t|t}(\cdot \mid y_1^t) \in \mathcal{M}$ implies $f_{t+1|t+1}(\cdot \mid y_1^{t+1}) \in \mathcal{M}$. In such a situation it is obviously sufficient to update the parameters in each step. In a linear state space model where the error distributions are finite mixtures of Gaussian distributions, it is easy to show that the family of finite Gaussian mixtures has this property (see Alspach and Sorensen (1972)). However, in each update the number of terms in the mixtures (and thus also the number of parameters) is at least doubled and thus the number of operations increases exponentially. So this is not feasible for exact computations, but it is the basis for an approximation, see Section 3.4.6.

Another example can be obtained by taking the linear state space model with Cauchy noises and as $\mathcal{M}$ the family of densities that can be written as finite linear combinations of functions of the form $(cx + d)/(b^2 + (x - a)^2)$. Note that if $c \neq 0$, this is not a density (it gets negative and it is not integrable), but due to special cancelation effects certain linear combinations are densities. In this case, the dimension of the parameters increases linearly, and thus computation is feasible also for longer series. But the cancelation effects require care in a numerical implementation. This example is due to Steyn (1996) where more cases of this kind can be found.

Finally, Ferrante and Vidoni (1998) have discussed a number of examples with additional arrows in the dependence graph such that exact updates are possible.

Despite all these efforts, the cases where exact computations are possible remain extremely limited.

3.4.5 Extended Kalman filter

The extended Kalman filter, see e.g. Anderson and Moore (1979), is an approximation for the conditional means $m_{t|s}$ and covariances $R_{t|s}$ in non-linear models (3.7)-(3.8) with known smooth functions g_t and h_t. It computes $m_{t|t-1}$, $R_{t|t-1}$ by using the standard Kalman filter with the linearized versions of (3.7)-(3.8):

$$
\begin{aligned}
X_t &= g_t(m_{t-1|t-1}, 0) + \frac{\partial}{\partial x} g_t(m_{t-1|t-1}, 0)(X_{t-1} - m_{t-1|t-1}) \\
&\quad + \frac{\partial}{\partial v} g_t(m_{t-1|t-1}, 0)V_t \\
Y_t &= h_t(m_{t|t-1}, 0)\frac{\partial}{\partial x} h_t(m_{t-1|t-1}, 0)(X_t - m_{t|t-1}) \\
&\quad + \frac{\partial}{\partial w} h_t(m_{t-1|t-1}, 0)W_t.
\end{aligned}
$$

Since it gives no information about the shape of the prediction and filtering densities, one usually assumes that they are Gaussian. Unfortunately, it is almost impossible to obtain information about the error of the approximation. Empirically, it turns out that this error can be large in many situations of interest, see e.g. Aidala (1979) or Weiss and Moore (1980).

3.4.6 Other approximate methods

Because the cases where computations can be done in closed form are rare and because the extended Kalman filter gives no information about the form of the conditional distributions and can be unstable, there is a need for other approximations. Kitagawa (1987) has developed numerical approximations. But in general, numerical integration is difficult in higher dimensions. This is aggravated here because a good choice of the knots for integration would already need the knowledge of how the filter densities look like.

Another approach for linear models with non-Gaussian noises has been via the use of mixture models, see e.g. Harrison and Stevens (1976) or Smith and West (1983). As was mentioned in Section 3.4.4, one then has for each component in the mixture a linear Kalman filter together with a recursion for the weights in the mixture. Since the number of components in $f_{t|t}$ increases exponentially with t, one has to reduce this number by some approximation. Usually one replaces a mixture of normals by a single normal such that the first two moments are matched. The algorithms in the literature differ by the method for choosing which components are collapsed. The most sophisticated methods proceed by collapsing similar components and by deleting components with small weights, but the implementation of this idea is rather delicate. Moreover, the restriction to linear state space models is severe.

For these reasons, attention in recent years has focused mainly on Monte Carlo methods. This is the subject of the next section.

3.5 Monte Carlo filtering and smoothing

Monte Carlo methods generate samples which are distributed according to the desired filtering or smoothing densities. Expectations are then approximated by sample averages, quantiles by corresponding order statistics, densities by kernel estimators etc.

3.5.1 Markov chain Monte Carlo: single updates

Using Markov chain Monte Carlo methods for state space models has been proposed by Carlin et al. (1992). The standard Gibbs sampler can be used to generate samples from the conditional distribution of x_0^T given y_1^T. It requires sampling from the so-called full conditionals, i.e. the density of X_t given $(x_s)_{s \neq t}$ and y_1^T. This is proportional to

$$a_t(x_{t-1}, x_t)a_{t+1}(x_t, x_{t+1})b_t(x_t, y_t).$$

Sampling from this conditional density can be done either directly or by using a rejection method. Alternatively, one can use the Metropolis-Hastings algorithm.

The disadvantage is that such a method is not recursive: When a new observation arrives, one has to start again from scratch. Also convergence of single update methods is usually slow because the conditional distributions are typically highly concentrated so that the chain moves too little in each update.

3.5.2 Markov chain Monte Carlo: multiple updates

It has been noted by Frühwirth-Schnatter (1994), Carter and Kohn (1994), Shephard (1994) and others that in some frequently used models one can update large blocks of the state variables in one step. This increases the speed of the Gibbs sampler often dramatically. It happens in situations where the state variable can be split in say two components $x_t = (x_{1,t}, x_{2,t})'$ and one can sample $x_{1,0}^T$ given $(x_{2,0}^T, y_1^T)$ and also $x_{2,0}^T$ given $(x_{1,0}^T, y_1^T)$. One then alternates between simulating from one of the two components while keeping the other component fixed.

For instance, Shephard (1994) consider a linear state space model (3.5)-(3.6) where the noises are scale mixtures of Gaussian distributions. Then we can introduce auxiliary i.i.d. variables $\lambda_t \sim G$, $\omega_t \sim H$ such that

$$V_t = \lambda_t \, \varepsilon_t, \quad W_t = \omega_t \, \eta_t$$

with ε_t and η_t standard Gaussian variables. We consider the enlarged state variable $z_t = (x_t', \lambda_t, \omega_t)'$. Conditional on $(\lambda_1^T, \omega_1^T)$, (x_0^T, y_1^T) is then a linear Gaussian state space model. In order to simulate x_0^T given $(y_1^T, \lambda_1^T, \omega_1^T)$, we compute the conditional means and variances of x_t given $(x_{t+1}, y_1^T, \lambda_1^T, \omega_1^T)$ according to (3.38)-(3.46) and generate Gaussian variables $x_T, x_{T-1} \ldots$ with that mean and variance backward in time. Moreover, given (x_0^T, y_1^T) the λ_t's and ω_t's are all independent and thus can be updated together in a straightforward way.

A similar procedure is possible if $x_{2,t}$ is discrete and conditioning on $x_{2,0}^T$ leads to a linear Gaussian state space model, see Carter and Kohn (1994). The conditional simulation of $x_{1,0}^T$ given $(x_{2,0}^T, y_1^T)$ proceeds as above. Conditional on $x_{1,0}^T$, $(x_{2,0}^T, y_1^T)$ is then a hidden Markov model. In order to simulate $x_{2,0}^T$ given $(x_{1,0}^T, y_1^T)$, we compute the conditional distribution of $x_{2,t}$ given $(x_{2,t+1}, x_{1,0}^T, y_1^T)$ according to (3.16) and (3.21) and generate a sample from the corresponding Markov chain backward in time.

An example is the following ion channel model which takes into account correlations of the noise and state dependent noise variance

$$Y_t = m(X_t) + C_t + \sigma(X_t)\eta_t.$$

Here (X_t) is a finite state Markov chain describing the state of the channel, (C_t) is a correlated (colored) noise modeled as a stationary Gaussian AR(1)-process and (η_t) is an independent Gaussian white noise with mean zero and variance one. The functions m and σ are assumed to be known. In this model, for given (x_0^T) we have a linear Gaussian state space model with observations $Y_t - m(x_t)$ and state variables C_t whereas for given (c_0^T) we have a hidden Markov model with observations $Y_t - c_t$ and state variables

X_t. It is straightforward to consider AR models of higher order for (C_t). In De Gunst et al. (1998) this method has been used in the slightly more complicated model

$$Y_t = \sum_{k=-r}^{r} \gamma_k m(X_{t+k}) + C_t + \sigma(X_t)\eta_t$$

which takes into account the blurring due to a filter in the recording process.

Even though multiple site updates are the solution for the problem of slow speed of convergence in many cases, they are not recursive and thus are not suitable for on-line implementation.

3.5.3 Recursive Monte Carlo filtering

Here we discuss methods to generate samples $(x_{t|t}^{(j)})$ according to the filtering density $f_{t|t}$ in a recursive way. The idea goes back a long time to Handschin and Mayne (1969) and Handschin (1970), but it has become popular only recently after Gordon et al. (1993), Isard and Blake (1996) and Kitagawa (1996). There is now an extensive literature, see Doucet (1998) and the book edited by Doucet et al. (2000). The method is often called the *particle filter*.

One starts by generating $(x_{0|0}^{(j)})$ according to the density $a_0(x)$ of X_0. Then the basic method proceeds by generating a sample $(x_{t|t}^{(j)})$ with density

$$g_{t|t}(x) \propto b_t(x, y_t) \sum_{i=1}^{N} a_t(x_{t-1|t-1}^{(i)}, x). \tag{3.52}$$

Comparing (3.52) with (3.16), we see that this amounts to replacing the integral $\int a_t(x, x_t) f_{t-1|t-1}(x \mid y_1^{t-1}) d\mu(x)$ by a sample average.

A compact description of this method which will be useful to study error propagation can be given using operator notation. If g is a density on $\mathcal{X}$ with respect to μ, then we denote by $E_N(g)$ the empirical distribution of a random sample of size N from g. In addition, we extend the operator notation introduced in Section 3.3.6 by considering A_t^* as acting on probability distributions η on $\mathcal{X}$ (not necessarily absolutely continuous) by

$$A_t^* \eta(x) = \int a_t(x', x) \eta(dx').$$

Then we have

$$g_{t|t} = B(A_t^* E_N(g_{t-1|t-1}), b_t(., y_t)). \tag{3.53}$$

This should be compared with the recursion (3.31).

In order to implement this method, we need to be able to generate samples according $g_{t|t}(x)$. If

$$c_t = \sup_x b_t(x, y_t)$$

is finite, we can combine the standard rejection algorithm with the standard algorithm for sampling from a mixture. This leads to the algorithm

1. Put $j = 1$.

2. Generate I uniform on $\{1, \ldots, N\}$ and X according to the distribution $a_t(x^{(I)}_{t-1|t-1}, x)d\mu(x)$.

3. Generate U uniform on $[0, c_t]$.

   If $U \leq b_t(X, y_t)$, put $x^{(j)}_{t|t} = X$ and $j = j + 1$.

4. If $j \leq N$, return to 2, otherwise stop.

This method produces an i.i.d. sample with exact distribution $g_{t|t}(x)d\mu(x)$. It has been discussed in Hürzeler and Künsch (1998). It can be slow if the "likelihood" $b_t(x, y_t)$ is in conflict with the "priors" $a_t(x^{(i)}_{t-1|t-1}, x)$ for the majority of indices i because then we have many rejects in step 3.

Most proposals discussed in the literature use approximate methods which are inspired by importance sampling and have been called Sampling Importance Resampling (SIR) by Rubin (1988). One generates first a sample $(z^{(j)})$ of size M according to a suitable proposal distribution $h(x)d\mu(x)$. Next, a resampling step selects from this the sample $(x^{(j)}_{t|t})$ of size N with probabilities

$$w_i \propto \frac{g_{t|t}(z^{(i)})}{h(z^{(i)})}.$$

The original proposal of a recursive Monte Carlo filter in Gordon et al. (1993) had

$$h(x) = \frac{1}{N} \sum_{i=1}^{N} a_t(x^{(i)}_{t-1|t-1}, x)$$

and $M = N$. This h is nothing else than an approximate prediction density.

The procedures discussed so far are illustrated in Figure 3.2. This figure is based on the model

$$X_t = X_{t-1} + V_t, \quad Y_t = X_t^2 + W_t$$

with $V_t \sim \mathcal{N}(0, 9)$ and $W_t \sim \mathcal{N}(0, 500)$. It is assumed that the filter density at time $t - 1$ is a mixture of two Gaussian densities. Sampling at time $t - 1$ is exact and $N = M = 100$ was used. One sees clearly that the Sampling Importance Resampling method produces ties in the filter sample whereas the basic method producing the filter samples according to (3.52) does not. The filter sample in (d) contains only 55 different values. These ties can be a problem if a small number of w_i's dominates the rest, that is if $f_{t|t}$ is substantially larger than $h(x)$ in some regions. Note that this typically happens when the "likelihood" is in conflict with the "priors" as described above, i.e. the two approaches suffer from the same problem. Choosing M for instance as $10N$ obviously alleviates the problem at the expense of more

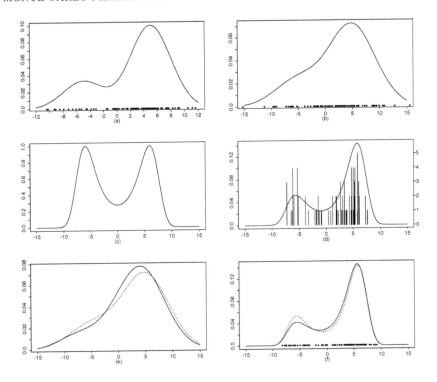

Figure 3.2 *Illustration of recursive Monte Carlo filters. (a) Filter density and filter sample at time $t-1$, (b) Prediction density and prediction sample at time t, (c) Likelihood of state at time t given observation at time t, (d) Filter density at time t and filter sample obtained by resampling the prediction sample of (b), (e) Approximate prediction density $N^{-1} \sum a_t(x_{t-1|t-1}^{(j)}, x)$ (solid) and exact prediction density (dotted) for comparison, (f) Approximate filter density $g_{t|t}$ (solid) and corresponding filter sample. Exact filter density (dotted) is given for comparison.*

work. The algorithm above can be considered as an adaptive method which gives a random M.

One can also try to overcome the basic problem by different choices for $h(z)$. Pitt and Shephard (1999) consider the case where $a_t(x_{t-1|t-1}^{(i)}, .)$ is concentrated around a value $m_{t|t-1}^{(i)}$ and then take

$$h(x) \propto \sum_{i=1}^{N} b_t(m_{t|t-1}^{(i)}, y_t) a_t(x_{t-1|t-1}^{(i)}, x).$$

This distribution generates automatically more values from those "priors" which are compatible with the "likelihood".

It has also been noted that the precision of the simulations can be improved by eliminating unnecessary randomness from the procedure. For instance, instead of choosing a random sample $(x_{t|t}^{(i)})$ from the $(z^{(j)})$ with probabilities (w_j), one can also choose a deterministic sample where each $z^{(i)}$ appears approximately Nw_i times (Higuchi, personal communication). A completely deterministic algorithm takes first each $z^{(i)}$ $[Nw_i]$ times where $[x]$ denotes the integer part of x. In a second step, one takes additional values $z^{(i)}$ once in decreasing order of $Nw_i - [Nw_i]$ until one has N values. In this way, values with very small w_i are never chosen which may lead to undesirable biases. An alternative method arranges the values $z^{(i)}$ on intervals of length $2\pi w_i$ on the unit circle in random order, spins a roulette wheel with N equidistant numbers and counts how many numbers lie in the interval of $z^{(i)}$, see e.g. Whitley (1994), p.67-68.

Another way to reduce unnecessary randomness is discussed in Carpenter et al. (1998). They note that

$$\sum_{i=1}^{M} w_i \psi(z^{(i)})$$

is always a more precise estimator of $\int \psi(x) g_{t|t}(x) d\mu(x)$ than

$$N^{-1} \sum_{i=1}^{N} \psi(x_{t|t}^{(i)})$$

and thus propose reweighting instead of resampling. But during the iterations, this leads to weights accumulating on only a few values and thus to imprecise Monte Carlo estimates in the long run. Ideally one would alternate between reweighting and resampling, but it seems difficult to decide when to do which of the two possibilities.

3.5.4 Recursive Monte Carlo smoothing

We assume that we have stored the filter samples $(x_{t|t}^{(j)})$ for $t = 1, \ldots, T$ and want to generate a sample from the distribution $p(x_0^T \mid y_1^T)$. Then according to Lemma 3.2 i) we have to generate $x_{t|T}^{(j)}$ given $x_{t+1|T}^{(j)}$ for $t = T-1, T-2, \ldots 0$ according to

$$\frac{a_{t+1}(x, x_{t+1|T}^{(j)}) f_{t|t}(x \mid y_1^t)}{\beta_j}$$

where $\beta_j = f_{t+1|t}(x_{t+1|T}^{(j)} \mid y_1^t)$ is just a normalization constant. Because $f_{t|t}$ is not available, we replace it by $g_{t|t}$ from (3.52). This means that we have to generate $x_{t|T}^{(j)}$ according to

$$g_{t|T}^{(j)}(x) \propto b_t(x, y_t)a_{t+1}(x, x_{t+1|T}^{(j)}) \sum_{i=1}^{N} a_t(x_{t-1|t-1}^{(i)}, x).$$

This has the same structure as (3.52). We thus can use the same combination of the rejection method with sampling from a mixture distribution. Assuming that

$$c_t^{(j)} = \sup_x b_t(x, y_t)a_{t+1}(x, x_{t+1|T}^{(j)})$$

is finite and can be computed, the algorithm is as follows

1. Put $j = 1$.

2. Generate I uniform on $\{1, \dots, N\}$ and X according to the distribution
   $a_t(x_{t-1|t-1}^{(I)}, x)d\mu(x)$.

3. Generate U uniform on $[0, c_t^{(j)}]$.
   If $U \leq b_t(X, y_t)a_{t+1}(X, x_{t+1|T}^{(j)})$, put $x_{t|T}^{(j)} = X$ and $j = j + 1$.

4. If $j \leq N$, return to 2, otherwise stop.

In a recursive implementation, if y_{T+1} becomes available, we have to compute only one additional filter sample $(x_{T+1|T+1}^{(j)})$, but we then have to regenerate the whole smoothing sample backward in time. This can be too time-consuming. An alternative is to generate $(x_{t|T+1}^{(j)})$ only for $t \geq T+1-m$ and to replace $x_{t|T+1}^{(j)}$ by $x_{t|T}^{(k)}$ for $t \leq T + 1 - m$ where

$$||x_{T+1-m|T+1}^{(j)} - x_{T+1-m|T}^{(k)}|| = \min_i ||x_{T+1-m|T+1}^{(j)} - x_{T+1-m|T}^{(i)}||.$$

This is justified by the Markov property and the fact that a new observation at time $T + 1$ has only a minor effect on states at times $t \leq T + 1 - m$ for a suitable m.

3.5.5 Examples

We take a few examples from Hürzeler (1998) and Hürzeler and Künsch (1998). First we look at two examples where state space models are used for smoothing noisy observations of a function h which is smooth except at a finite number of points. Our observations are

$$Y_t = h(t/T) + W_t$$

where the W_t's are independent standard Gaussian random variables. We model the unknown function values $h(t/T)$ as a random walk (3.11) or an integrated random walk (3.12) with heavy-tailed innovations. Outliers in the innovations correspond to places where h is not smooth. We estimate

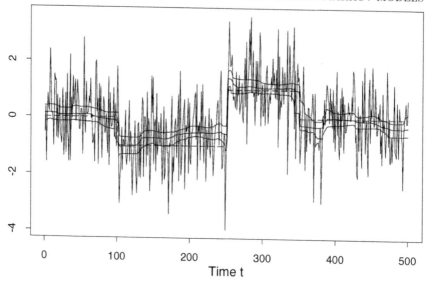

Figure 3.3 *Recursive Monte Carlo smoothing of a piecewise constant function. The true function (dashed), the observations (dotted) and the 10%-, 50%- and 90%-quantiles (solid) of the Monte Carlo smoother with $N = 1000$ are shown. From Hürzeler and Künsch (1998).*

$h(t/T)$ by a location functional (mean or median) of the smoothing density $f_{t|T}$.

In the first example, h is piecewise constant and T is 500. In this case, a random walk model for the state is appropriate. Following Kitagawa (1987) we use a t-distribution with 0.5 degrees of freedom for the innovations. The general behavior is shown in Figure 3.3. We see that the smoothing median is a good estimator even though the true function h is of course not a typical realization of a random walk. For this reason, the truth is often not within the 80% smoothing interval. In Figure 3.4 the Monte Carlo fluctuations in the estimated posterior median are shown near the first jump. They are large whenever the uncertainty about the state is large. But even in this difficult situation and with a small number N of replicates, the bias is minimal and the spread is acceptable.

In the second example, h is the so-called HeaviSine function of Donoho and Johnstone (1994), a sine function with two jumps, and $T = 2048$. Here an integrated random walk with Cauchy innovations is used for the states. The result is shown in Figure 3.5. The mean square error of the smoother mean is about 0.045. This is better than the values of 0.059 and 0.079 that can be obtained with two different wavelet thresholding methods, see Donoho and Johnstone (1994). Our method is far more computationally

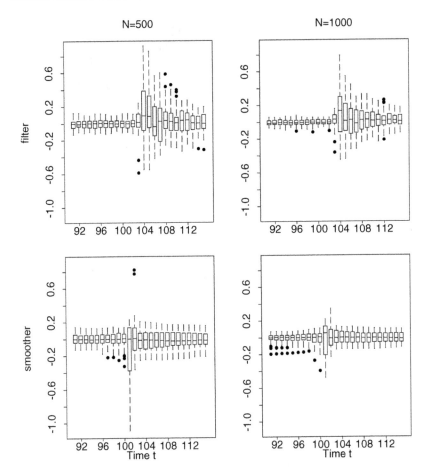

Figure 3.4 *Simulation errors in the recursive Monte Carlo filtering and smoothing median of a piecewise constant function. The error is defined with respect to the median using numerical integration as in Kitagawa (1987). Each boxplot is based on 200 repetitions of the Monte Carlo sampling whereas the data are fixed. Part of a large figure in Hürzeler and Künsch (1998).*

intensive than wavelets, but it can also handle more complex situations like irregularly spaced observations or blurred observations without much additional complications.

The third example considers the non-linear model

$$X_t = \alpha \cdot X_{t-1} + \beta \cdot \frac{X_{t-1}}{1 + X_{t-1}^2} + \gamma \cos(1.2t) + V_t \qquad (3.54)$$

$$Y_t = \frac{X_t^2}{20} + W_t \qquad (3.55)$$

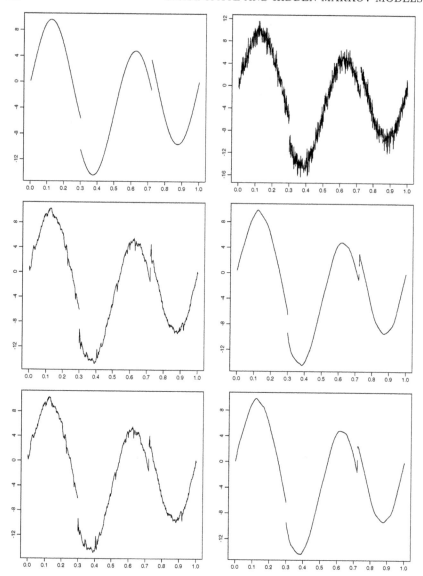

Figure 3.5 *Recursive Monte Carlo filtering and smoothing of a sine function with two jumps. First row: true function and noisy observations. Second and third rows: Two replicates of the mean of the Monte Carlo filter (left) and Monte Carlo smoother (right). From Hürzeler (1998).*

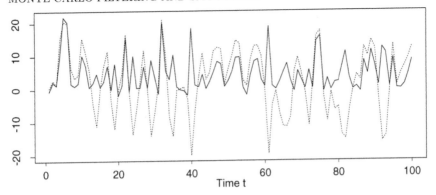

Figure 3.6 *Simulated series of the observations (solid line) and the state variables (dotted line) from the non-linear model (3.54) and (3.54). From Hürzeler and Künsch (1998).*

with $V_t \sim \mathcal{N}(0, \sigma_v^2)$ and $W_t \sim \mathcal{N}(0, \sigma_w^2)$ independent of each other. This model goes back to Andrade et al. (1978) and has been used as a kind of non-linear benchmark model, see Kitagawa (1987). The data of length $T = 100$ were generated according to this model with $\alpha = 0.5$, $\beta = 25$, $\gamma = 8$, $\sigma_v^2 = 10$ and $\sigma_w^2 = 1$ and are shown in Figure 3.6.

The behavior of the Monte Carlo filter based on $N = 1000$ replicates is shown in Figures 3.7 and 3.8 Because of the square in (3.55) information about the sign of x_t cannot come from y_t, but most come from the context. It turns out that in the filtering problem, i.e. if only observations from one side are available, there remains considerable uncertainty reflected in the wide difference between the 10%- and the 90%-quantiles. If observations from both sides are available, this uncertainty disappears. Figure 3.9 shows that the filtering density is in fact bimodal, whereas the smoothing density is unimodal.

3.5.6 Error propagation in the recursive Monte Carlo filter

We consider the simplest version of the recursive filter where in each step we generate the sample according to $g_{t|t}$ given in (3.52) instead of $f_{t|t}$. Obviously by doing so, the error gets propagated because $g_{t|t}$ depends on the sample generated in the previous step. This is most clearly visible by comparing the recursions (3.31) and (3.53). So it is natural to ask whether the error increases or remains stable and how it depends on the sample size N.

The results of Section 3.3.6 can be used to study this question. Using again operator notation, we can decompose the error $f_{t+1|t+1} - g_{t+1|t+1}$ at

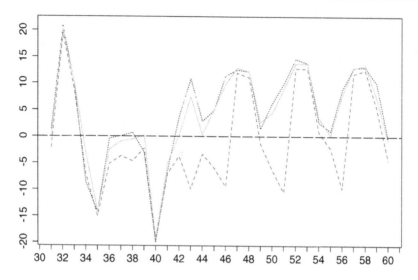

Figure 3.7 *Recursive Monte Carlo filtering in the non-linear model for t =* *31,...60. We show the unobserved state variable (solid), the median (dotted) and 10% and 90% quantiles (dashed) of the filter distribution. From Hürzeler and Künsch (1998).*

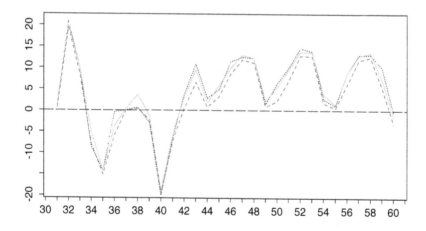

Figure 3.8 *Recursive Monte Carlo smoothing in the non-linear model for t =* *31,...60. We show the unobserved state variable (solid), the median (dotted) and 10% and 90% quantiles (dashed) of the filter distribution. From Hürzeler and Künsch (1998). From Hürzeler and Künsch (1998).*

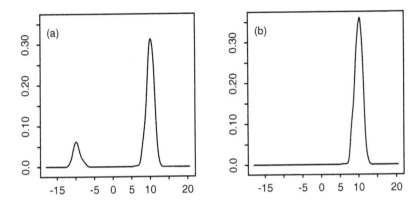

Figure 3.9 *Comparison of the filter (left) and smoother density (right) at $t = 46$ in the non-linear model. From Hürzeler and Künsch (1998).*

time $t + 1$ as

$$
\begin{aligned}
f_{t+1|t+1} &- g_{t+1|t+1} \\
&= B(A_{t+1}^* f_{t|t}, b_{t+1}(., y_{t+1})) - B(A_{t+1}^* g_{t|t}, b_{t+1}(., y_{t+1})) \\
&+ B(A_{t+1}^* g_{t|t}, b_{t+1}(., y_{t+1})) - B(A_{t+1}^* E_N(g_{t|t})), b_{t+1}(., y_{t+1}).
\end{aligned}
$$

Here the first difference represents the propagation of the error which was present already at time t and the second term represents the effect of the new error due to sampling at time t. If we try to control the L_1-norms of both terms separately, we face the same difficulty as in Section 3.3.6 because Bayes formula is not contracting. Looking at several transitions together leads to better results. Namely by (3.31) and (3.53), we obtain for any $k > 0$

$$
\begin{aligned}
f_{t+k|t+k} &- g_{t+k|t+k} \\
&= B(A_{t+k}^* \cdots B(A_{t+1}^* f_{t|t}, b_{t+1}), \ldots b_{t+k}) \\
&- B(A_{t+k}^* \cdots B(A_{t+1}^* g_{t|t}, b_{t+1}), \ldots b_{t+k}) \\
&+ \sum_{j=1}^{k} (B(A_{t+k}^* \cdots B(A_{t+j}^* g_{t+j-1|t+j-1}, b_{t+j}), \ldots b_{t+k}) \\
&\qquad - B(A_{t+k}^* \cdots B(A_{t+j}^* E_N(g_{t+j-1|t+j-1}), b_{t+j}), \ldots b_{t+k}))
\end{aligned}
$$

where b_s is shorthand for $b_s(., y_s)$. In addition to assumption **(A1)** we need a similar condition for the observation densities:

(A2) There exists a constant $C_b < \infty$ such that for all t and all x, x', y

$$\frac{b_t(x,y)}{b_t(x',y)} \leq C_b.$$

Using assumptions **(A1)** and **(A2)** and the results of Section 3.3.6, we obtain

$$\|f_{t+k|t+k} - g_{t+k|t+k}\|_1 \leq$$

$$C_a \left(1 - C_a^{-2}\right)^k \|f_{t|t} - g_{t|t}\|_1 + C_b \sum_{j=1}^{k} (C_b(1 - C_a^{-1}))^{k-j} \cdot$$

$$\|A_{t+j}^* E_N(g_{t+j-1|t+j-1}) - A_{t+j}^* g_{t+j-1|t+j-1}\|_1. \tag{3.56}$$

The following Lemma provides a bound for the L_1-norms in the last term on the right-hand side.

Lemma 3.10 *If the transition densities a_t satisfy Assumption **(A1)** from Section 3.3.6, then for any density g*

$$\mathbb{E}[\|A_t^* E_N(g) - A_t^* g\|_1] \leq (C_a/N)^{1/2}.$$

Proof. We denote the sample with density g by $(X^{(i)})$ and introduce

$$Z(x) = \sum_{i=1}^{N} (a_t(X^{(i)}, x) - A_t^* g(x)).$$

Then for an arbitrary x''

$$\begin{aligned}
(\mathbb{E}|Z(x)|)^2 &\leq \mathbb{E}[Z(x)^2] \\
&= N\left(\int a_t(x', x)^2 g(x')\mu(dx') - A_t^* g(x)^2\right) \\
&\leq N C_a a_t(x'', x) A_t^* g(x).
\end{aligned}$$

Therefore by Fubini and Schwarz' inequality

$$\begin{aligned}
\mathbb{E}[N^{-1} \int |Z(x)| d\mu(x)] &\leq (C_a/N)^{1/2} \int a_t(x'', x)^{1/2} (A_t^* g(x))^{1/2} d\mu(x) \\
&\leq (C_a/N)^{1/2}.
\end{aligned}$$

$\square$

Now we have all the necessary ingredients for proving our main result

Theorem 3.11 *Under assumptions **(A1)** and **(A2)**, to any $\epsilon > 0$, any $\eta > 0$ and any t there is an N such that for any y_1^t with probability at least $1 - \eta$*

$$\|f_{t|t} - g_{t|t}\|_1 \leq \epsilon.$$

(The probability refers to the generation of random samples in the Monte Carlo filter). The required sample size N grows like t^2.

Proof. By Lemma 3.10 and Markov's inequality it follows that with probability at least $1 - tC_a^{1/2}/(N^{1/2}\delta)$

$$||A_{s+1}^* E_N(g_{s|s}) - A_{s+1}^* g_{s|s}||_1 < \delta \text{ for all } 0 \leq s < t. \tag{3.57}$$

Now we choose first k such that

$$C_a \left(1 - C_a^{-2}\right)^k \leq \frac{1}{2},$$

then

$$\delta = \frac{\epsilon}{2C_b} \left(\sum_{j=0}^{k-1} (C_b(1 - C_a - 1))^j\right)^{-1}$$

and finally

$$N \geq \frac{t^2 C_a}{\delta^2 \eta^2}.$$

With these choices (3.57) is satisfied with probability at least $1 - \eta$, and if (3.57) holds, then by (3.56) $||f_{s|s} - g_{s|s}||_1 \leq \epsilon$ implies also $||f_{s+k|s+k} - g_{s+k|s+k}||_1 \leq \epsilon$. Iterating this completes the proof. $\square$

That the required sample size grows like t^2 is much better than exponential growth, but still not completely satisfactory. However, we believe that this is an artifact of our proof which uses Markov's inequality. With additional conditions we can obtain an exponential inequality which requires N to grow like $\log t$.

Related results on the convergence of the Monte Carlo filter can be found in Del Moral (1996) and Del Moral and Guionnet (1998, 1999).

3.6 Parameter estimation

Assume that the state transitions a_t depend on a parameter τ and the observation densities b_t on a parameter η. Both these parameters can be multidimensional. The combined parameter $(\tau', \eta')'$ will be denoted by θ.

3.6.1 Bayesian methods

For state space models, often the Bayesian viewpoint is adopted and prior distributions for the unknown parameters are introduced. In applications, a reasonable choice for these priors is not always easy, but we do not discuss this issue here. If one uses MCMC methods as described in Sections 3.5.1 and 3.5.2 to approximate the conditional distribution of the states given the data, then one can include updates on τ and η easily. Thus one can make simultaneous inferences on the states and the parameters. The graph describing the dependence between all variables contains two additional nodes for τ and η, and – assuming that the priors for τ and η are independent –

we have

$$p(\tau \mid x_0^T, y_1^T, \eta) = p(\tau \mid x_0^T) \propto p(\tau)a_0(x_0 \mid \tau) \prod_{t=1}^{T} a_t(x_{t-1}, x_t \mid \tau)$$

and

$$p(\eta \mid x_0^T, y_1^T, \tau) = p(\eta \mid x_0^T, y_1^T) \propto p(\eta) \prod_{t=1}^{T} b_t(x_t, y_t \mid \eta).$$

Exponential families for a_t and b_t with conjugate priors make the updating of τ and η particularly easy, but Metropolis updates can handle most other situations. This approach has been used in most published examples.

If one uses a recursive Monte Carlo filter as described in Section 3.5.3, one can proceed by including the parameters among the states x_t. The state transitions then become

$$\theta_t = \theta_{t-1}$$
$$x_t \mid (x_{t-1}, \theta_{t-1}) \sim a_t(x_{t-1}, x_t \mid \tau_{t-1})d\mu(x_t),$$

the initial distribution is

$$p(x_0, \theta_0) = p(\tau)p(\eta)a_0(x_0 \mid \tau)$$

and the observation transitions are

$$y_t \mid (x_t, \theta_t) \sim b_t(x_t, y_t \mid \eta_t)d\nu(y_t).$$

This has been discussed by Kitagawa (1998).

The main difficulty in this approach is that for all t $(\theta_{t|t}^{(i)})$ is a subsample of the prior sample $(\theta_{0|0}^{(i)})$. But the prior sample will have only few values in the region where the posterior is concentrated so that we do not obtain precise information about the location, spread and shape of the posterior. It thus seems advantageous to add a small noise to the transitions for θ_t:

$$\theta_t = \theta_{t-1} + V_t,$$

presumably with a decreasing variance for V_t in order not to discount information from early observations too much. How to choose these variances looks however difficult in general.

3.6.2 Monte Carlo likelihood approximations based on prediction samples

In this and the next section, we discuss Monte Carlo methods to approximate the likelihood function and the maximum likelihood estimator. Many ideas we will present originated in Geyer and Thompson (1992). The easiest way to approximate the likelihood function at a fixed value of the parameter starts from

$$
\begin{aligned}
L(\theta \mid y_1^T) &= p(y_1^T \mid \theta) = \prod_{t=1}^{T} p(y_t \mid y_1^{t-1}; \theta) \\
&= \prod_{t=1}^{T} \int b_t(x, y_t \mid \eta) f_{t|t-1}(x \mid y_1^{t-1}; \theta) d\mu(x) \qquad (3.58) \\
&= \prod_{t=1}^{T} \mathbb{E}[b_t(X_t, y_t \mid \eta) \mid y_1^{t-1}; \theta]. \qquad (3.59)
\end{aligned}
$$

If we have prediction samples $(x_{t|t-1}^{(j)})$ which are approximately distributed according to $f_{t|t-1}(x \mid y_1^{t-1}; \theta)$, we thus can approximate $\log L(\theta \mid y_1^T) - T \log N$ by

$$
\sum_{t=1}^{T} \log \left(\sum_{j=1}^{N} b_t(x_{t|t-1}^{(j)}, y_t \mid \eta) \right)
$$

The disadvantage of this formula is that if we need the likelihood at another value of θ, we have to generate prediction samples under a different model. Even if we use the same random numbers for all simulations, the likelihood function obtained in this way will be very noisy since different parameters will lead to different rejections in the generation of the samples. One thus has to apply some smoothing before attempting to find the maximum.

Using importance sampling, we can approximate the whole likelihood function based on a single set of prediction samples for a fixed parameter θ_0. If we define

$$
w_{t|t}(x; \theta, \theta_0) = \frac{f_{t|t}(x \mid y_1^t; \theta)}{f_{t|t}(x \mid y_1^t; \theta_0)},
$$

we can write

$$
L(\theta \mid y_1^T) = p(y_1^T \mid \theta) = \prod_{t=1}^{T} \mathbb{E}[b_t(X_t, y_t \mid \eta) w_{t|t}(X_t; \theta, \theta_0) \mid y_1^{t-1}; \theta_0] \quad (3.60)
$$

and approximate the factors on the left by sample means. However, the weights $w_{t|t}$ are not known in closed form. Using (3.14) and (3.16), we can write down recursions for $w_{t|t}$. The integrals in these recursions are then again approximated by sample averages, see Hürzeler (1998) and Hürzeler and Künsch (2000). The algorithm is quite computer intensive because its complexity is of the order $O(TN^2)$.

Although the equation (3.60) holds for all θ, the Monte Carlo approximation deteriorates if θ moves away from θ_0 because typically a few weights become dominant. One thus has to use an iterative procedure where one alternates between estimating the likelihood and maximizing it.

3.6.3 Monte Carlo approximations based on smoother samples

It is also possible to approximate the difference $\log L(\theta) - \log L(\theta_0)$ based on samples distributed according to the joint density of X_1^T given y_1^T for a fixed value of θ_0. By definition

$$\log L(\theta) = \log(p(y_1^T \mid \theta)) = \log(\int p(x_0^T, y_1^T \mid \theta) \prod_{t=0}^{T} d\mu(x_t))$$

which implies

$$
\begin{aligned}
\log L(\theta) - \log L(\theta_0) &= \log\left(\frac{p(y_1^T \mid \theta)}{p(y_1^T \mid \theta_0)}\right) \\
&= \log\left(\mathbb{E}\left[\frac{p(X_0^T, y_1^T \mid \theta)}{p(X_0^T, y_1^T \mid \theta_0)}\bigg| y_1^T; \theta_0\right]\right). \quad (3.61)
\end{aligned}
$$

The joint likelihood is given in (3.4) and the expectation can be approximated by a sample mean. This leads to

$$
\begin{aligned}
\log L(\theta) - \log L(\theta_0) &\approx \log\left(\frac{1}{N}\sum_{j=1}^{N}\frac{a_0(x_{0|T}^{(j)} \mid \tau)}{a_0(x_{0|T}^{(j)} \mid \tau_0)}\cdot\right. \\
&\left.\prod_{t=1}^{T}\frac{a_t(x_{t-1|T}^{(j)}, x_{t|T}^{(j)} \mid \tau)b_t(x_{t|T}^{(j)}, y_t \mid \eta)}{a_t(x_{t-1|T}^{(j)}, x_{t|T}^{(j)} \mid \tau_0)b_t(x_{t|T}^{(j)}, y_t \mid \eta_0)}\right).
\end{aligned}
$$

For the same reason as above, an iterative procedure is necessary in practice.

For later use we take in (3.61) $\theta = \theta_0 + h$ and let h tend to zero. Then it follows under suitable regularity conditions for exchanging differentiation and expectation

$$\frac{\partial}{\partial\theta}\log L(\theta) = \mathbb{E}\left[\frac{\partial}{\partial\theta}\log p(X_0^T, y_1^T \mid \theta) \mid y_1^T; \theta_0\right]. \quad (3.62)$$

A last possibility to find the maximum likelihood estimator is to use a stochastic version of the EM-algorithm. Such versions have been discussed by many people, e.g. by Wei and Tanner (1990) or Chan and Ledolter (1995). The E-step of the EM-algorithm consists in computing

$$
\begin{aligned}
Q(\theta \mid \theta_0) &= \mathbb{E}[\log p(X_0^T, y_1^T \mid \theta) \mid y_1^T, \theta_0] \\
&= \mathbb{E}[\log a_0(X_0 \mid \tau) \mid y_1^T, \theta_0] \\
&\quad + \sum_{t=1}^{T}\mathbb{E}[\log a_t(X_{t-1}, X_t \mid \tau) \mid y_1^T, \theta_0] \\
&\quad + \sum_{t=1}^{T}\mathbb{E}[\log b_t(X_t, y_t \mid \eta) \mid y_1^T, \theta_0].
\end{aligned}
$$

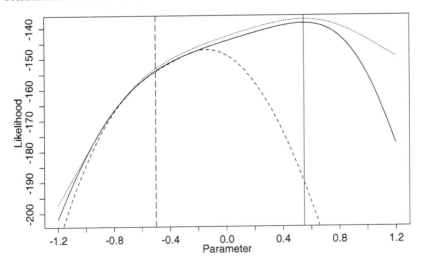

Figure 3.10 *AR(1) process with additive noise: the true likelihood function* $\log L(\theta)$ *(solid line), the filter approximation based on (3.60) (dotted line) and the smoother approximation based on (3.61) (dashed line) are drawn. The true MLE is* $\hat{\theta}_{ML} = 0.55$ *(solid vertical line) and the samples are produced for* $\theta_0 = -0.5$ *(dashed vertical line). From Hürzeler (1998).*

Clearly we can approximate the expectations by arithmetic means with the smoother samples. The M-step consists in maximizing $Q(\theta \mid \theta_0)$ with respect to θ in order to obtain the next θ_0.

The difficulty in implementing all these Monte Carlo maximum likelihood methods is in the allocation of computing resources. In the beginning, the Monte Carlo samples can be rather small because no precise approximation of $Q(\theta \mid \theta_0)$ or $\log L(\theta) - \log L(\theta_0)$ is needed to find a better θ_0. Later on in the iteration, this changes, and from some point on the iterates start to fluctuate randomly if the size of the smoother sample is kept fixed.

Stochastic EM algorithms in a more general setup have been discussed by Delyon et al. (1999).

3.6.4 Examples

We show some examples from Hürzeler (1998), see also Hürzeler and Künsch (2000). First we consider a Gaussian AR(1) process with added observation noise which becomes an ARMA(1,1) process:

$$X_t = \theta_0 \cdot X_{t-1} + V_t$$
$$Y_t = X_t + W_t$$

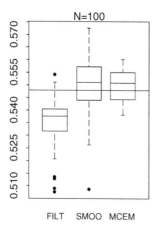
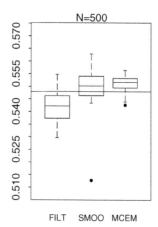

Figure 3.11 *AR(1) process with additive noise: Comparison of three different iterative methods to compute the maximum likelihood estimator and two Monte Carlo sample sizes. Each boxplot is based on 25 replicates of the Monte Carlo methods whereas the data are fixed. For a precise description of the methods, see text. From Hürzeler (1998).*

with $\theta_0 = 0.7$ and $V_t \sim \mathcal{N}(0, 0.7^2)$, $W_t \sim \mathcal{N}(0,1)$ independent of each other. The sample size is $T = 200$. We choose this example because the likelihood is available in closed form.

In Figure 3.10 we show the approximations of the likelihood function based on filtering (3.60) and on smoothing (3.61). We see that the filter approximation is quite near to the true likelihood function over a wide range of θ's. On the other hand, the approximation which is based on smoothing samples is really only a local approximation.

In Figure 3.11 we compare three different iterative methods to compute the maximum likelihood estimator and two Monte Carlo sample sizes. The first method uses the filter approximation of the likelihood based on (3.60), the second method uses the smoother approximation based on (3.61)and the third is the stochastic EM algorithm. All methods start with $\hat{\theta} = 0.7$. The first two methods use 6 iterations between approximation and maximization of the likelihood function and take the median of the last 4 values as the estimator. The EM-algorithm uses 21 iterations between the E- and the M-step and uses again the median of the last 4 values. We can see that the method using the filter approximation has a bias and that the stochas-

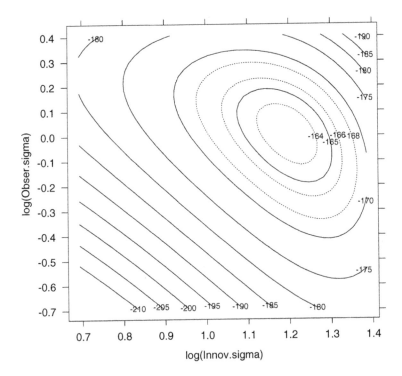

Figure 3.12 *Non-linear Model: Approximation of the (conditional) likelihood of the state and observation variances (with other parameters fixed at their true value). We have computed the approximation based on (3.58) at a regular grid of parameters and smoothed the resulting surface. From Hürzeler (1998).*

tic EM algorithm has the smallest fluctuations. Increasing the sample size reduces the fluctuations as expected. Note that a good behavior is achieved for small Monte Carlo sample sizes already.

Our second example is the non-linear model (3.54) and (3.55). We show in Figure 3.12 a smoothed version of the approximation of the likelihood function for the two variance parameters over a grid based on (3.58). Approximate maximum likelihood estimates can be found visually. Finally in Figure 3.13 we show the behavior of three different iterative approximation methods for the estimation of the regression parameters α, β and γ. The three methods are the same as in the previous example. A Monte Carlo sample size $N = 200$ was used. The starting values are $\hat{\alpha} = 0.9$, $\hat{\beta} = 18$ and $\hat{\gamma} = 20$ which is quite far from the true values 0.5, 25, and 8. We see that the stochastic EM algorithm converges slowly, but has the small-

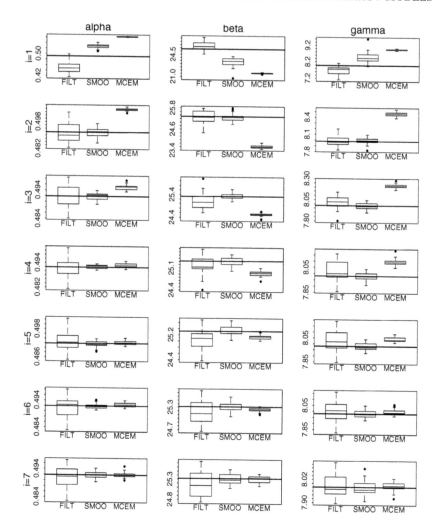

Figure 3.13 *Non-linear Model: Iterative methods for the MLE of regression pa-rameters. We show the sequences of approximations to the maximum likelihood estimators for the regression parameters* α, β *and* γ, *each row corresponding to one cycle in the iteration. The three methods are as in Figure 3.11. Each box plot consists of 25 replicates of the Monte Carlo simulations whereas the data are fixed. From Hürzeler (1998).*

est fluctuations. The other two converge quickly, but the one using filter approximations has large fluctuations and in one case also a bias.

3.6.5 Asymptotics of the maximum likelihood estimator

We summarize here recent work of Bickel et al. (1998) and Jensen and Petersen (1999) concerning the asymptotic properties of the maximum likelihood estimator $\hat{\theta}_T$ based on Y_1^T as T tends to infinity. We assume that the transitions a_t and b_t are the same for all t, that there is a stationary distribution for (X_t) and that the model is stationary.

The argument starts with the usual Taylor expansion of log likelihood at the true parameter θ_0. This implies that

$$-\sum_{t=1}^{T} \frac{\partial^2}{\partial\theta\partial\theta'} \log p(y_1^t \mid y_1^{t-1}, \theta_T^*)(\hat{\theta}_T - \theta_0) = \sum_{t=1}^{T} \frac{\partial}{\partial\theta} \log p(y_1^t \mid y_1^{t-1}, \theta_0)$$

where θ_T^* is between $\hat{\theta}_T$ and θ_0. The crucial point is to replace y_1^{t-1} by $y_{-\infty}^{t-1}$ in the conditioning on both sides. Then one can use the central limit theorem for martingale differences to show that the right-hand side divided by $\sqrt{T}$ is asymptotically normal with mean zero and covariance equal to the Fisher information

$$I(\theta_0) = \mathbb{E}[\frac{\partial}{\partial\theta} \log p(y_1^t \mid y_{-\infty}^{t-1}, \theta_0) \frac{\partial}{\partial\theta} \log p(y_1^t \mid y_{-\infty}^{t-1}, \theta_0)' \mid \theta_0].$$

Similarly, the ergodic theorem shows that

$$\frac{1}{T} \sum_{t=1}^{T} \frac{\partial^2}{\partial\theta\partial\theta'} \log p(y_1^t \mid y_{-\infty}^{t-1}, \theta_0)$$

converges almost surely to its expectation which is by a standard argument equal to $-I(\theta_0)$. An additional argument is needed to show that one may replace θ_T^* by θ_0.

Let us discuss the approach for showing that

$$\sum_{t=1}^{T} (\frac{\partial}{\partial\theta} \log p(y_1^t \mid y_1^{t-1}, \theta_0) - \frac{\partial}{\partial\theta} \log p(y_1^t \mid y_{-\infty}^{t-1}, \theta_0)) = o_P(\sqrt{T}). \quad (3.63)$$

Clearly

$$\log p(y_1^t \mid y_1^{t-1}, \theta_0) = \log p(y_1^t \mid \theta_0) - \log p(y_1^{t-1} \mid \theta_0),$$

and by (3.62) and (3.4)

$$\frac{\partial}{\partial\theta} \log p(y_1^t \mid \theta_0) = \mathbb{E}[\frac{\partial}{\partial\theta} \log a_0(X_0 \mid \theta_0) \mid y_1^t, \theta_0]$$
$$+ \sum_{s=1}^{t} \mathbb{E}[\frac{\partial}{\partial\theta} \log a(X_{s-1}, X_s \mid \theta_0) \mid y_1^t, \theta_0]$$

$$+ \sum_{s=1}^{t} \mathbb{E}[\frac{\partial}{\partial \theta} \log b(X_s, y_s \mid \theta_0) \mid y_1^t, \theta_0].$$

Hence the left-hand side of (3.63) contains the following terms

$$\mathbb{E}[h(X_{s-1}, X_s) \mid y_1^t, \theta_0] - \mathbb{E}[h(X_{s-1}, X_s) \mid y_1^{t-1}, \theta_0]$$
$$- \quad \mathbb{E}[h(X_{s-1}, X_s) \mid y_{-\infty}^t, \theta_0] + \mathbb{E}[h(X_{s-1}, X_s) \mid y_{-\infty}^{t-1}, \theta_0]$$

for some functions h and $1 \leq s \leq t$. For $s < t/2$ we pair the first term with the second and the third term with the fourth. On the other hand for $s > t$ we pair the first term with the third and the second term with the fourth. In this way the difference between the events we are conditioning on is always separated in time from the variables (X_s, X_{s-1}) whose expectations we compute. Similar arguments like those which have been used in Section 3.3.6 on forgetting the initial conditions can be used to bound these terms. For details, we refer the reader to Bickel et al. (1998) and Jensen and Petersen (1999). Part of the conditions needed to make the argument rigorous are similar to our **(A1)** and **(A2)**, but additional conditions for smoothness in θ are necessary.

3.7 Extensions of the model

In order to put the models and the methods discussed so far into a wider context, we discuss here two generalizations. The first considers spatial models instead of time series. Because there is no natural total order for spatial locations, defining Markov models is more complicated and recursive methods are no longer available. One has to use either approximations or Markov chain Monte Carlo even if the state space is finite. Historically, Markov chain Monte Carlo methods were developed in the spatial context (Metropolis et al. (1953), Geman and Geman (1984)) and found their way into time series models only afterwards (Carlin et al. (1992)).

In the second generalization, the states are not generated sequentially, but in a tree-like structure. Here, recursive methods are still available, but they have greater complexity.

3.7.1 Spatial models

In spatial models, the index t denotes a position in a finite index set L. For applications in image analysis L usually is the square lattice $\{1, \ldots T_1\} \times \{1, \ldots T_2\}$ of pixels. We assume that on L there is a neighborhood relation denoted by $\sim$ which is symmetric and not reflexive. This induces a graph structure on L, if all neighboring positions $t \sim s$ are joined by an edge. On the square lattice, each t has usually 4, 8 or 12 neighbors. A random field is then a collection of random variables $(X_t)_{t \in L}$ where each X_t takes values in a set $\mathcal{X}$ which we assume to be finite here. An element of $\mathcal{X}^L$ is called a *configuration* and a random field is called a *Markov random field* if the conditional distribution at one position given the rest only involves the values at the neighbors

$$p(x_t \mid x_s; s \neq t) = p(x_t \mid x_s; s \sim t). \qquad (3.64)$$

If we assume in addition that all configurations have positive probability, then by the Hammersley-Clifford theorem (see Lauritzen (1996), Theorem 3.9) we have the *Gibbs representation*

$$p(x) = \frac{1}{Z} \prod_{C \in \mathcal{C}} \exp(-\Phi_C(x_C)). \qquad (3.65)$$

Here $\mathcal{C}$ is the class of nonempty complete subsets of L (that is $C \in \mathcal{C}$ if any two different elements of C are neighbors), x_C is the restriction of the configuration $x = (x_t)_{t \in L}$ to $(x_t)_{t \in C}$, (Φ_C) is the so-called potential and Z is the normalization

$$Z = \sum_x \exp(-\sum_C \Phi_C(x_C)).$$

From the Gibbs representation (3.65), the computation of the conditional distribution of X_t given the rest is easy. We obtain

$$p(x_t \mid x_s; s \neq t) \propto p(x) \propto \prod_{C \in \mathcal{C}, C \ni t} \exp(-\Phi_C(x_C)) \qquad (3.66)$$

and the Markov property is obvious. The difficult part is the converse, i.e. getting from the $p(x_t \mid x_s; s \neq t)$ to (3.65).

In a *hidden Markov random field* we do not observe $(X_t)_{t \in L}$, but only variables $(Y_t)_{t \in L}$ which satisfy the assumption **(M2)** as before, i.e.

$$p(y \mid x) = \prod_{t \in L} b_t(x_t, y_t).$$

The main applications are in image restoration and segmentation where y is the observed noisy image and x is the clean image, see Geman and Geman (1984). If we want to account for some blurring in these applications, we have to extend the model by allowing Y_t to depend not only on X_t, but also on $X_s, s \sim t$. Other applications occur for instance in epidemiology, see Besag et al (1991).

The joint density of (x, y) is

$$p(x, y) = p(x)p(y \mid x) = \frac{1}{Z} \prod_C \exp(-\Phi_C(x_C)) \prod_{t \in L} b_t(x_t, y_t).$$

Because the conditional distribution $p(x \mid y)$ is proportional to $p(x, y)$ it also has a Gibbs representation and therefore $(X_t)_{t \in L}$ is also a Markov random field given the observations $(y_t)_{t \in L}$.

Unfortunately, spatial models are an order of magnitude more difficult to use than the time series models we have discussed so far. The following is an (incomplete) list of problems which do occur

1. Marginal probabilities $p(x_t)$ or $p(x_V)$ for $V \subset L$ and conditional probabilities $p(x_t \mid y)$ or $p(x_V \mid y)$ cannot be computed in closed form. The number of operations is exponential in the size of L and no effective reduction is known. This makes the computations prohibitive in almost all cases of interest.

2. One cannot simulate from $p(x)$ or $p(x \mid y)$ in a fixed number of steps, but has to use MCMC methods.

3. The X_t's may exhibit long-range dependence which means that the effect of how the field is defined at the boundary does not disappear even in the limit if L increases in all directions.

4. The joint distribution is not equal to the product of the conditional distributions for each x_t given the rest

$$\prod_{t \in L} p(x_t \mid x_s, s \sim t) \neq p(x).$$

5. The conditional distributions $p(x_t \mid x_s; s \sim t)$ cannot be chosen arbitrarily. In order to obtain a valid model, one has to use the Gibbs representation. Only the potential can be chosen arbitrarily.

6. If the potential contains unknown parameters τ, then the normalization Z also depends on τ, but $Z(\tau)$ cannot be evaluated in closed form. This makes maximum likelihood difficult even in the fully observed case when the X_t's are available.

7. Fully Bayesian inference on (x, τ, η) (where τ and η are the unknown parameters in the potential and b_t respectively) is difficult because

$$p(\tau \mid x, y, \eta) = p(\tau \mid x) \propto p(x \mid \tau) p(\tau)$$

also needs the normalization $Z(\tau)$.

8. Computation of the posterior mode (MAP)

$$\arg\max_x p(x \mid y) = \arg\max_x \left(-\sum_C \Phi_C(x_C) + \sum_t \log(b_t(x_t, y_t)) \right)$$

is made difficult because there are typically many local maxima.

We discuss here some possible methods how one can nevertheless overcome these difficulties. First, there is a number of methods which are simpler to use and avoid some of the computational problems. For instance one can use the so-called *pseudolikelihood*

$$\prod_{t \in L} p(x_t \mid x_s, s \sim t; \tau)$$

instead of the likelihood (Besag (1975)). Note that the normalizing constant in (3.66) involves only a summation over few terms and thus can be computed easily. The resulting pseudolikelihood estimator has been investigated extensively, and consistency and asymptotic normality has been proved (Comets (1992), Guyon and Künsch (1992)). It is less clear how one can justify the use of the pseudolikelihood instead of the likelihood in a Bayesian approach, e.g. when simulating from $p(\tau \mid x)$. Some empirical investigations on how well this works are reported in Rydén and Titterington (1998).

In physics, the *mean field approximation* is often used. There it is assumed that each X_t "sees" the mean value of its neighbors, but otherwise behaves independently. This means that one uses the approximation

$$\mathbb{E}[X_t] = \mathbb{E}[\mathbb{E}[X_t \mid X_s, s \sim t]] \approx \mathbb{E}[X_t \mid X_s = \mathbb{E}[X_s], s \sim t].$$

The right-hand side can be computed using (3.66). This gives a system of equations for the unknown quantities $\mathbb{E}[X_t]$ which can often be solved. For instance in a shift invariant case with periodic boundary conditions, one has $\mathbb{E}[X_t] = \text{constant}$ and the constant is determined by a single equation. The mean field approximation for the normalization Z is then

$$\begin{aligned}
Z &\approx \sum_x \exp(-\sum_t \sum_{C \ni t} \Phi_C(x_t \prod_{s \in C, s \neq t} \mathbb{E}[X_s])) \\
&= \prod_t \sum_{x_t} \exp(\sum_{C \ni t} \Phi_C(x_t \prod_{s \in C, s \neq t} \mathbb{E}[X_s])).
\end{aligned}$$

For a more detailed discussion of this approximation and its use in image analysis, see Zhang (1992).

Besag (1986) has proposed to compute instead of the posterior mode the so-called iterated conditional mode estimator where one visits the positions t in some sequential order and always computes

$$\hat{x}_t = \arg\max_{x_t} p(x_t \mid \hat{x}_s, s \sim t; y)$$

until nothing changes any more. The results depend on the starting value and the order of visits. It has been argued, that this may give better restorations than the MAP if started at $\arg\max_x p(y \mid x)$.

For the hidden case, when only (y_t) is available, Qian and Titterington (1992) have proposed a variation of the EM algorithm where the E-step is replaced by iterated conditional mode estimation and the M-step by a form of pseudo maximum likelihood:

$$\arg\max_{\tau, \eta} \sum_t \log p(y_t \mid \hat{x}_s, s \sim t; \tau, \eta).$$

The original feature of this proposal is the use of

$$p(y_t \mid \hat{x}_s, s \sim t; \tau, \eta) = \sum_{x_t} b_t(x_t, y_t \mid \eta) p(x_t \mid \hat{x}_s, s \sim t; \tau)$$

instead of the naive

$$p(\hat{x}_t \mid \hat{x}_s, s \sim t; \tau) b_t(\hat{x}_t, y_t \mid \eta).$$

The latter can easily give useless results because an extreme restoration, e.g. one where all $\hat{x}_t$ are equal, leads to an extreme $\hat{\tau}$ which then confirms the extreme restoration in the next E-step. In my own limited experience, this nice idea unfortunately did not perform particularly well. I conjecture that this is because an "optimal" and not an "average" restoration is used.

For image restoration purposes, it is often reasonable to assume η to be known because one can use known test images. One can then also determine τ off-line by ad-hoc methods, e.g. by considering which choices are good for some simple artificial images (see e.g. Künsch (1994)).

Finally, there are also a number of computing intensive proposals. For instance Geman and Geman (1984) have suggested to use *simulated annealing* to compute the posterior mode. A number of people have suggested to use MCMC to compute an approximate maximum likelihood estimator (Younes (1988), Geyer and Thompson (1992), Geyer (1991)). The starting point is

$$\frac{Z(\tau)}{Z(\tau_0)} = \mathbb{E}[\exp(-\sum_C (\Phi_C(X_C, \tau) - \Phi_C(X_C, \tau_0))) \mid \tau_0]$$

which is similar to (3.61). The methods then alternate between simulation at a fixed τ_0 in order to approximate the right hand side and maximization around that τ_0. Younes (1988) used incremental updates in the Monte Carlo approximation based on one additional simulated value and incremental changes in τ according to the Kiefer-Wolfowitz stochastic approximation. Geyer (1991) used approximations based on much longer simulations. These ideas can also be used in principle for the hidden case or for approximating the posterior for τ, but in my limited experience I had often difficulties to obtain convergence.

3.7.2 Stochastic context-free grammars

The purpose of this last section is to draw attention to a more general class of stochastic models than we have discussed before. It has a number of interesting applications in molecular biology (see e.g. Durbin et al. (1998)) and in computer vision (see e.g. Fu (1982) or Chou (1989)), but is usually not covered in the statistical literature. We make no attempt to give a comprehensive treatment.

Hidden Markov models have the basic feature that state values are generated sequentially. But this is not appropriate in some applications. For instance RNA sequences with the alphabet $\{A, C, G, U\}$ have a complicated secondary structure. They fold and form stems by pairing C with G and A with U. In between stems there are loops. This implies that there is an almost deterministic correspondence between different parts in the sequence which may be rather far apart. Also in some problems of computer vision, the dependence between constituents of objects is usually not sequential. For instance, characters are composed of vertical, horizontal and diagonal lines and different types of junctions which in turn are composed by small line segments which are composed by black pixels. In such cases, it is more natural to use a model where the states (secondary structure or image constituents respectively) are generated according to a Markovian branching tree model. It turns out that these models are related to Chomsky's (1956) theory of grammars.

A grammar consists of a finite set of nonterminal symbols denoted by capitals, another finite set of terminal symbols denoted by lower case symbols and production rules which describe how a string of symbols can be transformed into another string of symbols. One then starts with a nonterminal and applies the production rules successively until one ends up with a string of terminals. The main task for a given string of terminals is to find all starting symbols and sequences of production rules which produce that string (there maybe none or several ones). This is called *parsing*.

A stochastic grammar has probability distributions for the starting symbols and for the set of sequences of production rules. One then can rank the possible sequences of rules producing a given string according to their probability. In our terminology so far, terminals correspond to observations and nonterminals to states. Parsing is then nothing else than inference about the states given the observations.

In Chomsky's theory, there are four types of grammars called regular, context-free, context-sensitive and unrestricted. They differ by the type of production rules which are allowed. A regular grammar has only the following two types of productions

$$A \longrightarrow b, \quad A \longrightarrow bC$$

which can be applied to any nonterminal symbol in a string. The only possible strings in such a grammar are then $a_1 \ldots a_k$ and $a_1 \ldots a_k A$. For each nonterminal A we assume that there is a probability distribution on the set of allowed productions with A on the left. The next production depends only on the current string, and not on the history. Obviously, this corresponds to a slightly generalized hidden Markov models where the observation Y_t is the terminal symbol at position t and the state X_t is the nonterminal that occurs after the t-th application of the production rules. Then Y_t depends on the pair of states (X_{t-1}, X_t) .

In a context-free grammar of normal form, the allowed productions are

$$A \longrightarrow b, \quad A \longrightarrow BC$$

that can be applied to any nonterminal in a string. This generates a binary tree with nonterminals at the interior nodes, a nonterminal starting symbol at the root and terminals at the leaves. Each interior split of a node with symbol A into a left node B and a right node C corresponds to one application of the rule $A \longrightarrow BC$. The probabilities are assigned similarly as for regular grammars, namely at each node the splitting occurs independently of other splittings. This induces a Markovian structure on the tree. By the same idea one can produce not only sequences of terminals, but also two-dimensional arrays. For this it is sufficient to consider production rules $A \longrightarrow BC$ where the result BC is either arranged horizontally or vertically.

Inference about the generating tree for a given sequence (or array) of terminals can be made by the so-called inside-outside algorithm (Lari and Young, 1990), an analogue of the forward-backward recursions for hidden Markov models. In order to give an idea on how this algorithm works, let us consider an observed sequence of terminals y_1^T. For $1 \le s \le t \le T$ we denote the root of the subtree ending with y_s^t by $R(s,t)$ and we assume that we want to compute

$$\mathbb{P}[X_{R(s,t)} = A \mid y_1^T].$$

Because Y_s^t and (Y_1^{s-1}, Y_{t+1}^T) are conditionally independent given $x_{R(s,t)}$, we have

$$\mathbb{P}[X_{R(s,t)} = A, Y_1^T = y_1^T] = \alpha(s,t,A)(y_s^t)\beta(s,t,A)(y_1^{s-1}, y_{t+1}^T)$$

where

$$\alpha(s,t,A)(y_s^t) = \mathbb{P}[Y_s^t = y_s^t \mid X_{R(s,t)} = A]$$

and

$$\beta(s,t,A)(y_1^{s-1}, y_{t+1}^T) = \mathbb{P}[Y_1^{s-1} = y_1^{s-1}, Y_{t+1}^T = y_{t+1}^T, X_{R(s,t)} = A].$$

The inside algorithm computes the α's recursively, moving from the inside outwards, and the outside algorithm computes the β's, moving inwards from the outside. The inside recursion is easily seen to be

$$\alpha(s,t,A) = \sum_{u=s}^{t-1} \sum_{B,C} \alpha(s,u,B)\alpha(u+1,t,C)\mathbb{P}[A \to BC].$$

(We suppress the y-variables because they are assumed to be fixed.) As a byproduct we obtain also the likelihood of the observed sequence as

$$\mathbb{P}[Y_1^T = y_1^T] = \sum_A \alpha(1,T,A)\mathbb{P}[\text{Start} = A].$$

The outside algorithm is slightly more complicated. We leave it to the reader to check the following formula

$$
\begin{aligned}
\beta(s,t,A) \;=\; & \sum_{u=1}^{s-1}\sum_{B,C}\alpha(u,s-1,B)\beta(u,t,C)\mathbb{P}[C\to BA] \\
+\; & \sum_{u=t+1}^{T}\sum_{B,C}\alpha(t+1,u,B)\beta(s,u,C)\mathbb{P}[C\to AB]
\end{aligned}
$$

The inside-outside algorithm is needed if one wants to estimate production probabilities $\mathbb{P}[A\to BC]$ in the EM-framework. There is also an algorithm to compute the most likely tree (similar to the Viterbi algorithm), called Cocke-Younger-Kasami algorithm. For more details we refer the reader to the literature cited earlier in this section.

Context-sensitive grammars are more complicated. As the name says, the allowed productions depend not only on the symbol which is transformed, but also on neighboring symbols. Because of this additional dependence, even the assignment of a probability distribution to the set of allowed sequences of rules requires care.

Acknowledgments

I am grateful to Stu Geman for the collaboration on the paper Künsch et al. (1995). He pointed out to me the importance of the hidden Markov idea and its many applications. I am indebted to my former Ph. D. student Markus Hürzeler. It is by supervising his thesis and discussing with him that I became familiar with the Monte Carlo methods in this area. I thank him for the collaboration on this topic and for allowing me to use material from his thesis (Hürzeler (1998)) for these notes.

Finally I would like to thank Mathisca de Gunst and Barry Schouten for the collaboration on ion channel data, to Sake Vogelzang for Figure 1.1, to Genshiro Kitagawa and Tomoyuki Higuchi for interesting discussions and to Neil Shephard for some useful last minute comments.

3.8 References

Aidala, V. J. (1979) Kalman filter behavior in bearings-only tracking applications. *IEEE Trans. Aerospace Electronics Syst.*, **15**, 29–39.

Akaike, H. (1974) Markovian representation of stochastic processes and its application to the analysis of autoregressive and moving average processes. *Ann. Inst. Statist. Math.*, **26**, 363–387.

Alspach, D. L. and Sorensen, H. W. (1972) Nonlinear Bayesian estimation using Gaussian sum approximations. *IEEE Trans. Autom. Control*, **17**, 439–448.

Anderson, B. D. O. and Moore, J. B. (1979) *Optimal Filtering*. Prentice-Hall, London.

Andrade Netto, M. L., Gimeno, L. and Mendes, M.J. (1978) On the optimal and suboptimal nonlinear filtering problem for discrete-time systems. *IEEE Trans. Autom. Control*, **23**, 1062–1067.

Arapostathis, A. and Marcus, S. I. (1990) Analysis of an identification algorithm arising in the adaptive estimation of Markov chains. *Math. Control, Signals and Systems*, **3**, 1–29.

Ball, F. G. and Rice, J. A. (1992) Stochastic models for ion channels: introduction and bibliography. *Math. Biosci.*, **112**, 189–206.

Baum, L.E. and Petrie, T. (1966) Statistical inference for probabilistic functions of finite state Markov chains. *Ann. Math. Statist.*, **37**, 1554–1563.

Baum, L.E., Petrie, T., Soules, G. and Weiss, N. (1970) A maximization technique occurring in the statistical analysis of probabilistic functions of Markov chains. *Ann. Math. Statist.*, **41**, 164–171.

Besag, J. E. (1975) Statistical analysis of non-lattice data. *The Statistician*, **24**, 179–195.

Besag, J. (1986) On the statistical analysis of dirty pictures (with discussion). *J. Royal Statist. Soc.*, **48**, 259–279.

Besag, J., York, J. and Mollié, A. (1991) Bayesian image restoration, with two applications in spatial statistics (with discussion). *Ann. Inst. Statist. Math.*, **43**, 1–59.

Bickel, P.J., Ritov Y. and Ryden, T. (1998) Asymptotic normality of the maximum-likelihood estimator for general hidden Markov models. *Ann. Statist.*, **26**, 1614–1635.

Brockwell, P. J. and Davis, R. A. (1991) *Time Series: Theory and Methods*. 2nd ed., Springer, New York.

Carlin, B. P., Polson, N. G. and Stoffer, D. S. (1992) A Monte Carlo approach to nonnormal and nonlinear state-space modeling. *J. Amer. Statist. Assoc.*, **87**, 493–500.

Carpenter, J., Clifford, P. and Fearnhead, P. (1998) An improved particle filter for nonlinear problems. *Technical Report*, University of Oxford.

Carter, C. K. and Kohn, R. (1994) On Gibbs sampling for state space models. *Biometrika*, **81**, 541–553.

Chomsky, N. (1956) Three models for the description of language. *IRE Trans. Inform. Theory*, **2**, 113–124.

Chou, P. A. (1989) Recognition of equations using a two-dimensional stochastic context-free grammar. *Society of Photo Optical Instrumentation Engineers, Visual Communications and Image Processing IV* **1199**, 852–863.

Chaleyat-Maurel, M. and Michel, D. (1983) Des résultats de non existence de filtre de dimension finie. *C. R. Acad. Sci. Paris Sér. I Math.*, **296**, 933-936.

Chan, K. S. and Ledolter, J. (1995) A Monte Carlo EM estimation for time series involving counts. *J. Amer. Statist. Assoc.*, **90**, 242–252.

Chan, N. H. and Palma, W. (1998) State space modeling of long-memory processes. *Ann. Statist.*, **26**, 719–740.

Comets, F. (1992) On consistency of a class of estimators for exponential families of Markov random fields on the lattice. *Ann. Statist.*, **20**, 455–468.

De Gunst, M., Künsch, H. R. and Schouten, B. (1998) Statistical analysis of ion channel data using hidden Markov models with correlated noise and filtering. *Technical Report*, **WS-512**, Free University, Amsterdam.

Del Moral, P. (1996) Non-linear filtering: interacting particle resolution. *Markov Processes Related Fields*, **2**, 555–580.

Del Moral, P. and Guionnet, A. (1998) Large deviations for interacting particle systems. Applications to nonlinear filtering problems. *Stoch. Proc. Appl.*, **78**, 69–95.

Del Moral, P. and Guionnet, A. (1999) Central limit theorem for nonlinear filtering and interacting particle systems. *Ann. Appl. Probab.*, **9**, 275–297.

Delyon, B., Lavielle, M. and Moulines, E. (1999) Convergence of a stochastic approximation version of the EM algorithm. *Ann. Statist.*, **27**, 94–128.

Devroye, L. (1987) *A Course in Density Estimation*. Birkhäuser, Basel.

Dobrushin, R. L. (1956) Central limit theorem for non-stationary Markov chains I, II. *Theory Probab. Appl.* **1**, 65–80 and 329–383.

Donoho, D. L. and Johnstone, I. M. (1994) Ideal spatial adaptation by wavelet shrinkage. *Biometrika*, **81**, 425–455.

Douc, R., and Matias, C. (1999) Asymptotics of the maximum likelihood estimator for general hidden Markov models. *Preprint*, Université Paris-Sud, Orsay.

Doucet, A. (1998) On sequential simulation based methods for Bayesian filtering. *Technical Report*, Department of Electrical Engineering, University of Cambridge.

Doucet, A., de Freitas, J. F. G., and Gordon, N. J., eds. (2000) *Sequential Monte Carlo Methods in Practice*. Springer, New York, to appear.

Durbin, R., Eddy, S., Krogh, A. and Mitchison, G. (1998) *Biological Sequence Analysis*. Cambridge University Press, Cambridge.

Elliott, R. J., Aggoun, L. and Moore, J. B. (1995) *Hidden Markov Models: Estimation and Control*. Springer, New York.

Ferrante, M. and Vidoni, P. (1998) Finite dimensional filters for nonlinear stochastic difference equations with multiplicative noises. *Stoch. Proc. Appl.*, **77**, 69–81.

Frühwirth-Schnatter, S. (1994) Data augmentation and dynamic linear modeling. *J. Time Ser. Anal.*, **15**, 183–202.

Fu, K. S. (1982) *Syntactic Pattern Recognition and Applications*. Prentice-Hall, Englewood Cliffs, New Jersey.

Gallant, A. R. and Tauchen, G. (1998) Reprojecting partially observed systems with applications to interest rate diffusions. *J. Amer. Statist. Assoc.*, **93**, 10–24.

Geman, D. and Geman, S. (1984) Stochastic relaxation, Gibbs distributions and the Bayesian restoration of images. *IEEE Trans. Pattern Anal. Machine Intelligence*, **6**, 721–741.

Geyer, C. J. (1991) Markov chain Monte Carlo maximum likelihood. In *Computer Science and Statistics, Proceedings of the 23rd Symposium on the Interface*, Keramidas, E. M. and Kaufman, S. M., eds., 153–163. Interface Foundation of North America, Fairfax Station, Virginia.

Geyer, C. J. and Thompson, E. A. (1992) Constrained Monte Carlo maximum likelihood for dependent data (with discussion). *J. Royal Statist. Soc. B*, **54**, 657–699.

Ghil, M. and Ide, K. eds. (1997) Data assimilation in meteorology and oceanography: theory and practice. *J. Meteor. Soc. Japan, Special Issue*, **75**, **1B**, 111–496.

Goldman, N., Thorne, J. L. and Jones, D. T. (1996) Using evolutionary trees in protein secondary structure prediction and other comparative sequence analyses. *J. Molec. Biol.*, **263**, 196-208.

Gordon, N. J., Salmond, D. J. and Smith, A. F. M. (1993) Novel approach to nonlinear/non-Gaussian Bayesian state estimation *IEE proceedings F: Comm., Radar, Signal Processing*, **140**, 107–113.

Guyon, X. and Künsch, H. R. (1992) Asymptotic comparison of estimators in the Ising model. In *Stochastic Models, Statistical Methods, and Algorithms in Image Analysis*, Barone, P., Frigessi, A. and Piccioni, M., eds. Lecture Notes in Statistics, **74**, 177–198, Springer, New York.

Handschin, J. (1970) Monte Carlo techniques for prediction and filtering of nonlinear stochastic processes. *Automatica*, **6**, 555–563.

Handschin, J. E. and Mayne, D. Q. (1969) Monte Carlo techniques to estimate the conditional expectation in multi-stage non-linear filtering. *Intern. J. Control*, **9**, 547–559.

Hannan, E. J. and Deistler, M. (1988) *The Statistical Theory of Linear Systems*. Wiley, New York.

Harrison, P. J. and Stevens, C. F. (1976) Bayesian forecasting (with discussion). *J. Royal Statist. Soc. B*, **38**, 205–247.

Harvey, A. C. (1989) *Forecasting, Structural Time Series Models and the Kalman Filter*. Cambridge University Press, Cambridge.

Hughes, J. P. and Guttorp, P. (1999) A class of stochastic models for relating synoptic atmospheric patterns to regional hydrologic phenomena. *Water Resour. Res.*, **30**, 1535–1546.

Hughes, J. P., Guttorp, P. and Charles, S. P.(1999) A non-homogeneous hidden Markov model for precipitation. *J. Roy. Statist. Soc. C (Applied Statistics)*, **48**, 15–30.

Hürzeler, M. (1998) *Statistical Methods for General State-Space Models*. PhD Thesis Nr. 12674, ETH Zürich.

Hürzeler, M. and Künsch, H. R.(1998) Monte Carlo approximations for general state-space models. *J. Comp. and Graph. Statist.*, **7**, 175–193.

Hürzeler, M. and Künsch, H. R.(2000) Approximating and maximizing the likelihood for a general state space model. In *Sequential Monte Carlo Methods in*

Practice, Doucet, A., de Freitas, J. F. G., and Gordon, N. J., eds., Springer, New York, to appear.

Isard, M. and Blake, A. (1996) Contour tracking by stochastic propagation of conditional density. *Proceedings European Conference in Computer Vision, Cambridge*, **1**, 343–356.

Itô, H., Amari, S.-I. and Kobayashi, K. (1992) Identifiability of hidden Markov information sources and their minimum degrees of freedom. *IEEE Trans. Inform. Theory*, **38**, 324-333.

Jensen, J. L. and Petersen, N. V. (1999) Asymptotic normality of the maximum likelihood estimator in state space models. *Ann. Statist.*, **27**, 514–535.

Jones, R. H. (1980) Maximum likelihood fitting of ARMA models to time series with missing observations. *Technometrics*, **22**, 389–395.

Kalman, R. E. (1960) A new approach to linear filtering and prediction problems. *Trans. Amer. Soc. Mech. Eng., J. Basic Eng.*, **82**, 35–45.

Kalman, R. E., and Bucy, R. S. (1961) New results in linear filtering and prediction theory. *Trans. Amer. Soc. Mech. Eng., J. Basic Eng.*, **83**, 95–108.

Kitagawa, G. (1987) Non-Gaussian state-space modeling of nonstationary time series (with discussion). *J. Amer. Statist. Assoc.*, **82**, 1032–1063.

Kitagawa, G. (1996) Monte Carlo filter and smoother for non-Gaussian nonlinear state space models. *J. Comp. Graph. Statist.*, **5**, 1–25.

Kitagawa, G. (1998) A self-organizing state-space model. *J. Amer. Statist. Assoc.*, **93**, 1203–1215.

Kitagawa, G. and Gersch, W. (1996) *Smoothness Priors Analysis of Time Series.* Lecture Notes in Statistics **116**, Springer, New York.

Kunita, H. (1971) Asymptotic behavior of nonlinear filtering errors of Markov processes. *J. Multivariate Anal.*, **1**, 365–393.

Künsch, H. R. (1994) Robust priors for smoothing and image restoration. *Ann. Inst. Statist. Math.*, **46**, 1–19.

Künsch, H. R., Geman, S. and Kehagias, A. (1995) Hidden Markov random fields. *Ann. Appl. Probab.*, **5**, 577–602.

Lari, K. and Young, S. J. (1990) The estimation of stochastic context-free grammars using the inside-outside algorithm. *Comp. Speech Language*, **4**, 35–56.

Lauritzen, S. L. (1981) Time series analysis in 1880: A discussion of contributions made by T.N. Thiele. *International Statist. Review*, **49**, 319–331.

Lauritzen, S. L. (1996) *Graphical Models.* Oxford University Press, Oxford.

Le Cam, L. and Yang, G. L. (1990) *Asymptotics in Statistics: Some Basic Concepts.* Springer, New York.

LeGland, F. and Mevel, L. (2000) Exponential forgetting and geometric ergodicity in hidden Markov models. *Mathem. Control, Signals, Systems*, to appear.

Loève, M. (1978) *Probability Theory, 4th ed..* Springer, New York.

MacDonald, I. L. and Zucchini, W. (1997) *Hidden Markov and Other Models for Discrete-Valued Time Series.* Chapman and Hall, London.

Mammen, E. and van de Geer, S. (1997) Locally adaptive regression splines. *Ann. Statist.* **25**, 387–413.

Martin, R. D. and Thomson, D. J. (1982) Robust-resistant spectrum estimation. *Proc. IEEE*, **70**, 1097–1115.

Masreliez, C. J. (1975) Approximate non-Gaussian filtering with linear state and observation relations. *IEEE Trans. Autom. Control*, **20**, 361–371.

Meinhold, R. J. and Singpurwalla, N. D. (1983) Understanding the Kalman filter. *American Statistician*, **37**, 123–127.

Metropolis, N., Rosenbluth, A. W., Rosenbluth, M. N., Teller, A. H. and Teller, E. (1953) Equations of state calculations by fast computing machines. *Journal of Chemical Physics*, **21**, 1087–1091.

Muri, F. (1998) Modeling bacterial genomes using hidden Markov models. In *Compstat'98 Proceedings in Computational Statistics*, R. Payne and P. J. Green, eds., 89–100. Physica-Verlag, Heidelberg.

Pitt, M. K. and Shephard, N. (1999) Filtering via simulation: auxiliary particle filters. *J. Amer. Statist. Assoc.*, **94**, 590–599.

Plackett, R. L. (1950) Some theorems in least squares. *Biometrika*, **37**, 149–157.

Qian, W. and Titterington, D. M. (1992) Stochastic relaxations and EM algorithms for Markov random fields. *J. Statist. Computat. Simul.*, **40**, 55–69.

Rabiner, L. R. and Juang, B. H. (1993) *Fundamentals of Speech Recognition*. Prentice-Hall, Englewood Cliffs, New Jersey.

Rubin, D. (1988) Using the SIR algorithm to simulate posterior distributions. In *Bayesian Statistics 3*, Bernardo, J. M., DeGroot, M. H., Lindley, D. V. and Smith, A. F. M., eds., 395–402. Oxford University Press, Oxford.

Rydén, T. and Titterington, D. M. (1998) Computational Bayesian analysis of hidden Markov models. *J. Comput. Graph. Statist.*, **7**, 194–211.

Shephard, N. (1994) Partial non-Gaussian state space. *Biometrika*, **81**, 115–131.

Shephard, N. (1996) Statistical aspects of ARCH and stochastic volatility. In *Time Series Models with Econometric, Finance and Other Applications*, Cox, D. R., Hinkley, D. V. and Barndorff-Nielsen, O. E., eds., 1-67. Chapman and Hall, London.

Smith, A. F. M. and West, M. (1983) Monitoring renal transplants: An application of the multiprocess Kalman filter. *Biometrics*, **39**, 867-878.

Steyn, I., J. (1996) *State Space Models in Econometrics: A Field Guide*. PhD thesis, Department of Economics and Econometrics, Free University, Amsterdam.

Stolcke, A. and Omohundro, S. M. (1993) Hidden markov model induction by Bayesian model merging. In *Advances in Neural Information Processing Systems*, Hanson, S. J., Cowan, J. D. and Giles, C. L., eds., **5**, 11-18. Morgan Kaufman Publishers, San Mateo, CA.

Thompson, E. A. (2000) Monte Carlo methods on genetic structures. Chapter 4 of this volume.

Viterbi, A. J. (1967) Error bounds for convolutional codes and an asymptotically optimal decoding algorithm. *IEEE Trans. Inform. Theory*, **13**, 260–269.

Wei, G. C. G. and Tanner, M. A. (1990) A Monte Carlo implementation of the EM algorithm and the poor man's data augmentation algorithms. *J. Amer. Statist. Assoc.*, **85**, 699–704.

Wei, W. W. S. (1990) *Time Series Analysis: Univariate and Multivariate Methods*. Addison-Wesley, Redwood City, CA.

Weiss, H. and Moore, J. B. (1980) Improved extended Kalman filter design for passive tracking. *IEEE Trans. Autom. Control*, **25**, 807-811.

West, M. and Harrison, J. (1997) *Bayesian Forecasting and Dynamic Models*. 2nd ed., Springer, New York.

Whitley, D. (1994) A genetic algorithm tutorial. *Statistics and Computing*, **4**, 65-85.

Whittle, P. (1996) *Optimal Control: Basics and Beyond*. Wiley, Chichester, UK.

Younes, L. (1988) Estimation and annealing for Gibbsian fields. *Ann. Inst. Henri Poincaré*, **2**, 269–294.

Zucchini, W. and Guttorp, P. (1991) A hidden Markov model for space-time precipitation. *Water Resour. Res.*, **27**, 1917–23.

Zhang , J. (1992) The mean field theory in EM procedures for Markov random fields. *IEEE Trans. Signal Processing*, **40**, 2570-2583.

CHAPTER 4

Monte Carlo Methods on Genetic Structures

Elizabeth A. Thompson
University of Washington

4.1 Genetics, pedigrees, and structured systems

4.1.1 Introduction to Genetics

The *DNA* in the nuclei of cells of an individual is packaged into *chromosomes* each of which has a linear structure. There are 46 chromosomes in the nucleus of each normal human cell, 22 pairs of *autosomes* and a pair of *sex chromosomes*. In this chapter we will restrict attention to the autosomes, which contain the majority of the DNA coding for the proteins and affecting the characteristics of individuals. Similar approaches would apply to the sex chromosomes, and to the mitochondria which are packets of DNA in the cytoplasm of the cell and also contain functional genes, but the details differ. Of the two chromosomes of a pair, one derives from the DNA of the mother of the individual and the other derives from the DNA of the father.

Any small segment of the DNA of the chromosome is known as a *locus*. Typically, a locus used to refer to the segment of DNA coding for some functional protein, but it is now used to refer to any position characterized by a specific DNA sequence, or by specific forms of variation in the sequence. Those loci exhibiting observable variation in the DNA are *DNA marker loci*, and a *locus* simply indicates a particular position on a particular one of the pairs of chromosomes. The DNA at a locus may come in a variety of forms, or *alleles*. Since any individual has two chromosomes, they have two (possibly identical) alleles at each locus. The unordered pairs of alleles that an individual has is the individual's *genotype* at this locus. If the locus is one relating to a functional gene, the potentially observable characteristic of the individual is the *phenotype*.

For example, the DNA which codes for the antigens that determine an individual's *ABO* blood type is at a certain position on Chromosome 9.

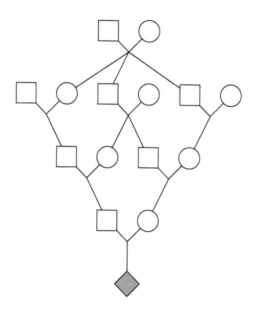

Figure 4.1 *The example pedigree from Goddard et al. (1996).*

This is a chromosome in mid-range size; chromosomes are numbered in approximately decreasing size order. This position is the *ABO locus*. There are three major alleles at the human *ABO locus*, *A*, *B*, and *O*, although these allelic types can be subdivided. There are thus six genotypes; *AA*, *AO*, *BB*, *BO*, *OO*, and *AB*. However there are only four phenotypes (*ABO* blood types), type-*A*, type-*B*, type-*O* and type-*AB*. Individuals with genotype *AA* or *AO* have type-*A* blood type; individuals with genotype *BB* or *BO* have type-*B* blood type. For each of the phenotypes *O* and *AB*, there is a single corresponding genotype.

A *pedigree* is a specification of the genealogical relationships among a set of individuals. A convenient form of this specification is to identify the father and the mother of each individual. Individuals at the top of the pedigree, whose parents are unspecified, are the *founders* of the pedigree; other individuals are *non-founders*. Relationships among individuals are defined relative to the specified pedigree; thus, by definition, the founders are unrelated. There are several alternative forms of graphical representation of a pedigree, but one of the most convenient is the *marriage node graph*. A marriage node represents any pairing of individuals who are parents of another individual in the pedigree. Lines downward from individuals connect them to each marriage node in which they are a parent, and lines downward from the marriage node connect to the offspring of this parent pair. Males are represented by squares and females by circles; a diamond indicates an individual of unspecified sex. An example pedigree is shown in Figure 4.1. In this pedigree, each individual has at most one marriage, but the pedigree is complicated by the three marriages between first cousins. This pedigree structure, which derives from an actual study (Goddard et al. 1996), will be used as an example throughout this chapter.

The DNA in an offspring is a copy of DNA in the individual's parents. For the DNA at a single locus, *Mendel's First Law* (1866) states that each individual has two "factors" (or *genes*) controlling a given characteristic, one being a copy of a corresponding gene in the father of the individual, the other a copy of a gene in the mother of the individual. Further, a copy of a randomly chosen one of the two is copied to each child, independently for different children, independently of genes contributed by the spouse. The word "gene" is overused in modern genetics, often referring to the locus ("the ABO gene"), or to an allele predisposing the individual to a particular disease or trait ("the cystic fibrosis gene"). Here we reserve the word "gene" for Mendel's original "factors"; the gene is the entity transmitted from parent to offspring.

The *gametes* are the sperm or egg cells, which combine to form an offspring individual. Each gamete carries only one of each pair of chromosomes. Thus of every chromosome pair in an offspring individual, one chromosome derives from the corresponding maternal chromosome pair and the other from the paternal chromosome pair. Thus, contrary to Mendel's second law Mendel (1866), there is dependence in the inheritance of genes at syntenic loci (that is, loci on the same chromosome). Such loci are said to be *linked*. If the maternal gene at some locus is known to be a copy of the mother's paternal gene, then the conditional probability that the individual's maternal gene at a linked locus also derives from the mother's father is greater than one half. The vector of alleles at loci on a chromosome is a *haplotype*, and a *multilocus genotype* is a pair of haplotypes. Note that the set of single-locus genotypes do not determine the multilocus genotype. The multilocus genotype includes a specification of *phase*; that is, which alleles (one at each locus) are on the same chromosome. For clarity, we refer to the observable set of single-locus genotypes at any set of DNA marker loci as marker phenotypes, even when these loci do not correspond to functional genes.

4.1.2 The conditional independence structures of genetics

In genetics, there are two key structures that underlie the patterns of dependence in data. There are the pedigree structure of relationships among individuals and the linear structure of a chromosome. Correspondingly there are two sets of latent variables, whose conditional independence structure is a key property facilitating modelling and analyses of genetic data.

The conditional independence structure of genotypes in a pedigree is straightforward; individuals receive copies of their parents' genes, and, together with their spouses, copy genes to their offspring. Considering the underlying genotypes $\mathbf{G} = \{G_{i,j}\}$ for all members i of a pedigree structure at all loci j, the probability of observed marker and/or trait phenotypes $\mathbf{Y}$ is

$$\Pr(\mathbf{Y}) \quad = \quad \sum_{\mathbf{G}} \Pr(\mathbf{Y} \mid \mathbf{G}) \Pr(\mathbf{G}) \tag{4.1}$$

We restrict attention here to models in which phenotypes of individuals are conditionally independent given their genotypes:

$$\Pr(\mathbf{Y}) = \sum_{\mathbf{G}} \left(\prod_{observed\ i} \Pr(Y_{i*} \mid G_{i*}) \right) \Pr(\mathbf{G}) \tag{4.2}$$

where Y_{i*} is the full set of phenotypic observations on individual i, and G_{i*} in the (multilocus) genotype of individual i. Assuming independence of the genotypes of founders, and of unrelated spouses, probability of genotypes $\mathbf{G}$ is

$$\Pr(\mathbf{G}) = \prod_{founders\ i} \Pr(G_{i*}) \prod_{nonfounders\ i} \Pr(G_{i*}|G_{M_i*}, G_{F_i*}) \tag{4.3}$$

where M_i and F_i are the parents of individual i.

These equations (4.1), (4.2) and (4.3), underlie computational approaches first developed by (Elston and Stewart 1971), and subsequently by others. They also indicate the fundamental elements of genetics models. First, a model is required for the genotypes, G_{i*} of founder individuals. The simplest such model is parametrized by allele frequencies, and postulates independence of the types of the two alleles at a given locus in a founder individual, and independence of genotypes at different loci. Second, a model is required for the transmission of genes from parents to offspring, providing $\Pr(G_{i*}|G_{M_i*}, G_{F_i*})$. These are models for the formation of gametes, *meiosis*, and *genetic linkage* . Third, a model is required to determine the probability of an individual's phenotype Y_{i*} given the genotype G_{i*}: that is, a specification of $\Pr(Y_{i*} \mid G_{i*})$ in equation (4.2). These probabilities are known as *penetrance probabilities*, and may depend on individual characteristics (age, sex) and environmental covariates (diet). More complex models might no longer assume the conditional independence of phenotypes given genotypes, and allow for shared familial environment (Cannings et al. 1980).

As an alternative to specifying the latent genotypes of all individuals at all loci, Mendelian inheritance may be characterized in terms of *meiosis indicators* $\mathbf{S} = \{S_{i,j}\}$, specifying the parental origins of genes at each locus

j in each meiosis i:

$$\begin{aligned} S_{i,j} \;=\; & 0 \quad \text{if copied gene at meiosis } i \text{ locus } j \\ & \qquad \text{is parent's maternal gene} \\ \;=\; & 1 \quad \text{if copied gene at meiosis } i \text{ locus } j \\ & \qquad \text{is parent's paternal gene.} \end{aligned} \tag{4.4}$$

We introduce the notation $S_{*j} = (S_{1,j}, \ldots, S_{m,j})$ for the set of all meiosis indicators on the pedigree at locus j; this vector is also known as an *inheritance vector* (Lander and Green 1987). Given ordered loci $j = 1, \ldots, L$, we define also $S_{i*} = (S_{i,1}, \ldots, S_{i,L})$ the vector of meiosis indicators for meiosis i. Mendel's First Law, section 4.1.1, then simply states that the indicators S_{i*} are independent, and $\Pr(S_{i,j} = 0) = \Pr(S_{i,j} = 1) = \frac{1}{2}$, for each j.

Mendel's Second Law (Mendel 1866) stated incorrectly that the S_{*j} are independent for different loci j. In fact, the components of the meiosis indicator vector for meiosis i, $S_{i*} = (S_{i,1}, \ldots, S_{iL})$, are dependent, for syntenic loci $j = 1, \ldots, L$. In genetic analyses of data on related individuals, absence of *genetic interference* is almost always assumed. This assumption is, at best, an approximation. Recent excellent accounts of modeling and computation under interference have been given by McPeek and Speed 1995, (Lin and Speed 1996) and (Browning 1999). It is also easy to extend the methods developed in section 4.3 to accommodate genetic interference (Thompson 1999), but we shall not consider this in this chapter. In the absence of interference, the components of S_{i*} have a simple conditional independence structure. If the loci are ordered $1, \ldots, L$ along a chromosome, then, given $S_{i,j}$, $(S_{i,1}, \ldots, S_{i,j-1})$ is independent of $(S_{i,j+1}, \ldots, S_{i,L})$. Or, (S_{*j}) is first-order Markov over j. In this case, the model for each S_{i*} is fully specified by Mendel's First Law, and by the $(L-1)$ *recombination frequencies* $\rho_1, \ldots, \rho_{L-1}$ where $\rho_j = \Pr(S_{i,j} \neq S_{i,j+1})$.

Now an alternative determination of the probability of data on related individuals in a pedigree is given by

$$\Pr(\mathbf{Y}) \;=\; \sum_{\mathbf{S}} \Pr(\mathbf{Y} \mid \mathbf{S}) \, \Pr(\mathbf{S}) \tag{4.5}$$

The space of latent variables $\mathbf{S}$, although large, is generally smaller than the space of latent genotypes $\mathbf{G}$ and has a simpler dependence structure. Often (although not always), data will be specific to a given locus. Let Y_{*j} now denote the data pertaining to locus j, so the full data pertaining to this chromosomal region is $\mathbf{Y} = (Y_{*1}, \ldots, Y_{*L})$, and

$$\Pr(\mathbf{Y} \mid \mathbf{S}) \;=\; \prod_{j} \Pr(Y_{*j} \mid S_{*j})$$

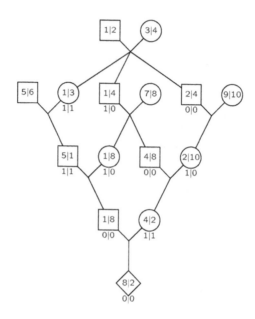

Figure 4.2 *Meiosis indicators S_{*j} determine descent of founder genes, and patterns of gene identity by descent, at any given locus j. The indicators $S_{i,j}$ are shown under the offspring individual, while the resulting labelled founder genes are shown within each individual.*

Further, since meioses i are independent, equation (4.5) becomes

$$\Pr(\mathbf{Y}) \;=\; \sum_{\mathbf{S}} \left(\prod_j \Pr(Y_{*j} \mid S_{*j}) \right) \left(\prod_i \Pr(S_{i*}) \right) \qquad (4.6)$$

However, determination of $\Pr(Y_{*j} \mid S_{*j})$ is non-trivial: Y_{*j} is all the data corresponding to locus j observed on an entire pedigree.

When Y_{*j} consist of single-locus genotypes for some members of the pedigree, efficient computation of $\Pr(Y_{*j} \mid S_{*j})$ is possible. First, S_{*j} determines which founder genes descend to which individuals, and hence which genes in observed individuals are descended from the same founder gene. Such genes are said to be *identical by descent* (IBD). We denote the pattern of gene identity by descent among genes of individuals observed at locus j by $\mathbf{B}(S_{*j})$, and we see

$$\Pr(Y_{*j} \mid S_{*j}) \;=\; \Pr(Y_{*j} \mid \mathbf{B}(S_{*j}))$$

Figure 4.2 gives an example, using the pedigree of Figure 4.1. The meiosis indicators shown under each individual are for the paternal and maternal meiosis to that individual. For easier visualization males are shown to the left and females to the right. For example it is seen that the paternal gene of the final individual is IBD with the maternal gene of his maternal grandfather, not because the gene descends directly (the mother does not carry this gene), but because both are copies of the same founder gene labelled "8". Now,

$$\Pr(Y_{*j} \mid \mathbf{B}(S_{*j})) \quad = \quad \sum_{\mathcal{A}(j)} \left(\prod_k q_j(a(k)) \right) \tag{4.7}$$

Here k denotes a founder gene, $a(k)$ its allelic type, and $q_j(a)$ the population frequency of allele a at locus j. Also, $\mathcal{A}(j)$ denotes an allocation of allelic types to the founder genes k at locus j, which, given S_{*j}, is consistent with the data Y_{*j}. This equation was given by Thompson (1974) who gave an example of ABO blood types on three individuals. The special case of two individuals (9 states $\mathbf{B}$) is discussed in Chapter 2 of Thompson (1986).

In general, efficient determination of all allocations $\mathcal{A}(j)$ at locus j compatible with data Y_{*j} is straightforward for genotypic data (for example, DNA marker phenotypes). An algorithm for this determination of is given by Sobel and Lange (1996). The implementation we use is due to Simon Heath (personal communication) and is described in more detail by (Thompson and Heath 1999). We use the same example pedigree, with the values of S_{*j} given in Figure 4.2, and assume five individuals observed with the genotypes shown in Figure 4.3(a). The method rests first on the fact that only founder genes having copies in observed individuals are constrained in allelic type: in our example, the genes labelled $\{1, 2, 4, 5, 8, 10\}$. Further two genes constrain each other's allelic type only when both are present in an observed individual. The *gene graph* (Figure 4.3(b)) connects all such pairs of genes. Allocation of allelic types may be considered separately for each component subgraph of connected genes. In our example, the genes $\{1, 5\}$ may be considered independently of $\{2, 4, 8, 10\}$. This assignment is readily accomplished, even on a much larger example. For given S_{*j} there are in general only 2, 1 or 0 possible assignments of allelic types to the genes of a component subgraph. For our example, there are two possible assignments for the first component and one for the second: $(a(1), a(5)) = (A, C)$ or (C, A) and $(a(2), a(4), a(8), a(10)) = (C, D, C, B)$. The algorithm can in principle be generalized to more complex phenotypes such as a quantitative trait measurement, using the conditional independence structure of the gene graph Figure 4.3(b), but the procedure becomes far more computationally intensive.

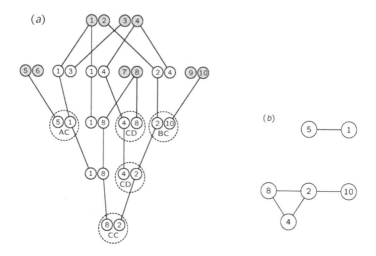

Figure 4.3 *Determination of probabilities* $\Pr(Y_{*j} \mid S_{*j})$. *The gene descent pattern is assumed to be that of Figure 4.2, and the pairs of genes are shown, rather than the individuals. Five individuals, shown as dashed circles, are assumed to be observed, with marker genotypes as indicated (see text for details). (a) Only genes present in observed individuals are constrained in type. (b) Two genes in a single observed individual are jointly constrained.*

4.1.3 "Peeling": sequential computation

We review here the algorithm given by Baum (1972) for the computation of the likelihood in a hidden Markov model. The procedure is general to any stochastic system with discrete-valued latent variables $\mathbf{X}_j$ with a first-order Markov structure, and outputs Y_j depending only on $\mathbf{X}_j$. However, for convenience, we retain the notation of section 4.1.2 with meiosis indicators $\mathbf{S}_{*j}$ and phenotypic data Y_{*j} for locus j, with loci ordered $j = 1, \ldots, L$ along a chromosome.

For data observations $\mathbf{Y} = (Y_{*j}, j = 1, \ldots, L)$, we want to compute $\Pr(\mathbf{Y})$. Due to the first-order Markov dependence of the S_{*j}, equation (4.5) can be written

$$
\begin{aligned}
\Pr(\mathbf{Y}) &= \sum_{\mathbf{S}} \Pr(\mathbf{S}, \mathbf{Y}) \\
&= \sum_{\mathbf{S}} \Pr(\mathbf{Y} \mid \mathbf{S}) \Pr(\mathbf{S})
\end{aligned}
$$

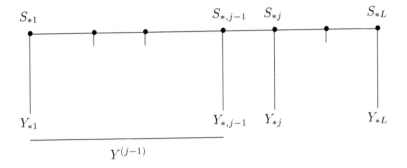

Figure 4.4 *The conditional independence structure of data, in the absence of genetic interference.*

$$= \sum_{\mathbf{s}} \left(\Pr(S_{*1}) \prod_{j=2}^{L} \Pr(S_{*j} \mid S_{*,j-1}) \right.$$

$$\left. \prod_{j=1}^{L} \Pr(Y_{*j} \mid S_{*j}) \right) \qquad (4.8)$$

The dependence structure is shown in Figure 4.4. For convenience we also define $Y^{(j)} = (Y_{*1}, \ldots, Y_{*j})$, the data determined by genes at loci along the chromosome up to and including locus j. Note $\mathbf{Y} = Y^{(L)}$.

Now define

$$R_j(s) \quad = \quad \Pr(Y_{*k}, k = (j+1), \ldots, L \mid S_{*j} = s)$$

with $R_L(s) = 1$ for all s, and note that

$$R_{j-1}(s) \quad = \quad \sum_{s^*} [\Pr(S_{*j} = s^* \mid S_{*,j-1} = s)$$

$$\Pr(Y_{*j} \mid S_{*j} = s^*) \, R_j(s^*)] \qquad (4.9)$$

while

$$\Pr(\mathbf{Y}) \quad = \quad \sum_{s^*} [\Pr(S_{*1} = s^*) \, \Pr(Y_{*1} \mid S_{*1} = s^*) R_1(s^*)]$$

Thus the L-dimensional sum (4.8) may be computed as a telescoping series of one-dimensional sums over the possible values s^* of each S_{*j} in turn, computed for each possible value s of $S_{*,j-1}$. Where each S_{*j} can take only a small number of possible values, this makes computation of the likelihood, $\Pr(\mathbf{Y})$, both feasible and practical. Computation is linear in the number

of marker loci L along a chromosome, so computation is practical even for
very large values of L.

In the case of meiosis indicators, the direction along a chromosome is
irrelevant and

$$\Pr(S_{*j} = s^* \mid S_{*,j-1} = s) \;\; = \;\; \Pr(S_{*,j-1} = s^* \mid S_{*j} = s)$$

However, for other hidden Marvov systems only the forward transitions
$\Pr(S_{*j} = s^* \mid S_{*,j-1} = s)$ may be readily available. Even so, it may be
desirable or even necessary to be able to effect the computation proceeding
either forwards or backwards. We will encounter an example where this is
the case in section 4.2.2. Equation (4.9) provides for peeling from L to 1,
using only the forward transition probabilities. Where only these forwards
transition probabilities are available, peeling in the direction from 1 to L is
also possible, but it is then more convenient to define the joint probability

$$\begin{aligned}
R_j^*(s) &= \Pr(Y_{*k}, k = 1, \ldots, j - 1, \; S_{*j} = s) \\
&= \Pr(Y^{(j-1)}, \; S_{*j} = s)
\end{aligned}$$

with $R_1(s) = \Pr(S_{*1} = s)$. Now equation (4.9) is replaced by

$$\begin{aligned}
R_{j+1}^*(s) &= \sum_{s^*} [\Pr(S_{*,j+1} = s \mid S_{*j} = s^*) \\
&\qquad\qquad \Pr(Y_{*j} \mid S_{*j} = s^*) \, R_j^*(s^*)] \qquad\qquad (4.10)
\end{aligned}$$

for $j = 1, 2, \ldots, L - 1$, with

$$\Pr(\mathbf{Y}) \;\; = \;\; \sum_{s^*} \Pr(Y_{*L} \mid S_{*L} = s^*) \, R_L^*(s^*)$$

We will return to these equations in section 4.2 in the context of likelihood
computations on the basis of data observed on members of a pedigree. We
note here only that efficient computation of the penetrance probabilities
$\Pr(Y_{*j} \mid S_{*j})$ (section 4.1.2) is key to the implementation.

4.1.4 The Baum algorithm for conditional probabilities

While the above method of likelihood computation was known to Baum
(1972), his primary aim was estimation of the transition probabilities of the
Markov chain, and of the probability relationship between input and output
(Baum and Petrie 1666, Baum et al. 1970). Here, these are the transition
probabilities $P(S_{*,j+1} = s \mid S_{*j} = s^*)$ and the penetrance probabilities
$P(Y_{*j} \mid S_{*j})$. If the latent variables $\mathbf{S}$ were observed, the sufficient statistics for estimation of these transition and penetrance parameters would be
simple functions of $\mathbf{Y}$ and $\mathbf{S}$. Thus, to estimate parameters of the model,
for example by using an EM algorithm (Dempster et al. 1977), one must
impute these functions of the underlying $\mathbf{S}$ conditional on $\mathbf{Y}$. Again, here

we use the notation of meiosis indicators of section 4.1.2, but the framework is general to any hidden Markov model.

Thus, the forward-backward algorithms of Baum et al. (1970) address *inter alia* the computation of marginal probabilities

$$Q_j(s) = \Pr(S_{*j} \mid \mathbf{Y}), \; j = 1, \ldots, L.$$

We define two functions

$$Q_j^\dagger(s) \;=\; \Pr(S_{*j} = s \mid Y^{(j)})$$
$$Q_{j+1}^*(s) \;=\; \Pr(S_{*,j+1} = s \mid Y^{(j)})$$

$Q_j^\dagger(\cdot)$ provides the imputation of S_{*j} given data $Y^{(j)}$ up to and including locus j, while $Q_{j+1}^*(\cdot)$ is the predictor of $S_{*,j+1}$ also given $Y^{(j)} = (Y_{*1}, \ldots, Y_{*j})$.

Then $Q_1^\dagger(s) = \Pr(S_{*1} = s \mid Y_{*1})$,

$$Q_{j+1}^*(s) \;=\; \Pr(S_{*,j+1} = s \mid Y^{(j)})$$
$$=\; \sum_{s^*} \Pr(S_{*,j+1} = s \mid S_{*j} = s^*) \, Q_j^\dagger(s^*) \qquad (4.11)$$

and

$$Q_{j+1}^\dagger(s) \;=\; \sum_{s^*} \Pr(S_{*,j+1} = s, \; S_{*j} = s^* \mid Y^{(j+1)})$$
$$=\; \frac{\sum_{s^*} \Pr(S_{*,j+1} = s, \; S_{*j} = s^*, \; Y_{*,j+1} \mid Y^{(j)})}{\Pr(Y_{*,j+1} \mid Y^{(j)})}$$
$$\propto\; \sum_{s^*} \Pr(S_{*,j+1} = s, \; S_{*j} = s^*, \; Y_{*,j+1} \mid Y^{(j)})$$
$$=\; \sum_{s^*} \Big(\Pr(Y_{*,j+1} \mid S_{*,j+1} = s)$$
$$\Pr(S_{*,j+1} = s \mid S_{*j} = s^*) \Pr(S_{*j} = s^* \mid Y^{(j)}) \Big)$$
$$=\; \Pr(Y_{*,j+1} \mid S_{*,j+1} = s)$$
$$\sum_{s^*} \Big(\Pr(S_{*,j+1} = s \mid S_{*j} = s^*) \, Q_j^\dagger(s^*) \Big). \qquad (4.12)$$

Provided $S_{*,j+1}$ takes only a limited number of values s, the probabilities may be normalized, giving each function $Q_j^\dagger(s)$, $j = 2, \ldots, L$, in turn, the final one being

$$Q_L(s) = Q_L^\dagger(s) = \Pr(S_{*L} = s \mid \mathbf{Y}) \qquad (4.13)$$

the desired distribution of S_{*L} given $\mathbf{Y}$.

Now we may proceed backwards to obtain $Q_j(\cdot)$ for

$j = L - 1, \ldots, 3, 2, 1$:

$$\begin{aligned}
&\Pr(S_{*,j-1} = s, S_{*j} = s^* \mid \mathbf{Y}) \\
&= \Pr(S_{*j} = s^* \mid \mathbf{Y}) \Pr(S_{*,j-1} = s \mid S_{*j} = s^*, \mathbf{Y}) \\
&= Q_j(s^*) \Pr(S_{*,j-1} = s \mid S_{*j} = s^*, Y^{(j-1)}) \\
&= Q_j(s^*) \frac{\Pr(S_{*j} = s^* \mid S_{*,j-1} = s)\Pr(S_{*,j-1} = s \mid Y^{(j-1)})}{\Pr(S_{*j} = s^* \mid Y^{(j-1)})} \\
&= Q_j(s^*) \Pr(S_{*j} = s^* \mid S_{*,j-1} = s) \frac{Q^{\dagger}_{j-1}(s)}{Q^{*}_j(s^*)} \qquad (4.14)
\end{aligned}$$

The second step uses conditional independence of $S_{*,j-1}$ and $Y_{*j}, \ldots, Y_{*L}$ given S_{*j}, and the third is an application of Bayes Theorem, using the conditional independence of S_{*j} and $Y^{(j-1)}$ given $S_{*,j-1}$. Note that this backward step involves both the forward probability function $Q^{\dagger}_{j-1}(\cdot)$ of equation (4.12) and the predictive probability $Q^{*}_j(\cdot)$ of equation (4.11). Now the marginal probabilities $Q_{j-1}(s) = \Pr(S_{*,j-1} = s \mid \mathbf{Y})$ are readily obtained by summing over s^*:

$$\begin{aligned}
Q_{j-1}(s) &= \Pr(S_{*,j-1} = s \mid \mathbf{Y}) \\
&= \sum_{s^*} \Pr(S_{*,j-1} = s, S_{*j} = s^* \mid \mathbf{Y}) \qquad (4.15)
\end{aligned}$$

In the context of time series, equation (4.11) is known as the predictor, and (4.12) as the filter, while the backward equations (4.14) is the smoother, incorporating all data $\mathbf{Y}$ into the estimation of each S_{*j}.

Finally, instead of computing the marginal distributions $Q_j(s)$, we may prefer a realization from the joint distribution $\Pr(\mathbf{S} \mid \mathbf{Y})$. The Baum algorithm provides this also. The forward computation is exactly as before (equation (4.12)). The backward computation is replaced by sampling. First, S_{*L} is sampled from $Q_L(\cdot)$ (equation (4.13). Then, similarly to equation (4.14), given a realization of $S_{*j} = s^*, S_{*,j+1}, \ldots, S_{*L}$

$$\begin{aligned}
&\Pr(S_{*,j-1} = s \mid S_{*j} = s^*, S_{*,j+1}, \ldots, S_{*L}, \mathbf{Y}) \\
&= \Pr(S_{*,j-1} = s \mid S_{*j} = s^*, Y^{(j-1)}) \\
&\propto \Pr(S_{*j} = s^* \mid S_{*,j-1} = s)Q^{\dagger}_{j-1}(s) \qquad (4.16)
\end{aligned}$$

again by a straightforward application of Bayes Theorem, where proportionality is w.r.t. s. Normalizing these probabilities, we can realize $S_{*,j-1}$. This is done for each $j = L, L - 1, \ldots, 4, 3, 2$ in turn, providing an overall realization $\mathbf{S} = (S_{*1}, \ldots, S_{*L})$ from $\Pr(\mathbf{S} \mid \mathbf{Y})$.

4.2 Computations on pedigrees

4.2.1 Peeling meiosis indicators

Phenotypic similarities among relatives result from the genes they share IBD (section 4.1.2) Among an ordered set of genes, any partition of the set can specify which subsets of the genes are IBD. Among a set of observed individuals, we denote this partition of their genes by $\mathbf{B}$, and refer to it as the *pattern* of gene identity by descent among the individuals. For a given locus j, and the m meioses of a pedigree, the meiosis indicators $\mathbf{S} = \{S_{i,j}; i = 1, ..., m\}$ of equation (4.4) determine the pattern, $\mathbf{B}$, of genes IBD in any currently observed set of individuals; $\mathbf{B} = \mathbf{B}(\mathbf{S})$. The probability of any data (i.e. observed characteristics of the individuals) depends on $\mathbf{S}$ only through $\mathbf{B}(\mathbf{S})$, and

$$\Pr(\mathbf{Y}) = \sum_{\mathbf{B}} \Pr(\mathbf{Y} \mid \mathbf{B}) \Pr(\mathbf{B})$$

$$= \sum_{\mathbf{S}} \Pr(\mathbf{Y} \mid \mathbf{B}(\mathbf{S})) \Pr(\mathbf{S}) \tag{4.17}$$

In partitioning the likelihood in this way, the "genetic model" is separated from the effects of genealogical and genetic structure. The probability of a given pattern of IBD $\mathbf{B}(S_{*j})$ at locus j depends only on the genealogical relationships among the observed individuals. At multiple linked loci, the probability $\Pr(\mathbf{B}(\mathbf{S}))$ depends also on the model for $\mathbf{S}$; that is, a model for the transmission of the genes at syntenic loci, such as the first-order Markov model of section 4.1.2. Given the gene identity pattern, $\mathbf{B}(\mathbf{S})$, the probability of data $\mathbf{Y}$ depends on the different types of genes, their frequencies, and how they affect observable phenotypes (see section 4.1.2). Thus, the passage of genes in pedigrees provides the connection between observable genetic characteristics and the pedigree structure, whether we are estimating relationships from genetic data, estimating the genetic basis of traits knowing the pedigree, or inferring the ancestry and descent of particular genes, knowing both the genetic model and the data.

Mendel's first law (section 4.1.1) provides that at a single locus j, the binary meiosis indicators $S_{i,j}$ are independent, each 0 or 1 with probability $\frac{1}{2}$. Under the assumption of absence of genetic interference (section 4.1.2), the model of first-order Markov dependence of S_{*j} along a chromosome gives rise to computational approaches based on "peeling" along a chromosome (section 4.1.3).

$$\Pr(\mathbf{Y}) = \sum_{\mathbf{S}} \left[\left(\prod_{j} \Pr(Y_{*j} \mid \mathbf{B}(S_{*j})) \right) \Pr(S_{*1}) \left(\prod_{j=2}^{L} \Pr(S_{*j} \mid S_{*,j-1}) \right) \right],$$

which is directly analogous to equation (4.8). Thus, the basic approach is a direct application of equation (4.9) or (4.10). When the phenotypes consist of single-locus genotypes of some individuals, the probabilities $\Pr(Y_{*j} \mid S_{*j})$ are computed using equation (4.7). In the context of genetic analysis, this computational approach was developed by Lander and Green (1987), and has been further developed by Kruglyak et al. (1996) and Kruglyak and Lander (1998).

However exact computation is limited to small pedigrees. If there are m meioses on the pedigree, then S_{*j} can take 2^m values. In moving along the chromosome, we must consider transitions from the 2^m values of S_{*j} to the 2^m values of $S_{*,j+1}$. The computation is thus of the order $4^m L$. In the pedigree of Figure 4.1. There are 10 non-founders, and thus in principle 20 meioses. However, where a founder has only one offspring in the pedigree, the meiosis from the founder to the offspring is irrelevant to probability computations and should be ignored. This reduces the number to $m = 18$. In fact, by taking advantage of symmetries in the binary indicators, an extra factor of $2 \times 2 = 4$ may be gained for each founder individual. In general, for a pedigree with n individuals, f of whom are founders, the effective number of meioses is $m = 2n - 3f$; this reduces the number for the pedigree of Figure 4.1 to $m = 15$ and within the computational bounds of the algorithm (Kruglyak et al. 96). However, the approach is limited to pedigrees of about this size or smaller.

Additionally, for each locus j, and for each value of S_{*j}, the probability $\Pr(Y_{*j} \mid \mathbf{B}(S_{*j}))$ must be computed. For DNA marker loci, at which single-locus genotypes are generally available, computation is straightforward for given S_{*j} using (equation (4.7)), but again this limits the number of S_{*j} that can be considered, and hence the size of the pedigree. For more complex trait models, where there is no 1-1 correspondence between single-locus genotypes and corresponding phenotypes, computation becomes much more intensive.

4.2.2 Peeling genotypes on pedigrees

The primary aim in computation of $\Pr(\mathbf{Y})$ on a pedigree is normally *segregation* or *linkage* analysis. Segregation analysis (Elston and Stewart 1971) is analysis of inheritance patterns and genotype-phenotype relationship at trait loci. Linkage analysis (Lathrop et al. 1984) involves analysis of the coinheritance or genetic linkage of genes at trait and/or marker loci. For segregation analysis, or for linkage analyses where trait loci are explicitly modeled, it is often more straightforward to use equation (4.2)

$$\Pr(\mathbf{Y}) = \sum_{\mathbf{G}} \left(\prod_{observed\ i} \Pr(Y_{i*} \mid G_{i*}) \right) \Pr(\mathbf{G})$$

in which the genotypes G_{i*} of individuals i are the latent variables. A pedigree structure provides a more complex dependence structure than the linear first-order Markov structure of transmission of genes at loci along chromosome. Nonetheless, the approach of section 4.1.3 can be applied, due to the conditional independence structure of **G** (equation (4.3)):

$$\Pr(\mathbf{G}) = \prod_{founders\ i} \Pr(G_{i*}) \prod_{nonfounders\ i} \Pr(G_{i*}|G_{M_i*}, G_{F_i*})$$

where M_i and F_i are the parents of individual i.

While each term of the product in (4.2) can be easily evaluated, the difficulty is in the sum over **G**. On a very small pedigree it may be possible to enumerate all possible genotypic configurations **G**, and to compute the sum directly. In other special cases, it may be possible to use a recursive algorithm to compute the gene identity pattern probabilities in the observed individuals, and hence to compute the marginal probability $\Pr(\mathbf{G})$ for these individuals alone. However, in general this is impractical. Independently, Hilden (1970, Elston and Stewart (1971), and Heuch and Li (1972) laid the foundations of the approach that made it possible to compute likelihoods of genetic models given data on large pedigrees. This approach underlies the successful mapping of many Mendelian traits between 1970 and 1990. However, computations are then limited to few loci, and to relatively simple pedigrees.

As in the first-order Markov case, the basic idea is simply one of efficient sequential summation. The number of terms in which a specific G_{i*}, the genotype of individual i, enters is limited to the penetrance term for i, $\Pr(Y_{i*} \mid G_{i*})$, and to transmission terms from the parents and to the offspring of individual i. Thus, performing a summation over the possible values of G_{i*} results in a function of (at worst) the genotypes of i's parents, spouses and offspring. Of course, this is only useful if implemented sensibly. By starting at the edges (top/bottom/side) of the pedigree, one limits the number of individuals whose genotypes must be considered jointly.

The graphical representation of a pedigree such as that of figure 4.1 has loops due to marriages between related individuals. There are also other causes of loops in a pedigree structure (Cannings et al. 1978). For a pedigree without loops, there always exists a sequence of nuclear families such that each is connected to the as yet unprocessed part of the pedigree via a single individual, the *pivot*. (In fact, there will be many alternative such sequences.) In this case, summation over the genotypes of the non-pivot members of each family leads to a function of only the pivot genotypes, which may be incorporated into the summation for that individual in due course. This sequential summation process has come to be known as "peeling", and the specification of the order of individuals (normally of nuclear families) in which summation will be carried out as the "peeling sequence". The process is clearly analogous to equations (4.9) and (4.10) of section 4.1.3, although the (unlooped) pedigree is now a tree structure, not a single chain. Elston and Stewart (1971) developed upwards (backwards) peeling (4.9) only, which limited the pedigree structures that could be considered. Using both upwards (4.9) and downwards (4.10) peeling, the procedure applies to any unlooped pedigree (Cannings et al. 1978). Cannings et al. (1978) named the approach *peeling*, and the functions $R_t(\cdot)$ *R-functions*. However, the idea of conditioning on individual genotypes and separating computations on conditionally independent branches of the pedigree, when computing probabilities on pedigrees, can be traced to Haldane and Smith (1947).

For a single locus, and unlooped pedigree, the limiting factor is the number of non-zero transmission probabilities $\Pr(G_{i*}|G_{M_i*}, G_{F_i*})$. Computation is of the order of K^3 where K is the potential number of genotypes of an individual, or k^6 where k is the number of alleles. The procedure applies also to linked loci: the genotype G_{i*} is the multilocus genotype of individual i. The transmission probabilities $\Pr(G_{i*}|G_{M_i,*}, G_{F_i,*})$ are then functions of the recombination frequencies between the loci. However, recall (section 4.1.1) that a multilocus genotype is an unordered pair of multilocus haplotypes. That is, it is a specification of not only the single-locus genotypes, but also phase information. If there are two diallelic loci, there are 4 haplotypes, and hence 10 genotypes; computation is feasible and fast. With more loci, K increases exponentially and computation rapidly becomes infeasible.

The Elston-Stewart approach was generalized to complex pedigrees and more complex penetrance models and traits by Cannings et al. (1978, 1980). (The Hilden (1970) approach also dealt, in principle, with arbitrarily complex pedigrees.) For a pedigree with loops, functions on the genotypes of a "cutset" of individuals may have to be considered. This is a set of individuals who divide a processed segment of pedigree from the unprocessed part. The processing therefore results in a function over the set of all possible genotype combinations for the individuals in the cutset. With K possible genotypes for each individual, there are K^n potential genotype combinations for n individuals. Even for a single autosomal diallelic locus, with $K = 3$, computation becomes infeasible unless n is small. A key component of the algorithm is thus the *peeling sequence*, the order in which summations will be performed over the genotypes of individuals of the pedigree. The objective of a good peeling sequence is to limit the cutset sizes as much as possible, and most importantly to limit the size of the maximal cutset.

Genotypic peeling has been an important approach in genetic analysis since 1971. On large though unlooped pedigrees, computations are possible for several loci jointly, which has been invaluable in linkage analyses of Mendelian traits. For a single locus, computation is feasible even on quite complex pedigrees, with multiple interconnected loops. However, on very complex pedigrees, with multiple intersecting loops, peeling becomes infeasible, particularly if there are more alleles, or more loci. Even for simple pedigrees, peeling becomes infeasible if there are multiple loci with multiple alleles; the number K of individual genotypes can be too large. Thus with the trend of the 1990s to have more and more data at multiple DNA markers, but often on fewer individuals, the approach of section 4.2.1 is increasingly used.

4.2.3 Importance sampling and Monte Carlo likelihood

For multiple linked genetic loci, on a large or complex pedigree, exact probabilities cannot be computed. Then, Monte Carlo estimation is an alternative way to compute sums, integrals, or expectations. To estimate a sum $\sum_{\mathbf{X}} g(\mathbf{X})$, note that

$$\sum_{\mathbf{X}} g(\mathbf{X}) = \sum_{\mathbf{X}} \frac{g(\mathbf{X})}{P^*(\mathbf{X})} P^*(\mathbf{X}) = E_{P^*} \left(\frac{g(\mathbf{X})}{P^*(\mathbf{X})} \right) \qquad (4.18)$$

where P^* is some distribution over the space of $\mathbf{X}$-values which assigns positive probability to every $\mathbf{X}$ for which $g(\mathbf{X}) > 0$. For convenience, we use here $\mathbf{X}$ to denote the variable over which summation or integration is required; in the genetic context, $\mathbf{X}$ might be the meiosis indicators $\mathbf{S}$ or latent genotypes $\mathbf{G}$. If $\mathbf{X}^{(1)}, ..., \mathbf{X}^{(N)}$ are realizations from the distribution $P^*(\cdot)$,

$$\frac{1}{N} \sum_{t=1}^{N} \left(\frac{g(\mathbf{X}^{(t)})}{P^*(\mathbf{X}^{(t)})} \right) \qquad (4.19)$$

is an unbiased estimator of the sum. Of course, it may not be an efficient estimator; in fact it may be a very inefficient. The art of Monte Carlo is finding good distributions to simulate from, and good ways of simulating from them, in order to get an estimator with small variance. Note this is not the standard statistical paradigm where variances of estimators refer to repeated sampling over the data random variables $\mathbf{Y}$. In our likelihood analyses the data $\mathbf{Y}$ are fixed, and we consider variances over the Monte Carlo sampling of $\mathbf{X}$.

Generally, the Monte Carlo variance of the estimator (4.19) will be small if $g(\mathbf{X})/P^*(\mathbf{X})$ is approximately constant. That is $P^*(\mathbf{X})$ should be approximately proportional to $g(\mathbf{X})$. However, if $P^*(\mathbf{X}) \propto g(\mathbf{X})$, $P^*(\mathbf{X}) = g(\mathbf{X})/\sum_{\mathbf{X}} g(\mathbf{X})$. If the normalizing constant $\sum_{\mathbf{X}} g(\mathbf{X})$ were known, the Monte Carlo would be pointless! (Hammersley and Handscomb 1964). The approximate proportionality is most important where $P^*(\mathbf{X})$ is small. Otherwise the estimator 4.19 is dominated by rare but very large contributions.

We can summarize the four requirements for $P^*(\cdot)$:
   (1) Realizations $\mathbf{X}^{(t)}$ from $P^*(\cdot)$ must be feasible.
   (2) Evaluation of $P^*(\mathbf{X}^{(t)})$ must be possible
   (3) $P^*(\mathbf{X})$ should be approximately proportional to the
       summand $g(\mathbf{X})$, and

(4) $P^*(\mathbf{X}) > 0$ if $g(\mathbf{X}) > 0$.

Note also that

$$\mathrm{E}_{P^*}(h(\mathbf{X})) = \sum_{\mathbf{X}} h(\mathbf{X})P^*(\mathbf{X})$$

$$= \mathrm{E}_{P^{**}}\left(h(\mathbf{X})\frac{P^*(\mathbf{X})}{P^{**}(\mathbf{X})}\right) \qquad (4.20)$$

Thus, in estimating an expectation under $P^*(\cdot)$, realization $\mathbf{X}^{(t)}$ may be sampled from any distribution $P^{**}(\cdot)$ and the integrand $h(\mathbf{X}^{(t)})$ reweighted by the factor $P^*(\mathbf{X}^{(t)})/P^{**}(\mathbf{X}^{(t)})$. This is useful when combining Monte Carlo realizations simulated under several different P^{**} to esimate an expectation under a single P^*. Thus the distributions used to realize latent variables in a Monte Carlo estimate may be chosen for computational simplicity as well as for Monte Carlo efficiency, although usually a balance must be struck between these criteria. In principle, reweighting is always possible, provided the support of $P^{**}(\cdot)$ includes that of $P^*(\cdot)$.

The simplest form of Monte Carlo is where we simulate independent, identically distributed (i.i.d.) realizations from some distribution. On pedigrees, the simplest distribution to simulate from is the prior distribution on genes or genotypes, which is done by "gene dropping". Labelled genes are assigned to the founders of the pedigree, and the meiosis indicators $\mathbf{S}$ are realized, determining the genes in current individuals (Figure 4.2). The required statistics relating to the resultant current genes are computed, or, if the statistics relate to genotypes or phenotypes, these may also be realized by assigning allelic types to founder genes. Such Monte Carlo estimates have been used by Edwards (1967) to estimate inbreeding coefficients, by MacCluer et al. (1986) to study the loss of genes in pedigrees of endangered species, and by Thompson et al. (1978) to study the potential power of a pedigree study.

A *simulation study* differs from a *Monte Carlo analysis*. Simulation studies are typically undertaken to discover empirically the distribution of a test statistic, or to assess the potential power of a study design. It involves the simulation of data random variables $\mathbf{Y}$ under a model $P_\theta(\mathbf{Y})$ of interest, indexed by parameters θ. In a Monte Carlo analysis, integrals, sums, or expectations are estimated by realizing random variables $\mathbf{S}$ from some distribution, P^*. The random variables $\mathbf{S}$ are not the data random variables $\mathbf{Y}$; normally, the data $\mathbf{Y}$ are fixed. The distribution P^* is simply a tool to provide estimates of the required expectations.

In practice, the difference may be slight. The probability distribution, $P^*(\cdot)$, used in a Monte Carlo estimation problem may often be closely related to the probability model $P_\theta(\cdot)$ underlying the data in a statistical problem. Conversely, the distribution used in a simulation study could be chosen arbitrarily, with reweighting (equation (4.20)) used to adjust the realizations to the distribution of interest. In a Monte Carlo analysis we shall normally simulate conditional on fixed data $\mathbf{Y}$, but in a simulation study it may sometimes also be desirable to simulate potential additional data conditional on the fixed values of partial data already obtained.

The likelihood of a genetic model indexed by parameters θ, given data on a pedigree, can be written

$$
\begin{aligned}
L(\theta) \;=\; P_\theta(\mathbf{Y}) \;&=\; \sum_{\mathbf{X}} P_\theta(\mathbf{X}, \mathbf{Y}) \\
&=\; \sum_{\mathbf{X}} P_\theta(\mathbf{Y} \mid \mathbf{X})\, P_\theta(\mathbf{X})
\end{aligned}
\tag{4.21}
$$

where $\mathbf{X}$ are latent variables, either the genotypes $\mathbf{G}$ (4.2) or the meiosis indicators $\mathbf{S}$ (4.6). This form lends itself to Monte Carlo estimation, when $\Pr(\mathbf{Y})$ cannot be computed exactly, due to the size or complexity of the pedigree and the number of linked genetic loci.

Thus, for fixed $\mathbf{Y}$,

$$
L(\theta) \;=\; \mathrm{E}_\theta(P_\theta(\mathbf{Y} \mid \mathbf{X}))
\tag{4.22}
$$

where the expectation is over $\mathbf{X}$ having the distribution $P_\theta(\mathbf{X})$. Where the variable $\mathbf{X}$ is the set of latent genotypes $\mathbf{G}$, this is the form given by Ott (1979). In principle, $L(\theta)$ could be estimated by simulating $\mathbf{G}$ from the prior genotype distribution under model θ and averaging the value of the penetrance probabilities $P_\theta(\mathbf{Y} \mid \mathbf{G})$ for the realized values of $\mathbf{G}$. This does not work well, except on very small pedigrees, since the realized $\mathbf{G}$ are almost certain to be inconsistent with data $\mathbf{Y}$, or at best to make infinitesimal contribution to the likelihood.

Better ideas normally involve some form of importance sampling, as in equations (4.18) and (4.20):

$$
L(\theta) \;=\; \mathrm{E}_{P^*}\left(\frac{P_\theta(\mathbf{X},\, \mathbf{Y})}{P^*(\mathbf{X})} \right)
\tag{4.23}
$$

Now realizations are made from $P^*(\mathbf{X})$, but this will only be effective if $P^*(\mathbf{X})$ is approximately proportional to $P_\theta(\mathbf{X},\, \mathbf{Y})$. An advantage of this approach is that a single set of realizations from $P^*(\mathbf{X})$ will provide a Monte Carlo estimate of $L(\theta)$ over a range of models θ. A disadvantage is that the range of models for which this estimation is effective is likely to be small, given the requirement that the single $P^*(\mathbf{X})$ must be approximately proportional to all the $P_\theta(\mathbf{X},\, \mathbf{Y})$.

The first use of importance sampling in this context is due to K. Lange in Ott (1979). In this case, $P^*(\mathbf{X})$ was taken to be $P_{\theta_0}(\mathbf{X})$ for some alternative, perhaps simpler, genetic model θ_0. However, this is no more effective than the original form (4.22) in addressing the need for $P^*(\mathbf{X})$ to be approximately proportional to $P_\theta(\mathbf{X},\, \mathbf{Y})$; again almost all realizations may be incompatible with $\mathbf{Y}$ or provide only infinitesimal contributions to the likelihood. Clearly the distribution $P^*(\mathbf{X})$ must be adapted to the data $\mathbf{Y}$. In fact,

$$P_\theta(\mathbf{X} \mid \mathbf{Y}) \propto P_\theta(\mathbf{X}, \mathbf{Y}). \qquad (4.24)$$

However the constant of proportionality is $L(\theta) = P_\theta(\mathbf{Y})$. If $P_\theta(\mathbf{X} \mid \mathbf{Y})$ can be computed, so also can $L(\theta) = P_\theta(\mathbf{Y})$. We return to Monte Carlo likelihood below, after first considering other forms of sampling or computation conditional upon data $\mathbf{Y}$.

4.2.4 Risk, Elods, conditional Elods, and sequential imputation

Risk probabilities

In analyses of data on a pedigree, under a model indexed by known values of the parameters θ, quantities of interest include the posterior genotype probabilities $P_\theta(G_{i*}|\mathbf{Y})$ for individuals i. These probabilities are known as *risk probabilities*, since the genotypes of interest are often those conferring a disease risk. In sections 4.1.3 and 4.1.4 we saw that, for a first-order Markov structure for latent variables S_{*j}, sequential computation of the likelihood $P_\theta(\mathbf{Y})$ using the functions

$$R_j(s) \;=\; P_\theta(Y_{*k}, k = (j+1), \ldots, L \mid S_{*j} = s)$$

had the same computational complexity as computation of conditional probabilities $Q_j(s) = P_\theta(S_{*j} = s \mid \mathbf{Y})$ using the two functions

$$\begin{aligned}
Q_j^\dagger(s) &\;=\; \Pr(S_{*j} = s \mid Y_{*1}, \ldots, Y_{*j}) \text{ and} \\
Q_{j+1}^*(s) &\;=\; \Pr(S_{*,j+1} = s \mid Y_{*1}, \ldots, Y_{*j})
\end{aligned}$$

The latter computation requires two passes along the chromosome (forward and backward), while the likelihood computation requires only one (forward or backward), but in both cases the computation is of order $4^m L$ where m is the number of meioses in the pedigree.

The same applies to latent variables G_{i*} on a pedigree structure. If $P_\theta(\mathbf{Y})$ can be computed, using the peeling method outlined in section 4.2.2, so also can the risk probabilities $P_\theta(G_{i*} \mid \mathbf{Y})$. This can be done by taking each individual i in turn, as the final individual L in a peeling sequence (equation (4.13)). However, it is more effectively accomplished by saving probabilities $R(G_{i*})$ (equation (4.9)), from peeling up the pedigree, and combining these with the functions $R^*(G_{i*})$ (equation (4.10)) obtained by progressing back down the pedigree. For example, if individual i divides the pedigree into two parts, the set $D(i)$ connected through his spouses and offspring, and the set $A(i)$ connected through his parents (including his siblings and their descendants), then in proceeding up the pedigree

$$R(G_{i*}) \;=\; P_\theta(\mathbf{Y}_{D(i)} \mid G_{i*})$$

while in proceeding down, relative to individual i,

$$R^*(G_{i*}) \;=\; P_\theta(\mathbf{Y}_{A(i)}, G_{i*})$$

so that

$$P_\theta(G_{i*} \mid \mathbf{Y}) \;\propto\; P_\theta(Y_{i*} \mid G_{i*})R(G_{i*})R^*(G_{i*})$$

and these probabilities may be normalized to give the required probabilities $P_\theta(G_{i*} \mid \mathbf{Y})$. This procedure of working back down the pedigree to obtain risk probabilities is sometimes known as *reverse peeling*. In the case where peeling always up the pedigree is computationally feasible, all risk probabilities on a large pedigree can be computed in two passes through the pedigree. Even on a complex pedigree, with multiple interconnecting loops, few passes through the pedigree are required to obtain all the marginal (over individuals i) conditional (on $\mathbf{Y}$) risk probabilities (Thompson 1981).

Elods

Simulation of data random variables $\mathbf{Y}$ is often undertaken as part of a power study. For example, simulation of latent genotypes $\mathbf{G}$ and resulting marker and trait phenotypes $\mathbf{Y}$ can be used to assess the power of a potential linkage study. Before the times of readily available genome-wide marker data, linkage detection was primarily a question analyzing the coinheritance of observed trait phenotypes $\mathbf{Y}_T$ and marker locus phenotypes $\mathbf{Y}_M$, for a single trait locus, T, and single marker locus, M. If the two loci are linked, the recombination frequency is $\rho < \frac{1}{2}$, while if they are unlinked inheritance is independent at the two loci ($\rho = \frac{1}{2}$). Thus we have the *lod score* (Morton55);

$$lod(\rho) \;=\; \log\left(\frac{P_\rho(\mathbf{Y}_M, \mathbf{Y}_T)}{P_{\rho=\frac{1}{2}}(\mathbf{Y}_M, \mathbf{Y}_T)} \right) \tag{4.25}$$

which is the logarithm of the likelihood ratio comparing the two hypotheses. The expected lod score is then

$$\begin{aligned}
Elod(\rho) \;=\;& \mathrm{E}_\rho(\log(P_\rho(\mathbf{Y}_M, \mathbf{Y}_T)) - \log(P_{\rho=\frac{1}{2}}(\mathbf{Y}_M, \mathbf{Y}_T))) \\
\;=\;& \mathrm{E}_\rho(\log(P_\rho(\mathbf{Y}_M, \mathbf{Y}_T)) - \\
& \log(P(\mathbf{Y}_M)) - \log(P(\mathbf{Y}_T)))
\end{aligned} \tag{4.26}$$

In advance of a study, one may compute the expected lod score to be obtained, given the sizes and counts of pedigree structures available. Thompson et al. (1978) first developed these *Elods* in the context of linkage analysis, and they have become quite widely used. In fact, Thompson et al. (1978) produced Monte Carlo estimates of the expectation in equation (4.26), by simulating the underlying trait and marker genotypes from

$P_\rho(\mathbf{G}_M, \mathbf{G}_T)$, and then the associated phenotypes, and then computing the lod score (4.25) for each realized set of phenotypes.

Elods conditional on trait data

As data at multiple DNA markers became potentially available, there was a rush to map Mendelian traits, using previously collected trait data. The *Elod* became an important tool in assessing whether there were sufficient trait data for probable linkage detection were the marker typing to be undertaken. One problem in using the *Elod* (4.26) is that the expectation is over both trait and marker phenotypes. Normally, however, there was already information on the trait phenotypes $\mathbf{Y}_T$ that would be available to researchers. Ploughman and Boehnke (1989) addressed this case. Given a single-locus trait model, and trait data $\mathbf{Y}_T$, it is possible to simulate the underlying inheritance patterns or genotypes, $\mathbf{G}_T$, at the trait locus. This is accomplished by a Monte Carlo version of reverse peeling analogous to that given by equation (4.16) in section 4.1.4. Once trait genotyps $\mathbf{G}_T$ are realized, conditional of the available trait data $\mathbf{Y}_T$, marker latent genotypes $\mathbf{G}_M$ and potentially observable marker phenotypes $\mathbf{Y}_M$ are readily obtained:

$$P_\rho(\mathbf{Y}_M, \mathbf{G}_M, \mathbf{G}_T \mid \mathbf{Y}_T) = P(\mathbf{Y}_M \mid \mathbf{G}_M)P_\rho(\mathbf{G}_M \mid \mathbf{G}_T)$$
$$P(\mathbf{G}_T \mid \mathbf{Y}_T) \qquad (4.27)$$

The dependence structure here is a special case of that shown in Figure 4.4. The combined realizations $(\mathbf{Y}_T, \mathbf{Y}_M)$ may be used to estimate a *Elod*, conditional upon the fixed $\mathbf{Y}_T$, These conditional *Elods* have become an essential tool in applied studies, particularly during the 1980s when many Mendelian traits were mapped, and marker typing remained the most expensive component of studies.

Monte Carlo likelihood by sequential imputation

We turn now to a use of reverse peeling in the Monte Carlo estimation of likelihoods. Recall that efficient Monte Carlo estimation of the likelihood $L(\theta) = P_\theta(\mathbf{Y})$ will result from sampling latent genotypes $\mathbf{G}$ from a distribution $P^*(\mathbf{G})$ close to proportional to the joint probability $P_\theta(\mathbf{G}, \mathbf{Y})$

$$P^*(\mathbf{G}) \approx P_\theta(\mathbf{G} \mid \mathbf{Y}) \propto P_\theta(\mathbf{G}, \mathbf{Y})$$

(equation (4.24)). The following approach to choice of $P^*(\mathbf{G})$ is due to Kong et al. (1994) and Irwin et al. (1994).

Suppose, as before, there are data at L genetic loci (say a disease and $L-1$ markers) on a chromosome, and assume absence of genetic interference. Let Y_{*j} again denote the data for locus j and G_{*j} the underlying genotypes at that locus for all members of the pedigree. Note that genotypes G_{*j} satisfy the same first-order Markov dependence over loci as do the meiosis indicators S_{*j} (Figure 4.4). For any specified θ_0 of interest, a realization G_{*j}^* is obtained for each locus in turn from the distribution

$$P_{\theta_0}(G_{*j} \mid G_{*1}^*, \ldots G_{*,j-1}^*, Y_{*1}, \ldots, Y_{*,j-1}, Y_{*j}) =$$
$$P_{\theta_0}(G_{*j} \mid G^{*(j-1)}, Y^{(j)})$$

where as in section 4.1.3, $Y^{(j)} = (Y_{*1}, \ldots, Y_{*j})$, $G^{(j)}$ is analogously defined, and θ indexes the genetic model. Predictive weights w_j are computed:

$$w_j = P_{\theta_0}(Y_{*j} \mid Y^{(j-1)}, G^{*(j-1)})$$

Now

$$P_{\theta_0}(G_{*j} \mid G^{(j-1)}, Y^{(j)}) = \frac{P_{\theta_0}(G_{*j}, Y_{*j} \mid G^{*(j-1)}, Y^{(j-1)})}{P_{\theta_0}(Y_{*j} \mid G^{(j-1)}, Y^{(j-1)})}$$
$$= \frac{P_{\theta_0}(G_{*j}, Y_{*j} \mid G^{*(j-1)}, Y^{(j-1)})}{w_j}$$

Thus the joint simulation distribution for $\mathbf{G}^* = (G_{*1}^*, \ldots, G_{*L}^*)$ is

$$P^*(\mathbf{G}^*) = \frac{P_{\theta_0}(\mathbf{Y}, \mathbf{G}^*)}{W_L(\mathbf{G}^*)}$$

where $W_L(\mathbf{G}^*) = \prod_{j=1}^{L} w_j$. Thus

$$\mathrm{E}_{P^*}(W_L(\mathbf{G}^*)) = \sum_{\mathbf{G}^*} W_L(\mathbf{G}^*) P^*(\mathbf{G}^*) = P_{\theta_0}(\mathbf{Y}) \qquad (4.28)$$

Thus a Monte Carlo estimate of $L(\theta_0) = P_{\theta_0}(\mathbf{Y})$ is given by the mean value of $W_L(\mathbf{G}^*)$, over repeated independent repetitions of the sequential imputation process. Repeating the process for different trait locus positions on the chromosome, one obtains an estimated likelihood curve for the location of the trait locus.

In genetic analyses, conditional expectations, given the data, with respect to some particular model $P_{\theta_0}(\cdot)$ are often needed. These address such questions as: In which meioses and at what locations are the recombinations? Who should be sampled to obtain most additional information about the trait model or trait locus position? Where are the biggest uncertainties in underlying marker genotypes? How would it affect inferences to reduce such uncertainty? In principle, such expectations can be readily estimated, using the sequential imputation probability distribution P^* and computed weights W_L:

$$\mathrm{E}_{\theta_0}(g(\mathbf{G}, \mathbf{Y}) \mid \mathbf{Y}) = \sum_{\mathbf{G}} g(\mathbf{G}, \mathbf{Y}) P_{\theta_0}(\mathbf{G} \mid \mathbf{Y})$$
$$= \sum_{\mathbf{G}} g(\mathbf{G}, \mathbf{Y}) \frac{P^*(\mathbf{G}) W_L(\mathbf{G})}{P_{\theta_0}(\mathbf{Y})}$$

$$= \frac{\mathrm{E}_{P^*}(g(\mathbf{G}, \mathbf{Y})W_L(\mathbf{G}))}{P_{\theta_0}(\mathbf{Y})}$$

The normalising factor $P_{\theta_0}(\mathbf{Y})$ is the unknown likelihood. Equation (4.28) provides a Monte Carlo estimate of $P_{\theta_0}(\mathbf{Y})$, so that

$$\mathrm{E}_{\theta_0}(g(\mathbf{G}, \mathbf{Y}) \mid \mathbf{Y}) = \frac{\mathrm{E}_{P^*}(g(\mathbf{G}, \mathbf{Y})W_L(\mathbf{G}))}{\mathrm{E}_{P^*}(W_L(\mathbf{G}))}.$$

4.2.5 Monte Carlo likelihood ratio estimation

Monte Carlo likelihood ratios

Recall again (equation (4.24)) that, since, for phenotypic data $\mathbf{Y}$

$$L(\theta) \quad = \quad P_\theta(\mathbf{Y}) \quad = \quad \sum_{\mathbf{X}} P_\theta(\mathbf{Y}, \mathbf{X})$$

where latent variables $\mathbf{X}$ are genotypes $\mathbf{G}$ or meiosis indicators $\mathbf{S}$, efficient Monte Carlo estimation of $L(\theta)$ will result from sampling from a distribution $P^*(\mathbf{X})$ close to proportional to the joint probability $P_\theta(\mathbf{Y}, \mathbf{X})$

$$P^*(\mathbf{X}) \approx P_\theta(\mathbf{X} \mid \mathbf{Y}) \propto P_\theta(\mathbf{Y}, \mathbf{X})$$

One possible choice is thus to simulate, by methods as yet unspecified, not from $P_\theta(\mathbf{X} \mid \mathbf{Y})$ but from $P_{\theta_0}(\mathbf{X} \mid \mathbf{Y})$, where $\theta_0 \approx \theta$. Then

$$
\begin{aligned}
P_\theta(\mathbf{Y}) &= \sum_{\mathbf{X}} P_\theta(\mathbf{Y}, \mathbf{X}) = \sum_{\mathbf{X}} \frac{P_\theta(\mathbf{Y}, \mathbf{X})}{P_{\theta_0}(\mathbf{X} \mid \mathbf{Y})} P_{\theta_0}(\mathbf{X} \mid \mathbf{Y}) \\
&= \mathrm{E}_{\theta_0}\left(\frac{P_\theta(\mathbf{Y}, \mathbf{X})}{P_{\theta_0}(\mathbf{X} \mid \mathbf{Y})} \ \Big| \ \mathbf{Y} \right) \\
&= P_{\theta_0}(\mathbf{Y}) \, \mathrm{E}_{\theta_0}\left(\frac{P_\theta(\mathbf{Y}, \mathbf{X})}{P_{\theta_0}(\mathbf{Y}, \mathbf{X})} \ \Big| \ \mathbf{Y} \right)
\end{aligned}
$$

Hence in genetic analysis, or in any missing-data context, we have the key formula of Thompson and Guo (1991)

$$\frac{L(\theta)}{L(\theta_0)} = \frac{P_\theta(\mathbf{Y})}{P_{\theta_0}(\mathbf{Y})} = \mathrm{E}_{\theta_0}\left(\frac{P_\theta(\mathbf{Y}, \mathbf{X})}{P_{\theta_0}(\mathbf{Y}, \mathbf{X})} \ \Big| \ \mathbf{Y} \right) \qquad (4.29)$$

In this expectation, $\mathbf{X}$ is the random variable, $\mathbf{Y}$ is fixed. The distribution of $\mathbf{X}$ is $P_{\theta_0}(\cdot \mid \mathbf{Y})$. If $\mathbf{X}^{(t)}$, $t = 1, ..., N$, are realized from this distribution then the likelihood ratio can be estimated by

$$\frac{1}{N} \sum_{t=1}^{N} \left(\frac{P_\theta(\mathbf{Y}, \mathbf{X}^{(t)})}{P_{\theta_0}(\mathbf{Y}, \mathbf{X}^{(t)})} \right)$$

In section 4.3 we shall see how Markov chain Monte Carlo (MCMC) can be used to realize $\mathbf{X}$ from $P_{\theta_0}(\cdot \mid \mathbf{Y})$.

Simulation at a single model θ_0 provides an estimate of the relative likelihood $L(\theta)/L(\theta_0)$ as a function of θ. This will be a satisfactory estimator only for those θ close to θ_0; specifically, for those θ for which $P_{\theta_0}(\mathbf{X}|\mathbf{Y})$ is close to proportional to $P_{\theta_0}(\mathbf{Y}, \mathbf{X})$. Sometimes, primary interest is in the shape of the likelihood surface in the neighbourhood of some specific point, such as the maximum likelihood estimate (MLE). In this case, preliminary MCMC runs and likelihood ratio function estimates can be used to obtain a ballpark value of the MLE. Alternatively, Monte Carlo EM can be used (Guo and Thompson 1992). Once a ballpark estimate is found, one very large MCMC run can provide an accurate estimate of the MLE and of the likelihood in the region. However, this approach has limitations. One may be interested in the likelihood surface, or in log-likelihood differences, over large regions in the parameter space. Or, the large MCMC run may reveal that one's initial estimate was not sufficiently close to the MLE, and additional large runs may be necessary. It is desirable to find a method that combines realizations from all the runs, and provides an estimate of the likelihood surface over a range of parameter values.

Monte Carlo likelihood surfaces

One way of combining realizations from different MCMC samplers was provided by Geyer (1991a). MCMC samplers are run at many models, covering the range of interest, say at $\theta_0, \theta_1, ..., \theta_K$. The sets of N_j realizations from $P_{\theta_j}(\mathbf{X}|\mathbf{Y})$, $j = 0, 1, \ldots, K$, give a combined set of realizations from

$$P^*(\mathbf{X}) \;=\; \frac{1}{\sum_j N_j} \sum_{j=0}^{K} N_j P_{\theta_j}(\mathbf{X}|\mathbf{Y})$$

$$\;=\; \frac{1}{\sum_j N_j} \sum_{j=0}^{K} N_j P_{\theta_j}(\mathbf{Y}, \mathbf{X})/L(\theta_j)$$

and writing the likelihood estimation formula as an expectation w.r.t. this $P^*()$

$$L(\theta_j) \;=\; \mathrm{E}_{P^*} \left(\frac{P_{\theta_j}(\mathbf{Y}, \mathbf{X})}{P^*(\mathbf{X})} \right)$$

Now, although we have a sample from P^*, the denominator $P^*(\mathbf{X})$ cannot be explicitly computed, since it depends on the unknown $L(\theta_j)$, but we have the implicit Monte Carlo estimating equations

$$L(\theta_j) \;=\; \sum_{\mathbf{X}^*} \left(\frac{P_{\theta_j}(\mathbf{Y}, \mathbf{X}^*)}{\sum_{l=0}^{K} N_l P_{\theta_l}(\mathbf{Y}, \mathbf{X}^*)/L(\theta_l)} \right)$$

$$\;=\; \sum_{\mathbf{X}^*} \left(\sum_{l=0}^{K} N_l \frac{P_{\theta_l}(\mathbf{Y}, \mathbf{X}^*)}{P_{\theta_j}(\mathbf{Y}, \mathbf{X}^*)} \frac{1}{L(\theta_l)} \right)^{-1} \qquad (4.30)$$

for $j = 0, ..., K$, where the sum is over the total set of realizations $\mathbf{X}^*$. These equations determine only the relative values of $L(\theta_j)$, but can be solved iteratively for these relative values. For example, one may iterate equation (4.30) directly, renormalizing after each cycle, to keep one value, say $L(\theta_0)$ fixed ($=1$). This iterative procedure is globally convergent to the unique solution of equation (4.30). Once the relative values of $L(\theta_j)$ are found, then, for any other value of θ in the range spanned by the set of θ_j, $L(\theta)$ can be estimated by

$$L(\theta) = \sum_{\mathbf{X}^*} \left(\sum_{l=0}^{K} N_l \frac{P_{\theta_l}(\mathbf{Y}, \mathbf{X}^*)}{P_{\theta}(\mathbf{Y}, \mathbf{X}^*)} \frac{1}{L(\theta_l)} \right)^{-1} \tag{4.31}$$

where the sum is over the same total set of realizations as before. (Again, the estimate is relative to $L(\theta_0) = 1$.) Geyer (1991a) named this method "reverse logistic regression".

There are two requirements for this approach to be an effective solution to the likelihood estimation problem. First, each sampler $P_{\theta_j}(\mathbf{X}|\mathbf{Y})$, $j = 0, \ldots, K$ must cover well that part of the space of $\mathbf{X}$-values that has high total probability mass under that probability distribution — for an MCMC sampler this is a non-trivial consideration (section 4.3). Second, even if the separate samplers are behaving "well", in this sense, we need good "overlap" between adjacent models, for the mixture estimates to be effective. The posterior probability that a particular observation $\mathbf{X}$ derived from the sample P_{θ_j} is

$$\frac{N_j P_{\theta_j}(\mathbf{X}|\mathbf{Y})}{\sum_{l=0}^{K} N_l P_{\theta_l}(\mathbf{X}|\mathbf{Y})}$$

For every j, the values of these probabilities should not be too close to 1 for too large a proportion of the sampled $\mathbf{X}$-values. Thus adjacent parameter values θ_j must be chosen not too far apart, where distance is measured in terms of the probabilities of the $\mathbf{X}$-values generated.

Other difficulties with using the reverse logistic regression method concern computational resources. Either the realized $\mathbf{X}^*$, or at least the values $P_{\theta}(\mathbf{Y}, \mathbf{X}^*)$ for each θ of interest, must be saved, in order for equations (4.30) and (4.31) to be implemented. This can demand massive amounts of storage. An alternative is to use block averages of the ratios of $P_{\theta_j}(\mathbf{Y}, \mathbf{X})/P_{\theta_l}(\mathbf{Y}, \mathbf{X})$ in equation (4.30) (Thompson 1994b). In the extreme case, this block might be the average over a full run of the sampler at a given θ_j. Let

$$R_j(\theta_l, \theta_j) = N_j^{-1} \sum_{\mathbf{X}^{*(j)}} \frac{P_{\theta_l}(\mathbf{Y}, \mathbf{X}^{*(j)})}{P_{\theta_j}(\mathbf{Y}, \mathbf{X}^{*(j)})}$$

be the likelihood ratio estimate of $L(\theta_l)/L(\theta_j)$ from realizations $\mathbf{X}^{*(j)}$ at θ_j. Here the chosen values of l may vary with j. We define $R_j(\theta_l, \theta_j)$ to be 0 if $L(\theta_l)/L(\theta_j)$ is not estimated from realizations under model θ_j. At a minumum, for each j, values for $R_j(\theta_l, \theta j)$ should be computed for the values θ_l adjacent to θ_j. Then the estimating equation (4.30) becomes

$$L(\theta_j) = \left(\sum_{l=0}^{K} R_j(\theta_l, \theta j) \frac{1}{L(\theta_l)} \right)^{-1} \qquad (4.32)$$

Writing $\nu_j = 1/L(\theta_j)$, $R_{jl} = R_j(\theta_l, \theta_j)$, $\boldsymbol{\nu} = (\nu_j)$, and $\mathbf{R} = (R_{jl})$, equation (4.32) becomes

$$\boldsymbol{\nu} = \mathbf{R}\boldsymbol{\nu}$$

That is, the vector of ν_j-values is a right eigenvector of the matrix $\mathbf{R}$. Asymptotically, for large Monte Carlo runs, each computed R_{jl}-value converges to $L(\theta_l)/L(\theta_j) = \nu_j/\nu_l$. Thus, if, for each j, R_{jl} is evaluated for s other θ_l values, then each evaluated $R_{jl}\nu_l$ is approximately ν_j, and the corresponding eigenvalue should be s. This provides one check on the performance of the method.

4.3 MCMC methods for Multilocus Genetic Data

4.3.1 Monte Carlo estimation of location score curves

In modern genetic analysis, a primary goal is the localization of trait genes. Genetic markers have been mapped throughout the genome at a scale suitable for multipoint linkage analysis. While there may be uncertainties about marker locations, or other aspects of the marker model such as allele frequencies, these are normally assumed known. Here we denote the marker model parameters by Γ_M. For a complex trait, the trait model parameters, β, are also unknown. While in some analyses, joint maximization of the likelihood with respect to trait model β and trait locus position γ may be attempted, often a *location score curve* (Lathrop et al. 1984, Lange 1997) is computed for fixed β. The likelihood (or a profile likelihood) is evaluated as a function of a hypothesized trait-locus location γ, against a fixed marker map Γ_M. The overall model is indexed by parameter $\theta = (\beta, \gamma, \Gamma_M)$. As before, the likelihood is

$$L(\theta) = P_\theta(\mathbf{Y}) = \sum_{\mathbf{X}} P_\theta(\mathbf{Y} \mid \mathbf{X}) \, P_\theta(\mathbf{X}) \qquad (4.33)$$

(equation (4.21)) which may take the form (4.1) if $\mathbf{X} = \mathbf{G}$, the underlying genotypes, or (4.6) if $\mathbf{X} = \mathbf{S}$, the inheritance patterns of genes. For computations with multiple marker loci, the Lander-Green paradigm (4.6) is more natural and more effective, but exact computation is limited to small pedigrees.

As in the discussion of *Elods* (equations (4.26) and (4.27)), for convenience we partition the data $\mathbf{Y}$ into the trait data $\mathbf{Y}_T$ and marker data $\mathbf{Y}_M$. The corresponding latent variables are partitioned into $\mathbf{X}_T$ and $\mathbf{X}_M$. Monte Carlo estimation of the location score curve is always feasible. The form that follows directly from equation (4.29) is

$$\frac{L(\beta,\ \gamma_1,\ \Gamma_M)}{L(\beta,\ \gamma_0,\ \Gamma_M)} =$$

$$\mathrm{E}_{\theta_0}\left(\frac{P_{\theta_1}(\mathbf{Y}_T,\mathbf{Y}_M\mid\mathbf{X}_T,\mathbf{X}_M)P_{\theta_1}(\mathbf{X}_T,\mathbf{X}_M)}{P_{\theta_0}(\mathbf{Y}_T,\mathbf{Y}_M\mid\mathbf{X}_T,\mathbf{X}_M)P_{\theta_0}(\mathbf{X}_T,\mathbf{X}_M)}\ \middle|\ \mathbf{Y}_T,\mathbf{Y}_M\right)$$

for two trait locus positions γ_1 and γ_0. Noting the fact that only the position of the trait locus differs between numerator and denominator, the above equation reduces to

$$\frac{L(\beta,\ \gamma_1,\ \Gamma_M)}{L(\beta,\ \gamma_0,\ \Gamma_M)} = \mathrm{E}_{\theta_0}\left(\frac{P_{\gamma_1}(\mathbf{X}_T\mid\mathbf{X}_M)}{P_{\gamma_0}(\mathbf{X}_T\mid\mathbf{X}_M)}\ \middle|\ \mathbf{Y}_T,\mathbf{Y}_M\right) \qquad (4.34)$$

Thus only the conditional probability of trait-locus latent variables given marker-loci latent variables appears explicitly in the estimator. Although realization of the latent variables is complex, and requires MCMC methods, computation of the estimate from the realizations is generally very straightforward (Thompson and Guo 1991).

One practical difficulty of the above approach is accurate estimation of log-likelihood differences for trait locations in different marker intervals. The likelihood-ratio estimate (4.29) works well in comparing locations within an interval, and in principle the mixtures method (4.30) facilitates estimation between intervals. However, in practice, values of $\mathbf{X}_T$ realized at γ_0 may have very small probabilities under γ_1 if there is a marker locus between the two positions. Additionally, the usual objective is to estimate the lod-score relative to the base-point in which the trait locus position γ is not within the marker map Γ_M – the trait locus is unlinked. Again this can be accomplished by using the mixtures method (4.30), but several intervening positions γ, linked to but not within the marker map, may be required for effective estimation (Thompson 1994b). The procedure becomes computationally quite intensive.

The difficulties in Monte Carlo estimation of location score curves are due to the explicit modelling and sampling of the trait locus variables. An alternative procedure is due to Lange and Sobel (1991), who, using our current notation, write the likelihood in the form

$$\begin{aligned} L(\beta,\gamma,\Gamma_M) &= P_{\beta,\gamma,\Gamma_M}(\mathbf{Y}_M,\mathbf{Y}_T) \\ &\propto P_{\beta,\gamma,\Gamma_M}(\mathbf{Y}_T\mid\mathbf{Y}_M) \\ &= \sum_{\mathbf{X}_M}P_{\beta,\gamma}(\mathbf{Y}_T\mid\mathbf{X}_M)P_{\Gamma_M}(\mathbf{X}_M\mid\mathbf{Y}_M) \qquad (4.35) \\ &= \mathrm{E}_{\Gamma_M}(P_{\beta,\gamma}(\mathbf{Y}_T\mid\mathbf{X}_M)\mid\mathbf{Y}_M). \qquad (4.36) \end{aligned}$$

Now latent variables $\mathbf{X}_M$ are sampled from their conditional distribution given the marker data $\mathbf{Y}_M$. Provided exact computation of $P_{\beta,\gamma}(\mathbf{Y}_T \mid \mathbf{X}_M)$ is possible for alternative trait models (β) or locations (γ), we have a Monte Carlo estimate of $L(\beta, \gamma, \Gamma_M)$, while comparison to the unlinked base-point requires only $P_\beta(\mathbf{Y}_T)$. Since Γ_M is fixed the Monte Carlo requires only a single set of realizations $\mathbf{X}_M^{(t)}$. The disadvantage is that $P_{\beta,\gamma}(\mathbf{Y}_T \mid \mathbf{X}_M^{(t)})$ must be computed for each such realization; this requires a single-locus peeling computation for the trait-locus data under the trait model. Further, this computation must be done, not only for each realization $\mathbf{X}_M^{(t)}$, but also for each β and γ at which a likelihood estimate is required.

In many cases, however, the gains outweigh the costs, except when the simulation distribution $P_{\Gamma_M}(\mathbf{X}_M \mid \mathbf{Y}_M)$ is not close to proportional to the ideal importance-sampling target distribution $P_{\beta,\gamma,\Gamma_M}(\mathbf{X}_M \mid \mathbf{Y}_M, \mathbf{Y}_T)$. This is particularly so for models (trait locations) γ which are not close to the truth, and for a trait which provides substantial information about the inheritance patterns of genes at the underlying trait locus, and hence also at linked marker loci. In fact, the cases where the Monte Carlo estimate based on equation (4.36) performs poorly are precisely those in which the likelihood ratio estimator (4.34) also has difficulties. There continue to be interesting open questions in the estimation of multilocus linkage likelihoods.

Finally, consider again equation (4.35), and the sampling of $\mathbf{X}_M$ from its conditional distribution given the marker data $\mathbf{Y}_M$, particularly for the case $\mathbf{X}_M = \mathbf{S}_M$, the meiosis indicators at the marker loci. Regardless of the trait model, provided it has some genetic component, related affected individuals, or related individuals exhibiting extreme trait values, will share genes IBD at trait loci with some increased probability. Hence also they will share genes IBD with increased probability at marker loci linked to those trait loci. In so-called "non-parametric" computations for linkage detection, marker data on a pedigree are analyzed to detect regions of the genome in which there is evidence for excess gene IBD among affected individuals, or individuals exhibiting extreme trait values. Such regions provide evidence for linkage. The Monte Carlo sampling of $\mathbf{S}_M$ given marker data $\mathbf{Y}_M$ provides direct estimates of posterior probabilities of patterns of gene IBD $\mathbf{J}(\mathbf{S}_M)$. Obtaining the exact marginal probabilities of S_{*j} given the full data $\mathbf{Y}$, using the Baum algorithm (equation (4.15)), was discussed in section 4.1.4. Here we have only Monte Carlo estimates of these probabilities, but have estimates of $P_\theta(\mathbf{S} \mid \mathbf{Y})$ which can be used to score events jointly over loci. Moreover, using MCMC, we will have these realizations for larger pedigrees in which the exact computations are infeasible. For the case where only marker data are considered, many of the problems of the Monte Carlo estimation procedures are much reduced, provided a good MCMC sampler is used (see sections 4.3.4, 4.3.5). The statistical problem becomes one of development of appropriate test statistics, to detect linkage on the basis of estimated posterior IBD probabilities.

4.3.2 Markov chain Monte Carlo

To implement effective Monte Carlo estimates of location score curves (4.34), or of any likelihoods based on multilocus genetic data on extended pedigrees (equation (4.29)), some form of sampling of latent variables $\mathbf{X}$ conditional upon data $\mathbf{Y}$ is required (equation 4.24). As before, we use $\mathbf{X}$ to denote generic latent variables which in practice will be underlying genotypes, $\mathbf{G}$, or meiosis indicators, $\mathbf{S}$. Now (equation 4.24)

$$P_\theta(\mathbf{X} \mid \mathbf{Y}) = \frac{P_\theta(\mathbf{X}, \mathbf{Y})}{P_\theta(\mathbf{Y})} \propto P_\theta(\mathbf{X}, \mathbf{Y})$$

as a distribution over $\mathbf{X}$ for fixed $\mathbf{Y}$. The objective of Monte Carlo estimation is to obtain the likelihood $P_\theta(\mathbf{Y})$; if it were available Monte Carlo would be unnecessary. Thus, whereas the joint probability $P_\theta(\mathbf{X}, \mathbf{Y})$ is easily computed, the conditional probability $P_\theta(\mathbf{X} \mid \mathbf{Y})$ is not. Markov chain Monte Carlo (MCMC) is a method designed for obtaining realizations from a distribution, such as $P_\theta(\mathbf{X} \mid \mathbf{Y})$, known only up to a normalizing constant.

We review very briefly the Metropolis-Hastings class of algorithms (Hastings 1970), in the context of generating dependent realizations from a conditional probability distribution $P_\theta(\mathbf{X} \mid \mathbf{Y})$ on a space $\mathcal{X}$, where $P_\theta(\mathbf{Y})$ cannot be computed. For each $\mathbf{X}$ in $\mathcal{X}$ a "proposal distribution" $q(\cdot; \mathbf{X})$ is defined. Then, if the process is now at $\mathbf{X}$ the next value is generated as follows:

(1) Generate $\mathbf{X}^*$ from the proposal distribution $q(\cdot; \mathbf{X})$
(2) Compute the Hastings ratio

$$h(\mathbf{X}^*; \mathbf{X}) = \frac{q(\mathbf{X}; \mathbf{X}^*)P_\theta(\mathbf{X}^* \mid \mathbf{Y})}{q(\mathbf{X}^*; \mathbf{X})P_\theta(\mathbf{X} \mid \mathbf{Y})}$$

Note that $h(\cdot; \cdot)$ depends only on the ratio of densities $P_\theta(\cdot \mid \mathbf{Y})$, so that the normalizing constant $P_\theta(\mathbf{Y})$ need not be computed.
(3) With probability $a = \min(1, h(\mathbf{X}^*; \mathbf{X}))$ the process moves to $\mathbf{X}^*$ and with probability $(1 - a)$ it remains at $\mathbf{X}$.

Thus, a first-order Markov chain on $\mathcal{X}$ is defined. Consider any two points $\mathbf{X}$ and $\mathbf{X}^*$ in $\mathcal{X}$, and assume, without loss of generality, that

$$q(\mathbf{X}; \mathbf{X}^*)P_\theta(\mathbf{X}^* \mid \mathbf{Y}) > q(\mathbf{X}^*; \mathbf{X})P_\theta(\mathbf{X} \mid \mathbf{Y}),$$

and consider the transition probability $\pi(\mathbf{X} \to \mathbf{X}^*)$. Since the Hastings ratio $h(\mathbf{X}^*; \mathbf{X}) > 1$ this is simply the proposal probability $q(\mathbf{X}^*; \mathbf{X})$. In the reverse direction

$$\pi(\mathbf{X}^* \to \mathbf{X}) = q(\mathbf{X}; \mathbf{X}^*)h(\mathbf{X}; \mathbf{X}^*)$$

$$= q(\mathbf{X}; \mathbf{X}^*) \frac{q(\mathbf{X}^*; \mathbf{X}) P_\theta(\mathbf{X} \mid \mathbf{Y})}{q(\mathbf{X}; \mathbf{X}^*) P_\theta(\mathbf{X}^* \mid \mathbf{Y})}$$

$$= \pi(\mathbf{X} \to \mathbf{X}^*) \frac{P_\theta(\mathbf{X} \mid \mathbf{Y})}{P_\theta(\mathbf{X}^* \mid \mathbf{Y})}.$$

Provided the relevant probabilities are non-zero, detailed balance obtains for the distribution $P_\theta(\cdot \mid \mathbf{Y})$, which is thus an equilibrium distribution of the Markov chain. Provided the transition probabilities define an irreducible chain, this is the unique equilibrium distribution. After the chain has been run for "long enough", each realization has the target distribution $P_\theta(\cdot \mid \mathbf{Y})$, and continuing to run the chain provides a sequence of (dependent) realizations from this target distribution.

The algorithm of Metropolis et al. (1953) is a special case; if $q(\mathbf{X}^*; \mathbf{X}) = q(\mathbf{X}; \mathbf{X}^*)$ the Hastings ratio reduces to the odds ratio of the proposal state $\mathbf{X}^*$ versus the current state $\mathbf{X}$. The Gibbs sampler (Geman and Geman 1984) is also a special case, in which $\mathbf{X}^*$ differs from $\mathbf{X}$ in only a subset of the components $\mathbf{X}_I = \{X_i; i \in I\}$, say. If $\mathbf{X}_{-I}$ denotes the remaining components

$$q(\mathbf{X}^*; \mathbf{X}) = 0 \text{ if } \mathbf{X}^*_{-I} \neq \mathbf{X}_{-I}$$

$$q(\mathbf{X}^*; \mathbf{X}) = P_\theta(X_I^* \mid \mathbf{X}_{-I}, \mathbf{Y}) \text{ if } \mathbf{X}^*_{-I} = \mathbf{X}_{-I} \qquad (4.37)$$

Then, if $\mathbf{X}^*_{-I} = \mathbf{X}_{-I}$,

$$h(\mathbf{X}^*; \mathbf{X}) = \frac{P_\theta(\mathbf{X}_I \mid \mathbf{X}^*_{-I}, \mathbf{Y}) P_\theta(\mathbf{X}^* \mid \mathbf{Y})}{P_\theta(\mathbf{X}_I^* \mid \mathbf{X}_{-I}, \mathbf{Y}) P_\theta(\mathbf{X} \mid \mathbf{Y})}$$

$$= \frac{P_\theta(\mathbf{X} \mid \mathbf{Y}) P_\theta(\mathbf{X}^* \mid \mathbf{Y}) P_\theta(\mathbf{X}_{-I} \mid \mathbf{Y})}{P_\theta(\mathbf{X}^* \mid \mathbf{Y}) P_\theta(\mathbf{X} \mid \mathbf{Y}) P_\theta(\mathbf{X}^*_{-I} \mid \mathbf{Y})} = 1$$

Since $a \equiv \min(1, h) \equiv 1$, in the Gibbs sampler there is no rejection step, but $\mathbf{X}^* = \mathbf{X}$ is of course possible. If I consists of a single component, or even only a few components, changes in $\mathbf{X}$ are necessarily small. In principle, I may consist of any part of $\mathbf{X}$ provided the full conditional distribution $P_\theta(\mathbf{X}_I \mid \mathbf{X}_{-I}, \mathbf{Y})$ is available. The choice of I is thus a balance between good mixing of the sampler and computational feasibility. Reversible-jump MCMC (Green 1995) is an extension to the basic Metropolis algorithm, in which the dimension of $\mathbf{X}$ may change, and a Jacobian of the transformation must thus be included in the Hastings ratio. Although reversible-jump MCMC has been used in the development of genetic analysis methods (Heath 1997, Browning 1999), we shall not cover these methods in this chapter.

In order for the time-average over the chain to converge to the expectation under the equilibrium distribution, the ergodic theorem must apply. For a discrete state space, irreducibility of the Markov chain is required. Although it is important to check that an MCMC sampler is irreducible, the mixing properties of the chain are of far greater practical importance.

Any chain can be made irreducible, using Metropolis rejection. Consider the occasional use of a proposal distribution which does not depend on the current $\mathbf{X}$, and puts positive probability on every $\mathbf{X}^*$ in $\mathcal{X}$ for which $P_\theta(\mathbf{X}^*|\mathbf{Y})$ is positive. This proposal will be accepted with positive but infinitesimal probability. Nothing has been achieved in practice, but the sampler is irreducible. Occasional use of a well-designed "restart" proposal distribution is often useful; this is one of several key ideas in improving samplers, and in assessing performance. However, such "restarts" should be considered as a tool for improving mixing, not for achieving irreducibility in an otherwise reducible sampler.

Convergence is an issue related to mixing; there are two (related) convergence issues which often get confused. One is convergence of the marginal distribution of each configuration $\mathbf{X}^{(t)}$ at step t to the equilibrium distribution of the Markov chain as t becomes large. The other relates to the convergence of a time-average over the chain to the expectation of the function under the equilibrium distribution. Both depend on the mixing properties of the Markov chain. The first can (in principle) be addressed by burn-in (discarding enough realizations at the start of the run), but it is seldom the major problem, in practice. The second issue remains even if the Markov chain starts in the equilibrium distribution. This is a much more major problem; all relevant parts of the space must be sampled by realizations from a single Markov chain. Starting the chain in different parts of the space $\mathcal{X}$ is no help if we have no idea how to weight the realizations from the different starts. (see Geyer 1992 for more discussion).

There are the usual sorts of Central Limit Theorems for Monte Carlo estimators, but we refer readers to the theoretical literature for these. On the other hand, estimation of the standard deviations of Monte Carlo estimators of expectations is a practical necessity. There are several approaches which are easily implemented, but assessment of the performance of the method in a particular instance is often hard. Again, Geyer 1992 is a good reference. One of the simplest methods of estimating Monte Carlo variances is by using batch means (Hastings 1970). The realizations are divided into sufficiently large batches so that the batch means are "almost independent", and the variance of independent batch means is related to the variance of the overall mean (the estimator of the expectation). Autocorrelation between the batches can be tested for. This method is quite effective if the sampler is mixing well, but otherwise can severely underestimate Monte Carlo variance. However, the same is true of other variance estimators also, and the batch-means approach is very easily implemented.

Although, after convergence, each time-step of the Markov chain provides a realization from the target distribution, successive realizations may be highly correlated. Particularly if the use of the realizations is computationally intensive, or storage is limited, it may be desirable to subsample the chain at a *spacing* of some number of time-steps. Effective variance esti-

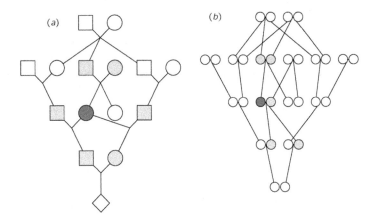

Figure 4.5 *The neighbourhood for a genotypic Gibbs sampler on a pedigree: (a) the individual neighbourhood, and (b) the gene neighbourhood.*

mation is also important in deciding the the spacing to be used in sampling realizations from an MCMC. The optimal spacing is the one that achieves minimum computational cost for given variance of the resulting estimator. This optimal value depends on the relative costs of generating the samples and of evaluating the contribution to the estimator at the realised values, but is seldom large.

4.3.3 Single-site updating methods

As in other areas of application, the earliest MCMC samplers that were used to realize latent variables on pedigrees conditional on phenotypic data were mainly single-site updating methods. The proposed changes to the latent variable configurations were thus very small. Lange an Matthysse (1989) used as their latent variables both the genotypes and inheritance patterns of genes, and used a Metropolis algorithm to propose changes. Sheehan (1990) and Thompson and Guo (1991) used a Gibbs sampling approach, using the genotypes as the latent variables, while Thompson (1994c,1994b) used a Metropolis algorithm to update a single meiosis indicator $S_{i,j}$ for meiosis i and locus j.

Unfortunately, in genetic examples the constraints on genotypes **G** or meiosis indicators **S** imposed by Mendelian segregation and discrete marker phenotypes mean that any proposal that makes multiple changes to the current value of **G** or **S** has a high probability of proposing a configuration inconsistent with the data **Y**. By contrast, although proposed changes are small, single-site updates are easily proposed and often accepted. The genes and heritable effects in an individual are determined by those in his parents, and jointly with those in his spouse, influence those in his offspring (figure 4.5(a)). This neighbourhood structure means that a single-site Gibbs sampler is easy to implement. Each genetic effect in each individual is successively updated, conditional upon the remainder. Specifically, where genotypes **G** are the latent variables, underlying genotypes for both trait and marker loci are sampled individual by individual and locus by locus.

For certain data configurations, the single-site genotypic Gibbs sampler is not irreducible when a locus is multiallelic. However, theoretical irreducibility can always be easily achieved. The practical problem is failure of the sampler to mix adequately. This can be a problem on large pedigrees even for diallelic loci, particularly if underlying genotypes are highly constrained (but not determined) by the data. The reducibility of the Gibbs sampler for genetic loci with more than two alleles was first addressed by Sheehan and Thomas (1993), in the context of a single-genotype Gibbs sampler. Their method used modification of either the segregation probabilities or the penetrance probabilities, so that the sampler was no longer irreducible. For example, modifying the penetrances

$$P^*(Y_i \mid G_i) \quad = P_\theta(Y_i \mid G_i) \quad \text{if } P_\theta(Y_i \mid G_i) > 0$$
$$P^*(Y_i \mid G_i) \quad \quad = \quad c \quad \quad \text{if } P_\theta(Y_i \mid G_i) = 0$$

Then

$$\frac{P_\theta(\mathbf{G}, \mathbf{Y})}{P^*(\mathbf{G}, \mathbf{Y})} \quad = \quad 1 \text{ if } P_\theta(\mathbf{Y} \mid \mathbf{G}) > 0$$
$$= \quad 0 \text{ if } P_\theta(\mathbf{Y} \mid \mathbf{G}) = 0$$

Thus no reweighting is required in order for the realizations to represent the distribution of genotypes under the true genetic model. All realizations consistent with the true model have equal weight; those inconsistent with it are just dropped from the output sample. Lin et al. (1993) used similar penetrance modifications to achieve irreducibility, but used Metropolis-coupled samplers (Geyer 1991b), coupling a sampler under the true model to samplers which were not only irreducible, but also moved more quickly around the space. Rather than a uniform penetrance modification, only the changes necessary to achieve irreducibility are made. The expansion of the space that is sampled is therefore limited.

Several methods for more efficient sampling of the space of feasible underlying genotype configurations have been developed. Some of these are due to Shili Lin (Lin et al. 1993,1994a). Others are due to Eric Sobel Sobel and Lange (1993) and to Charles Geyer (Geyer and Thompson 1995). We briefly outline here only the methods of Lin et al.(1993, 1994a), directed specifically towards sampling of genotypes at polymorphic marker loci where there are many unsampled individuals in the pedigree. These methods use a form of "heated proposals", resulting in samplers that move around the space of genotypic configurations far more effectively.

For a single-site update to component G_i, the proposal distribution for the Gibbs sampler (equation (4.37)) is

$$
\begin{aligned}
q_i(\mathbf{G}^*; \mathbf{G}) &= P_\theta(G_i^* \mid \mathbf{G}_{-i}, \mathbf{Y}) \text{ for component } i \\
G_j^* &= G_j \text{ for } j \neq i \quad \text{or } \mathbf{G}_{-i}^* = \mathbf{G}_{-i}
\end{aligned}
$$

This full conditional distribution for G_i is easily computed, but only small changes to $\mathbf{G}$ are possible at each step. On the other hand, the full conditionals for larger blocks of components $\mathbf{G}_I = \{G_i; i \in I\}$ are more computationally intensive or even infeasible. One possibility is to base a Metropolis-Hastings sampler on this local conditional distribution for the single component G_i, but in a way that enhances movement around the space. The method of Lin et al. (1994a) "flattens" the proposal distribution in a manner similar to simulated annealing, using a "temperature" parameter T:

$$
\begin{aligned}
q_i(\mathbf{G}^*; \mathbf{G}) &\propto (P_\theta(G_i^* \mid \mathbf{G}_{-i}, \mathbf{Y}))^{1/T} \text{ for component } i \\
G_j^* &= G_j \text{ for } j \neq i, \quad \text{or } \mathbf{G}_{-i}^* = \mathbf{G}_{-i}
\end{aligned}
$$

The Hastings ratio is then

$$
\begin{aligned}
h(\mathbf{G}^*; \mathbf{G}) &= \frac{q(\mathbf{G}; \mathbf{G}^*) P_\theta(\mathbf{G}^* \mid \mathbf{Y})}{q(\mathbf{G}^*; \mathbf{G}) P_\theta(\mathbf{G} \mid \mathbf{Y})} \\
&= \frac{(P_\theta(G_i \mid \mathbf{G}_{-i}^*, \mathbf{Y}))^{1/T} P_\theta(\mathbf{G}^* \mid \mathbf{Y})}{(P_\theta(G_i^* \mid \mathbf{G}_{-i}, \mathbf{Y}))^{1/T} P_\theta(\mathbf{G} \mid \mathbf{Y})} \\
&= \frac{(P_\theta(\mathbf{G} \mid \mathbf{Y}))^{1/T} P_\theta(\mathbf{G}^* \mid \mathbf{Y}) (P_\theta(\mathbf{G}_{-i} \mid \mathbf{Y}))^{1/T}}{(P_\theta(\mathbf{G}^* \mid \mathbf{Y}))^{1/T} P_\theta(\mathbf{G} \mid \mathbf{Y}) (P_\theta(\mathbf{G}_{-i}^* \mid \mathbf{Y}))^{1/T}} \\
&= \frac{(P_\theta(\mathbf{G}^* \mid \mathbf{Y}))^{1-1/T}}{(P_\theta(\mathbf{G} \mid , \mathbf{Y}))^{1-1/T}} \\
&= \frac{(P_\theta(G_i^* \mid \mathbf{G}_{-i}, \mathbf{Y}))^{1-1/T}}{(P_\theta(G_i \mid \mathbf{G}_{-i}, \mathbf{Y}))^{1-1/T}}
\end{aligned}
$$

using, in several steps, the fact that $G_{-i} = G_{-i}^*$. The Hastings ratio is thus as easily computed as the local conditionals $P_\theta(G_i \mid \mathbf{G}_{-i}, \mathbf{Y})$. An interesting feature of this system is that, with $T > 1$ the probability of change in $\mathbf{G}$ is reduced from that for the Gibbs sampler, $T = 1$ (C. Jennison, pers.comm. 1992). However, because this increases the probability that the sampler remains in low-probability states, it increases the overall probability of a succession of changes that moves $\mathbf{G}$ to a different part of the space. The probabilities of single-step changes are not necessarily indicative of overall performance of the sampler, particularly in high-dimensional spaces.

Under the assumption that S_{*j} are first-order Markov over loci j, the single-site meiosis indicator sampler is also easily implemented (Thompson

1994c). Since $S_{i,j}$ is binary, a Metropolis algorithm is natural. A meiosis i and locus j are selected at random, and a change from $S_{i,j} = s$ to $S_{i,j} = (1-s)$ is proposed. This proposal changes only the recombinant/non-recombinant status in the two intervals adjoining locus j, and the conditional probability of marker data at locus j:

$$
\begin{aligned}
h &= \frac{P_\theta(\mathbf{Y} \mid \mathbf{S}^*) P_\theta(\mathbf{S}^*)}{P_\theta(\mathbf{Y} \mid \mathbf{S}) P_\theta(\mathbf{S})} \\
&= \frac{P_\theta(Y_{*j} \mid S_{*j}^*) P_\theta(S_{i,j}^* \mid S_{i,j-1}, S_{i,j+1})}{P_\theta(Y_{*j} \mid S_{*j}) P_\theta(S_{i,j} \mid S_{i,j-1}, S_{i,j+1})} \\
&= \frac{P_\theta(Y_{*j} \mid S_{*j}^*)}{P_\theta(Y_{*j} \mid S_{*j})} \left(\frac{\rho_{j-1}}{1 - \rho_{j-1}}\right)^{T_{j-1}} \left(\frac{\rho_j}{1 - \rho_j}\right)^{T_j}
\end{aligned}
$$

where $T_{j-1} = (|S_{i,j-1} - s| - |S_{i,j-1} - 1 + s|)$ is the indicator of whether the proposal places ($T_{j-1} = +1$) or removes ($T_{j-1} = -1$) a recombination between locus $j - 1$ and j, $\rho_{j-1} = \Pr(S_{i,j-1} \neq S_{i,j})$ is the recombination frequency between locus $j - 1$ and locus j, T_j and ρ_j are similarly defined for the interval j to $j + 1$, and $\rho_0 = \rho_L = \frac{1}{2}$. The first term in the Hastings ratio h is given by equation (4.7) and is easily computed by the methods of section 4.1.2 provided there are not too many data Y_{*j} on the pedigree. Generally, the space of latent variables is smaller for $\mathbf{S}$ than for $\mathbf{G}$, and hence MCMC more effective. The sampler may not be irreducible (Sobel and Lange 1996), but there are many fewer constraints than with a genotypic sampler and irreducibility is often provable on a locus-by-locus basis (Thompson 1994c, Thompson and Heath 1999). Note that, provided recombination frequencies between adjacent loci are strictly positive, irreducibility is a single-locus issue.

4.3.4 Block updating; combining exact and MC computation

A major difficulty with MCMC methods is to ensure proper mixing of the samplers, and hence efficient Monte Carlo estimation. On large pedigrees, with models or data involving multiple linked loci, single-variable MCMC updating methods are not effective. Some approaches to improving mixing involve some combination of exact and Monte Carlo computation. One straightforward idea is simply to compute exactly on those parts of the pedigree on which this is possible (Thompson 1991). The results from peeling peripheral parts of the pedigree enter as potentials on nodes of the remaining core (Geyer and Thompson 1995). Another early approach was to use an integrated ("Rao-Blackwellized") estimator, integrating some of the latent variables out of the numerator and denominator of equation (4.29). Since this equation applies for arbitrary latent variables $\mathbf{X} = (\mathbf{Z}^{(1)}, \mathbf{Z}^{(2)})$ it will apply with any reduced set of variables $\mathbf{Z}^{(1)}$, provided the relevant "complete-data" likelihood

$$P_\theta(\mathbf{Y}, \mathbf{Z}^{(1)}) \;=\; \sum_{\mathbf{Z}^{(2)}} P_\theta(\mathbf{Y}, \mathbf{Z}^{(1)}, \mathbf{Z}^{(2)})$$

can be computed. Thompson (1994) used this idea to improve Monte Carlo likelihood analysis of genetic models with several latent heritable components.

Recently a variety of joint-updating schemes have been developed. For example, Jensen et al. (1995) update genotypes of blocks of individuals jointly at several loci. Heath (1997) and Thompson and Heath (1999) use the meiosis indicators $\mathbf{S} = \{S_{i,j}\}$. Heath (1997) updates jointly the components of S_{*j}, the indicators at a single locus j: the *L-sampler*. Thompson and Heath (1999) update jointly the components of S_{i*}, the meiosis indicators for all loci in a single meiosis i: the *M-sampler*. All these MCMC methods provide, directly or indirectly, realizations of the descent of genes in pedigrees and the genotypes of individuals, and hence Monte Carlo estimates of likelihoods for linkage and segregation analysis, and the probabilities of gene identity by descent and haplotype sharing conditional on observed trait and marker data $\mathbf{Y}$. In a Bayesian framework, in which the segregation and linkage parameters of genetic models are assigned prior probability distributions, the same methods provide posterior probabilities of linkage and trait gene effects and locations.

In the locus-by-locus sampler (L-sampler) first developed by Kong (1991a), all genotypes $G_{*j} = \{G_{i,j}\}$ at a single locus j are updated conditionally upon those at neighboring loci. Computationally the approach is analogous to the sequential imputation method of section 4.2.4, except that sampling is from the full conditional of G_{*j}. Heath (1997) has further developed the L-sampler, and widened its scope, using S_{*j} rather than G_{*j}. Because of the structure, this full conditional distribution

$$P_\theta(S_{*j} \mid \mathbf{S}_{-*j}, \mathbf{Y}) \;=\; P_\theta(S_{*j} \mid S_{*,j-1}, S_{*,j+1}, Y_{*j})$$

depends only on current values of $S_{*,j-1}$ and $S_{*,j+1}$ and data Y_{*j}. Thus, the calculation of $P_\theta(S_{*j} \mid \mathbf{S}_{-*j}, \mathbf{Y})$ is a single-locus peeling computation analogous to those of section 4.2.2, and is often feasible. The developments of Heath (1997) are in the context of Bayesian analyses of quantitative traits, under models of several loci contributing additively to the trait value. His approach uses a variety of improved sampling and computational ideas, including more efficient peeling algorithms, integrated proposal distributions (Besag et al. 1995) and reversible jump MCMC (Green 1995). However, the output are realizations of putative trait loci from a Bayesian posterior; no likelihood or lod score is obtained. One great advantage of the L-sampler is that it is irreducible, provided only that recombination probabilities between adjacent loci are strictly positive. Moreover, this MCMC sampling is a great improvement over single-site methods. However, when there are multiple tightly linked marker loci, mixing can be poor.

4.3.5 Tightly-linked loci: the M-sampler

The single-site $(S_{i,j})$ or single-locus (S_{*j}) update has mixing problems when loci are tightly linked. An alternative form of block-updating is to update all loci in a given meiosis (S_{i*}). The M-sampler is a whole-meiosis Gibbs sampler (Thompson and Heath 1999) for S_{i*}. To implement this we must compute

$$\Pr(S_{i*} \mid \{S_{k*}, k \neq i\}, \mathbf{Y})$$

As previously (section 4.1.3), we suppose that the marker data $\mathbf{Y}$ can be partitioned into data relating to each locus $j = 1, 2, ..., L$, and that the loci are numbered in order along the chromosome. Then

$$\mathbf{Y} = (Y_{*1}, ..., Y_{*L}).$$

As in section 4.1.3, let

$$Y^{(j)} = (Y_{*1}, ..., Y_{*j}), \quad \text{so } \mathbf{Y} = Y^{(L)}.$$

We have seen in section 4.1.2 that $\Pr(Y_{*j} \mid S_{*j})$ can be easily computed.
    Now define

$$Q_j^\dagger(s) = \Pr(S_{i,j} = s \mid \{S_{k*}, k \neq i\}, Y^{(j)}) \qquad (4.38)$$

for $s = 0, 1$. Note that this function $Q_j^\dagger(\cdot)$ is analogous, but not identical, to the function $Q_j^\dagger(\cdot)$ of section 4.1.4. There the probability was jointly for all components of S_{*j}, conditional on $Y^{(j)}$; here the probability is for $S_{i,j}$ conditioning additionally on indicators at other meioses $\{S_{k*}, k \neq i\}$. Since meiosis indicators S_{i*} are *a priori* independent over i, and become dependent only through conditioning on the data Y_{*j}, $Q_j^\dagger(s)$ is the probability for the meiosis indicator $S_{i,j}$, given the data $Y^{(j)}$ and other $(k \neq i)$ meiosis indicators at loci up to and including locus j. Thus, by analogy with section 4.1.4, $Q_j^\dagger(s)$ may be computed sequentially just as in equation (4.12). The only difference is that now, rather than considering all 2^m possible values of S_{*j}, we consider only values of the single binary indicator $S_{i,j}$, conditioning on the remainder $(k \neq i)$ which remain fixed. In meiosis i, there is no recombination between locus $(j-1)$ and locus j if the value $(s = 0, 1)$ of $S_{i,j}$ is the same as at locus $(j-1)$, and there is recombination if the values differ. That is

$$Q_1^\dagger(s) \propto \Pr(Y_{*1} \mid S_{*1})$$

and

$$Q_j^\dagger(s) \propto \Pr(Y_{*j} \mid S_{*j}) (Q_{j-1}^\dagger(s)(1 - \rho_{j-1})$$

$$+ Q_{j-1}^{\dagger}(1-s)\rho_{j-1}) \qquad (4.39)$$

for $j = 2, ..., L$. In this equation, S_{*j} takes the current value at meioses k other than i, and the value s for meiosis i, and where ρ_{j-1} is the recombination frequency between locus $j-1$ and locus j. Thus we may compute (4.39) for each j in turn, working forwards sequentially along the chromosome.

Finally we have computed

$$Q_L^{\dagger}(s) = \Pr(S_{i,L} = s \mid \{S_{k*}, k \neq i\}, \mathbf{Y} = Y^{(L)})$$

and thus $S_{i,L}$ may be sampled from this desired conditional distribution. Suppose now each $S_{i,l}$ has been successively sampled from the required distribution for $l = L, L - 1, ..., j + 1, j$. Then

$$\Pr(S_{i,j-1} = s \mid \{S_{k*}, k \neq i\}, \{S_{i,l}, l = j, ..., L\}, \mathbf{Y})$$
$$\propto Q_{j-1}^{\dagger}(s) \, (T_j \, \rho_{j-1} \, + \, (1 - T_j)(1 - \rho_{j-1})) \qquad (4.40)$$

where $T_j = |S_{i,j} - s|$ is the indicator of recombination in the interval $j - 1$ to j. Thus we may work backwards along the chromosome, sampling each $S_{i,j}$ in turn $(j = L, ..., 1)$, obtaining overall a joint realization of $S_{i,j}$, $j = 1, ..., L$ from its full conditional distribution given $\{S_{k*}, k \neq i\}$ and $\mathbf{Y}$. Again, this is directly analogous to the equation (4.16) of section 4.1.4.

The ways in which MCMC samplers can be extended, combined, and improved, are almost limitless. The M-sampler does not suffer poor mixing due to tightly linked loci, but can mix poorly where there are extended ancestral paths of descent in a pedigree. Additionally, the M-sampler may not be irreducible. Combination of the L-sampler and M-sampler (Heath and Thompson 1997) can help to avoid these problems. Whereas the L-sampler is often the more computationally intensive, and seems to take longer to achieve stable probability estimates, the M-sampler may simply fail to sample the part of the space containing the majority of the probability mass. Combining the two samplers, say in the ratio of 10 M-iterations to 1 L-iteration, can achieve more robust and reliable results with higher Monte Carlo precision.

Another improvement of the M-sampler is to update S_{i*} for several meioses i jointly. Analogously, the L-sampler might update jointly S_{*j} for several loci j. For the M-sampler this is computationally straightforward, even for joint updating of as many as 10 meioses ($2^{10} = 1024$) —the difficulty is in deciding the appropriate meioses to combine. For the L-sampler, on a complex pedigree, usually no more than two or three loci can be updated jointly.

Throughout this chapter we have ignored the fact that genetic maps differ between males and females: the order of loci is the same, but the recombination frequencies can differ quite widely. Linkage analysis computations should accommodate different values of recombination frequencies for males and females. For the M-sampler this is particularly straightforward, since each meiosis is in a male or in a female. The M-sampler can also incorporate more general meiosis models, including genetic interference, by using a Metropolis-Hastings acceptance/rejection step (Thompson, 1999).

4.4 Conclusion

Over the last ten years, genetic data have become increasingly complex. On the one hand, most simple Mendelian traits have been mapped, and studies of complex traits have become the major challenge. To resolve the genetic basis of these traits, and to localize the contributing genes, analysis of data on extended pedigrees is desirable. On the other hand, data are increasingly available at more and more linked DNA marker loci; only by using these data effectively will it be possible to detect and localize the genes contributing to genetically complex human diseases. Even with the increasing speeds of current computers, exact computation of probabilities and likelihoods is often infeasible.

Monte Carlo methods in general, and Markov chain Monte Carlo (MCMC) in particular, provide a solution in some cases, permitting accurate estimation of probabilities of data at multiple genetic loci observed on a subset of the members of an extended pedigree. Earlier MCMC methods had limited success, due to poor mixing and the many constraints inherent in the descent of genes in pedigrees. More recent methods have solved several of the earlier problems, and the range of genetic models which can used to analyze data using MCMC computations is increasingly rapidly. In this chapter we have focussed on a single trait locus and multiple linked genetic markers. However, the trait model can be extended, for example to oligogenic quantitative traits (Heath 1997), and so also can the model for the transmission of genes, to accommodate different genetic maps in males and females and genetic interference (Thompson 1999). Nevertheless there remain several challenges in developing effective and computationally feasible MCMC methods, particularly in the estimation of likelihood ratios and lod scores.

Acknowledgment

I am very grateful to Dr. David Balding and Dr. Sharon Browning for many helpful and detailed comments on an earlier draft. Their comments have greatly increased the clarity the text. Parts of this chapter derive from work with Dr. Simon Heath; specifically, figures 4.1, 4.2 and 4.3 are reproduced from Thompson and Heath (1999) and are due to Dr. Heath. The writing of this chapter was supported in part by NIH grant GM-46255.

4.5 References

Baum LE (1972) An inequality and associated maximization technique in statistical estimation for probabilistic functions on Markov processes. In O Shisha, ed., *Inequalities-III; Proceedings of the Third Symposium on Inequalities. University of California Los Angeles, 1969*, 1–8. Academic Press, New York.

Baum LE, Petrie T (1966) Statistical inference for probabilistic functions of finite state Markov chains. *Annals of Mathematical Statistics* 37:1554–1563.

Baum LE, Petrie T, Soules G, Weiss N (1970) A maximization technique occurring in the statistical analysis of probabilistic functions on Markov chains. *Annals of Mathematical Statistics* 41:164–171.

Besag JE, Green P, Higdon D, Mengerson K (1995) Bayesian computation and stochastic systems. *Statistical Science* 10:3–66.

Browning S (1999) Monte Carlo Likelihood Calculation for Identity by Descent Data. Ph.D. thesis, Department of Statistics, University of Washington.

Cannings C, Thompson EA, Skolnick MH (1978) Probability functions on complex pedigrees. *Advances of Applied Probability* 10:26–61.

— (1980) Pedigree analysis of complex models. In J Mielke, M Crawford, eds., *Current Developments in Anthropological Genetics*, 251–298. Plenum Press, New York.

Dempster AP, Laird NM, Rubin DB (1977) Maximum likelihood from incomplete data via the EM algorithm (with discussion). *Journal of the Royal Statistical Society* B 39:1–37.

Edwards AWF (1967) Automatic construction of genealogies from phenotypic information (AUTOKIN). *Bulletin of the European Society of Human Genetics* 1:42–43.

Elston RC, Stewart J (1971) A general model for the analysis of pedigree data. *Human Heredity* 21:523–542.

Geman S, Geman D (1984) Stochastic relaxation, Gibbs distributions, and the Bayesian restoration of images. *IEEE Transactions on Pattern Analysis and Machine Intelligence* 6:721–741.

Geyer CJ (1991a) Markov chain Monte Carlo maximum likelihood. In EM Keramidas, SM Kaufman, eds., *Computer Science and Statistics: Proceedings of the 23rd Symposium on the Interface*, 156–163. Interface Foundation of North America, Fairfax Station, VA.

— (1991b) Reweighting Monte Carlo Mixtures. Technical Report 568, School of Statistics, University of Minnesota.

— (1992) Practical Markov chain Monte Carlo (with discussion). *Statistical Science* 7:473–511.

Geyer CJ, Thompson EA (1995) Annealing Markov chain Monte Carlo with applications to ancestral inference. *Journal of the American Statistical Association* 90:909–920.

Goddard KA, Yu CE, Oshima J, Miki T, Nakura J, Piussan C, Martin GM, et al. (1996) Toward localization of the Werner syndrome gene by linkage disequilibrium and ancestral haplotyping: lessons learned from analysis of

35 chromosome 8p11.1-21.1 markers. *American Journal of Human Genetics* 58:1286–1302.

Green PJ (1995) Reversible jump Markov chain Monte Carlo computation and Bayesian model determination. *Biometrika* 82:711–732.

Guo SW, Thompson EA (1992) A Monte Carlo method for combined segregation and linkage analysis. *American Journal of Human Genetics* 51:1111–1126.

Haldane JBS, Smith CAB (1947) A new estimate of the linkage between the genes for colour-blindness and haemolphilia in man. *Annals of Eugenics* 14:10–31.

Hammersley JM, Handscomb DC (1964) *Monte Carlo Methods*. Methuen and Co., London.

Hastings WK (1970) Monte Carlo sampling methods using Markov chains and their applications. *Biometrika* 57:97–109.

Heath S, Thompson EA (1997) MCMC samplers for multilocus analyses on complex pedigrees. *American Journal of Human Genetics* 61:A278.

Heath SC (1997) Markov chain Monte Carlo segregation and linkage analysis for oligogenic models. *American Journal of Human Genetics* 61:748–760.

Heuch I, Li FMH (1972) PEDIG—A computer program for calculation of genotype probabilities, using phenotypic information. *Clinical Genetics* 3:501–504.

Hilden J (1970) GENEX—An algebraic approach to pedigree probability calculus. *Clinical Genetics* 1:319–348.

Irwin M, Cox N, Kong A (1994) Sequential imputation for multilocus linkage analysis. *Proceedings of the National Academy of Sciences USA* 91:11684–11688.

Jensen CS, Kjaerulff U, Kong A (1995) Blocking Gibbs sampling in very large probabilistic expert systems. *International Journal of Human Computer Studies* 42:647–666.

Kong A (1991) Analysis of pedigree data using methods combining peeling and Gibbs sampling. In EM Keramidas, SM Kaufman, eds., *Computer Science and Statistics: Proceedings of the 23rd Symposium on the Interface*, 379–385. Interface Foundation of North America, Fairfax Station, VA.

Kong A, Liu J, Wong WH (1994) Sequential imputations and Bayesian missing data problems. *Journal of the American Statistical Association* 89:278–288.

Kruglyak L, Daly MJ, Reeve-Daly MP, Lander ES (1996) Parametric and nonparametric linkage analysis: a unified multipoint approach. *American Journal of Human Genetics* 58:1347–1363.

Kruglyak L, Lander ES (1998) Faster multipoint linkage analysis using Fourier transforms. *Journal of Computational Biology* 5:1–7.

Lander ES, Green P (1987) Construction of multilocus genetic linkage maps in humans. *Proceedings of the National Academy of Sciences USA* 84:2363–2367.

Lange K (1997) Mathematical and Statistical Methods for Genetic Analysis. Statistics for Biology and Health. Springer Verlag, New York.

Lange K, Matthysse S (1989) Simulation of pedigree genotypes by random walks. *American Journal of Human Genetics* 45:959–970.

Lange K, Sobel E (1991) A random walk method for computing genetic location scores. *American Journal of Human Genetics* 49:1320–1334.

Lathrop GM, Lalouel JM, Julier C., Ott J (1984) Strategies for multilocus linkage analysis in humans. *Proceedings of the National Academy of Sciences USA* 81:3443–3446.

Lin S, Speed TP (1996) Incorporating crossover interference into pedigree analysis using the χ^2 model. *Human Heredity* 46:315–322.

Lin S, Thompson EA, Wijsman EM (1993) Achieving irreducibility of the Markov chain Monte Carlo method applied to pedigree data. *IMA Journal of Mathematics Applied in Medicine & Biology* 10:1–17.

— (1994) An algorithm for Monte Carlo estimation of genotype probabilities on complex pedigrees. *Annals of Human Genetics* 58:343–357.

MacCluer JW, VandeBerg JL, Read B, Ryder OA (1986) Pedigree analysis by computer simulation. *Zoo Biology* 5:147–160.

McPeek MS, Speed TP (1995) Modeling interference in genetic recombination. *Genetics* 139:1031–1044.

Mendel G (1866) Experiments in plant hybridisation. In JH Bennett, ed., *English Translation and Commentary by R. A. Fisher*. Oliver and Boyd, Edinburgh, 1965.

Metropolis N, Rosenbluth AW, Rosenbluth MN, Teller AH, Teller E (1953) Equations of state calculations by fast computing machines. *Journal of Chemical Physics* 21:1087–1092.

Morton NE (1955) Sequential tests for the detection of linkage. *American Journal of Human Genetics* 7:277–318.

Ott J (1979) Maximum likelihood estimation by counting methods under polygenic and mixed models in human pedigrees. *American Journal of Human Genetics* 31:161–175.

Ploughman LM, Boehnke M (1989) Estimating the power of a proposed linkage study for a complex genetic trait. *American Journal of Human Genetics* 44:543–551.

Sheehan NA (1990) Genetic Restoration on Complex Pedigrees. Ph.D. thesis, Department of Statistics, University of Washington.

Sheehan NA, Thomas AW (1993) On the irreducibility of a Markov chain defined on a space of genotype configurations by a sampling scheme. *Biometrics* 49:163–175.

Sobel E, Lange K (1993) Metropolis sampling in pedigree analysis. *Statistical Methods in Medical Research* 2:263–282.

— (1996) Descent graphs in pedigree analysis: Applications to haplotyping, location scores, and marker-sharing statistics. *American Journal of Human Genetics* 58:1323–1337.

Thompson EA (1974) Gene identities and multiple relationships. *Biometrics* 30:667–680

— (1981) Pedigree analysis of Hodgkin's disease in a Newfoundland genealogy. *Annals of Human Genetics* 45:279–292.

— (1986) *Pedigree Analysis in Human Genetics*. Johns Hopkins University Press, Baltimore.

— (1991) Probabilities on complex pedigrees: the Gibbs sampler approach. In EM Keramidas, SM Kaufman, eds., *Computer Science and Statistics: Proceedings of the 23rd Symposium on the Interface*, 321–328. Interface Foundation of North America, Fairfax Station, VA.

— (1994a) Monte Carlo estimation of multilocus autozygosity probabilities. In J Sall, A Lehman, eds., *Proceedings of the 1994 Interface Conference*, 498–506. Fairfax Station, VA.

— (1994b) Monte Carlo likelihood in genetic mapping. *Statistical Science* 9:355–366.

— (1994c) Monte Carlo likelihood in the genetic mapping of complex traits. *Philosophical Transactions of the Royal Society of London* B 344:345–351.

— (1999) MCMC estimation of multi-locus genome sharing and multipoint gene location scores. *International Statistical Review*, in press.

Thompson EA, Guo SW (1991) Evaluation of likelihood ratios for complex genetic models. *IMA Journal of Mathematics Applied in Medicine & Biology* 8:149–169.

Thompson EA, Heath SC (1999) Estimation of conditional multilocus gene identity among relatives. In F Seillier-Moiseiwitsch, ed., *Statistics in Molecular Biology and Genetics: Selected Proceedings of a 1997 Joint AMS-IMS-SIAM Summer Conference on Statistics in Molecular Biology*, IMS Lecture Note–Monograph Series Volume 33, 95–113. Institute of Mathematical Statistics, Hayward, CA.

Thompson EA, Kravitz K, Hill J, Skolnick MH (1978) Linkage and the power of a pedigree structure. In NE Morton, ed., *Genetic Epidemiology*, 247–253. Academic Press.

CHAPTER 5

Renormalization
of Interacting Diffusions

Frank den Hollander

University of Nijmegen

This chapter is a mini-review of some recent results on the large space-time behavior of infinite systems of interacting diffusions. Although the techniques that have been used come from stochastic analysis, the results are driven by renormalization ideas from statistical physics, centered around the notion of universality.

5.1 Introduction

Statistical physics is the bridge between the microscopic world of interacting particles and the macroscopic world of physical phenomena. Its goal is to describe the behavior on large space-time scales of stochastic systems with a large number of locally interacting components and to explain the different types of phenomena occurring in such systems. A central task of statistical physics is to explain *universality*, i.e., the experimentally observed fact that for many systems the large scale behavior is to a certain extent independent of the precise nature of the local interaction. One way to approach this task is via the method of *renormalization*, which will be demonstrated here on a specific class of models.

The fact that on large space-time scales stochastic systems show universality should come as no surprise. Intuitively one expects "averaging principles" similar in spirit to the law of large numbers to be in force. However, it turns out to be very hard to make this intuition precise in concrete examples, even though it should be quite general. [1]

Renormalization is the idea to look at the system along a *sequence of increasing space-time scales* and to show that *block averages* along this sequence have a dynamics that converges to a simple limiting dynamics

---

[1] A well-known example is the behavior of fluids, which is described by the so-called hydrodynamic equations. These equations are the same for all fluids, except for certain transport coefficients, like shear and bulk viscosity, thermal conductivity, etc.

as one moves up in the hierarchy of scales. Each step up in this hierarchy is associated with a *renormalization transformation*, acting on appropriate quantities, in which some information on the previous scale is averaged out. In this way the iterates of the renormalization transformation form the link between the microscopic and the macroscopic world. The universal limiting dynamics arises as the attracting fixed point or attracting orbit of the renormalization transformation. [2]

In what follows we will make the above intuition precise for a class of models of mutually attracting diffusions, namely, we will identify explicitly all the objects appearing in italics above.

5.2 The model

We consider a collection of diffusions

$$\{X_i(t): t \geq 0\}_{i \in \mathcal{L}} \tag{5.1}$$

indexed by a *lattice* $\mathcal{L}$ and taking values in a *state space* $\mathcal{S}$ that is a closed convex subset of $\mathbb{R}$. A typical choice is

$$\mathcal{L} = \mathbb{Z}^N \ (N \in \mathbb{N}), \quad \mathcal{S} = \mathbb{R}. \tag{5.2}$$

The diffusions evolve according to the following system of coupled SDE's:

$$dX_i(t) = \sum_{j \in \mathcal{L}} a(j - i)[X_j(t) - X_i(t)] \, dt + \sqrt{b(X_i(t))} \, dW_i(t) \tag{5.3}$$

$(i \in \mathcal{L}, t \geq 0)$.

Here, $a(\cdot)$ is a nonnegative function on $\mathcal{L}$ controlling the interacting strength between different components, $b(\cdot)$ is a nonnegative function on $\mathcal{S}$ controlling the diffusion rate of single components, and $\{W_i(t): t \geq 0\}_{i \in \mathcal{L}}$ is an i.i.d. collection of standard Brownian motions on $\mathcal{S}$. As the initial condition for (5.3) we choose the constant configuration:

$$X_i(0) = \theta \quad \forall i \in \mathcal{L} \quad \text{for some } \theta \in \mathcal{S}. \tag{5.4}$$

The right-hand side of (5.3) consists of two parts:

- an *interactive* part, where different components attract each other with a strength depending on their relative location;

- a *noninteractive* part, where each component follows an autonomous Brownian motion with a local diffusion rate.

---

[2] This short introduction focuses on renormalization in *non-equilibrium* statistical physics. Renormalization is also successfully applied in *equilibrium* statistical physics, for instance, to study Gibbs measures, percolation, etc. Moreover, various types of renormalization transformations other than block averages are possible, like majority rules, dilution rules, etc.

What makes the system in (5.3) interesting is that these parts compete: the first tends to make the configuration flat, the second tends to make it noisy.

For the system in (5.3) to make sense, some restrictions must be placed on $a(\cdot)$ and $b(\cdot)$. For instance, we must require that $0 < \sum_{i \in \mathcal{L}} a(i) < \infty$. Furthermore, in order for (5.3) to have a unique strong solution we must require that $b(\cdot)$ is Lipschitz continuous on $\mathcal{S}$. In addition, $b(\cdot)$ should behave properly near the boundary of $\mathcal{S}$, in order for the components not to leave $\mathcal{S}$ in a finite time.

5.3 Interpretation of the model

The system in (5.3) is a model for population dynamics, where we imagine that each site of $\mathcal{L}$ carries a large number of individuals.

Suppose that $\mathcal{S} = [0, 1]$. Then we have a model where each individual can be of two "types": $X_i(t)$ denotes the fraction of individuals at site i at time t of type 1 and $1 - X_i(t)$ the fraction of individuals of type 2. The states at different sites change due to two random mechanisms:

- *migration*: individuals move randomly between different sites;

- *resampling*: individuals reproduce randomly at single sites.

The system in (5.3) is the *continuum limit* where the number of individuals per site tends to infinity. The migration is controlled by $a(\cdot)$, the resampling by $b(\cdot)$. For more background the reader is referred to Sawyer and Felsenstein (1983), Ethier and Kurtz (1986) Chapter 10, Swart (1999) Chapter 1.

Suppose that $\mathcal{S} = [0, \infty)$. Then we have a model where each individual can have a certain "mass": $X_i(t)$ describes the mass of the population at site i at time t. Again, the states at different sites change due to migration and resampling.

5.4 Block averages and renormalization

As explained in Section 5.1, we want to look at the system in (5.3) along a sequence of increasing space-time scales. To that end we define the following *block averages*. For $n \in \mathbb{N}_0$, let

$$X_i^{(n)}(t) = \frac{1}{|B_i^{(n)}|} \sum_{j \in B_i^{(n)}} X_j(s_n t) \qquad (i \in \mathcal{L}, t \geq 0) \tag{5.5}$$

with

$$B_i^{(n)} = \{j \in \mathcal{L} : \|j - i\| \leq n\}. \tag{5.6}$$

This is the average of the components in a block of radius n centered at $i \in \mathcal{L}$, with time scaled up by a factor s_n. For the distance $\| \cdot \|$ we may choose some appropriate norm on $\mathcal{L}$, e.g. the Euclidean norm in case

$\mathcal{L} = \mathbb{Z}^N \, (N \in \mathbb{N})$. The function $n \mapsto s_n$ is called the *space-time scaling relation* and needs to be chosen appropriately depending on the model. Typically s_n increases with n, e.g. $s_n = |B_0^{(n)}|$.

If we fix n and substitute (5.5) into (5.3), then we find that the collection

$$\{X_i^{(n)}(t): t \geq 0\}_{i \in \mathcal{L}} \tag{5.7}$$

again satisfies a system of coupled SDE's, but with a more complicated structure. As n increases this system becomes gradually more involved, because as the blocks get larger more information is summed out. However, the intuitive idea is that in the limit as $n \to \infty$ the behavior of the system should simplify because of some underlying "averaging principle". This leads us to the following:

- PROGRAM: Show that, as $n \to \infty$,

$$\{X_0^{(n)}(t): t \geq 0\} \implies \{Y(t): t \geq 0\}, \tag{5.8}$$

and identify the limiting process explicitly.

Here, $\implies$ denotes weak convergence on path space, and 0 is chosen as a typical center of the n-blocks. The process $\{Y(t): t \geq 0\}$ is obviously important, because it describes our system in (5.3) on a large space-time scale: the limit $n \to \infty$ effectuates the link between the microscopic and the macroscopic level of description. Typically we may expect this limiting process to display universality properties: it should be to a certain extent independent of the particular choice of the coefficients $a(\cdot)$ and $b(\cdot)$ in (5.3).

In the sequel we will demonstrate that the above program can indeed be carried out successfully in a number of cases, with interesting conclusions.

5.5 The hierarchical lattice

It would be nice if we could work with the lattice $\mathcal{L} = \mathbb{Z}^N \, (N \in \mathbb{N})$. Unfortunately, this is at present too difficult. However, a complete story can be told when

$$\mathcal{L} = \Omega_N = \text{ the } N\text{-dimensional } \textit{hierarchical lattice} \tag{5.9}$$

with N large. The hierarchical lattice is defined as

$$\Omega_N = \left\{ i = (i_1, i_2, \ldots): i_k \in \{0, 1, \ldots, N-1\}, i_k \neq 0 \text{ finitely often} \right\}. \tag{5.10}$$

Think of Ω_N as the collection of infinite telephone numbers on N symbols ending with all 0's. With componentwise addition modulo N, Ω_N is a countable group. On Ω_N we have a natural norm:

$$\|i - j\| = \inf\{l \in \mathbb{N}_0: i_k = j_k \, \forall k > l\}, \tag{5.11}$$

called the *hierarchical distance*. Note that for each site in Ω_N the number of sites within distance l is $N^l \, (l \in \mathbb{N}_0)$.

One way to visualize Ω_N is through the following picture:

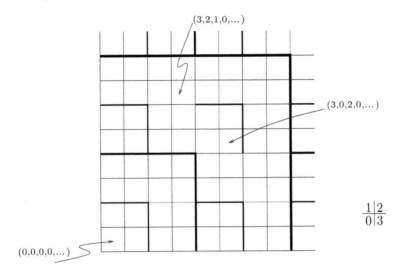

Figure 5.1 *The hierarchical lattice Ω_N for $N = 4$.*

For instance, site $i = (3,0,2,0,\ldots)$ is situated south-east in the 1-block around i, this 1-block is situated south-west in the 2-block around i, and this 2-block is situated north-east in the 3-block around i. This way of drawing the picture does justice to the hierarchical distance.

At first sight the choice $\mathcal{L} = \Omega_N$ may seem a bit artificial. However, this choice arises naturally in an application from genetics. What Ω_N does is organize the sites according to their "genetic information". The population is ordered into families, clans, villages, etc. Site i contains all individuals that are in family i_1 in clan i_2 in village i_3, etc. Individuals migrate between different sites, i.e., change family and/or clan and/or village, etc., and resample at single sites. See Sawyer and Felsenstein (1983).

For the state space S we will consider the following three choices:

$$S = [0,1], \ [0,\infty), \ \mathbb{R}. \tag{5.12}$$

For the interaction function $a(\cdot)$ on Ω_N we pick

$$a(0) = 0, \qquad a(i) = \sum_{l \geq \|i\|} c_l N^{1-2l} \qquad (i \neq 0), \tag{5.13}$$

where $c_1, c_2, \ldots \in (0,\infty)$ are constants that are assumed to satisfy

$$\sum_{l \in \mathbb{N}} \frac{1}{c_l} = \infty. \tag{5.14}$$

We also need to assume that $\sum_{l \in \mathbb{N}} c_l N^{-l} < \infty$, as can be seen from (5.16) below. Since $|\{i \in \Omega_N: \|i\| \le l\}| = N^l$ $(l \in \mathbb{N})$, what (5.13) says is that the total interaction strength between a given component and all the components at a given hierarchical distance decays by a factor N^{-1} per unit in the hierarchical distance. The reason for (5.14) will become clear later on. [3]

For the diffusion function $b(\cdot)$ on $\mathcal{S}$ we assume that it satisfies the following conditions:

- $\mathcal{S} = [0,1]$:

    (i) $b(0) = b(1) = 0$
    (ii) $b(x) > 0 \; \forall x \in (0,1)$
    (iii) b is Lipschitz on $[0,1]$.

- $\mathcal{S} = [0,\infty)$:

    (i) $b(0) = 0$
    (ii) $b(x) > 0 \; \forall x > 0$
    (iii) b is locally Lipschitz on $[0,\infty)$
    (iv) $b(x) = o(x^2) \; (x \to \infty)$.

- $\mathcal{S} = \mathbb{R}$:

    (i) b is bounded away from 0
    (ii) b is locally Lipschitz on $\mathbb{R}$
    (iii) $b(x) = o(x^2) \; (|x| \to \infty)$.

The class of such functions will be denoted by $\mathcal{B}$ in each of the three cases. For the space-time scaling relation we pick

$$s_n = N^n = |\{i \in \Omega_N: \|i\| \le n\}|. \qquad (5.15)$$

To exhibit the N-dependence, we write the components of (5.3) as $X_i^N(t)$, and the n-blocks in (5.5) as $X_i^{N,(n)}(t)$. In terms of the block averages the system in (5.3) with initial condition in (5.4) can be rewritten as

$$dX_i^N(t) = \sum_{n \ge 1} c_n N^{1-n} \left[X_i^{N,(n)}(t/N^n) - X_i^N(t) \right] dt + \sqrt{b(X_i^N(t))} \, dW_i(t)$$

$$X_i^N(0) = \theta \qquad (i \in \Omega_N, t \ge 0).$$

$$(5.16)$$

This representation shows that each component is linearly attracted to a weighted average of the block averages around itself, such that the weight

---

[3] Equation (5.14) is precisely the condition under which the random walk on Ω_N with jump rate $a(j-i)$ between sites i and j given by (5.13) is recurrent. This fact plays an important role in the universality properties. See also the last paragraph of Section 5.9.

decreases by a factor N^{-1} per unit in the block hierarchy. It also explains why we chose the interaction function $a(\cdot)$ to be of the form given in (5.13).

Apart from $\mathcal{S}$ and N, there are three parameters appearing in (5.16): θ, b and $(c_l)_{l \in \mathbb{N}}$. These parameters will play a key role in the renormalization approach that is described next.

5.6 The renormalization transformation

Let us have a closer look at equation (5.16) for $i = 0$ and see what happens when we pass to the limit $N \to \infty$, at least heuristically. Clearly, only the term with $n = 1$ survives. Moreover, the term $X_0^{N,(1)}(t/N)$ converges to θ for all $t \geq 0$. We therefore find that

$$\{X_0^N(t) \colon t \geq 0\} \Longrightarrow \{Y(t) \colon t \geq 0\} \tag{5.17}$$

with the right-hand side given by the SDE

$$dY(t) = c_1[\theta - Y(t)]\, dt + \sqrt{b(Y(t))}\, dW(t), \qquad Y(0) = \theta. \tag{5.18}$$

In other words, in the limit as $N \to \infty$ the single components decouple and follow a simple diffusion equation parametrized by θ, b and c_1.

It turns out that something similar happens with the n-blocks for $n \geq 1$, as is seen from the following result due to Dawson and Greven (1993a, 1993b, 1993c, 1996).

Theorem 5.1 *Fix* $\theta \in \mathcal{S}$, $b \in \mathcal{B}$, $(c_l)_{l \in \mathbb{N}} \subset (0, \infty)$ *and* $n \in \mathbb{N}_0$. *Then, as* $N \to \infty$,

$$\{X_0^{N,(n)}(t) \colon t \geq 0\} \Longrightarrow \{Y^{\theta, F^{(n)}b, c_{n+1}}(t) \colon t \geq 0\}, \tag{5.19}$$

where $\{Y^{y,b,c}(t) \colon t \geq 0\}$ *is the unique strong solution of the SDE*

$$dY(t) = c[y - Y(t)]\, dt + \sqrt{b(Y(t))}\, dW(t), \qquad Y(0) = y. \tag{5.20}$$

In (5.19), $F^{(n)} = F_{c_n} \circ \cdots \circ F_{c_1}$ *is the* n-*th iterate of a renormalization transformation* F_c, *acting on* b, *which is identified in Theorem 5.2.*

Theorem 5.2 *Fix* $c \in (0, \infty)$. *Then* F_c *acts on* $b \in \mathcal{B}$ *as follows:*

$$(Fb)(y) = \int_{\mathcal{S}} b(x)\nu^{y,b,c}(dx) \qquad (y \in \mathcal{S}). \tag{5.21}$$

Here, $\nu^{y,b,c}$ *is the unique equilibrium associated with* (5.20), *which is given by the formula*

$$\nu^{y,b,c}(dx) = \frac{\mu^{y,b,c}(x)dx}{\int_{\mathcal{S}} \mu^{y,b,c}(x)dx}, \tag{5.22}$$

where

$$\mu^{y,b,c}(x) = \frac{1}{b(x)} \exp\left[-\int_y^x \frac{c(z-y)}{b(z)}dz \right]. \tag{5.23}$$

The formula in (5.23) is the solution (modulo normalization) of the equation

$$c(y - x)\mu(x) = \frac{\partial}{\partial x}[b(x)\mu(x)], \tag{5.24}$$

which expresses the equilibrium property.

Here is a heuristic explanation of what is behind Theorems 5.1 and 5.2. By summing (5.16) over the components in a 1-block and using the special properties of the hierarchical distance, we get the system of SDE's:

$$dX_i^{N,(1)}(t) = \sum_{n \geq 2} c_n N^{2-n} \left[X_i^{N,(n)}(t/N^{n-1}) - X_i^{N,(1)}(t) \right] dt$$

$$+ \frac{1}{\sqrt{N}} \sum_{j \in \Omega_N: \, \|j-i\| \leq 1} \sqrt{b(X_j^N(Nt))} \, dW_j(t) \tag{5.25}$$

$$X_i^N(0) = \theta \qquad (i \in \Omega_N, t \geq 0).$$

Here we have scaled up time by a factor N, both in the 1-block and in the Brownian motions (note the factor $1/\sqrt{N}$). Three observations are now crucial:

– If a 1-block has value y, then each of the N components in this block is linearly attracted towards a value that is approximately y, since the attraction towards the values of the n-blocks with $n \geq 2$ is weak when N is large (recall (5.16)). This linear attraction is indicated in Fig. 2 by springs.

– On a time scale smaller than N each of the components in the 1-block approximately reaches the equilibrium ν^{y,b,c_1} (recall (5.17–5.18)).

– On the time scale N the value of the n-blocks with $n \geq 2$ is still very close to the initial value θ.

By combining these three observations we see that the diffusion rate of a 1-block at value y is given by $(F_{c_1} b)(y)$, the *average diffusion rate of a component* in the equilibrium ν^{y,b,c_1}, and that the drift is towards the initial value θ with drift strength c_2. This is precisely what (5.19-5.20) says for $n = 1$.

The same type of argument applies when we go from n-blocks to $(n+1)$-blocks, which is why the iterates $F^{(n)}$ appear in (5.19).

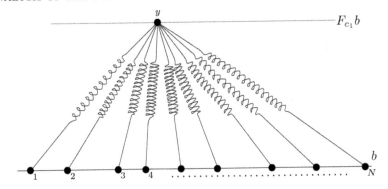

$$F_{c_1} b$$

$$b$$

Figure 5.2 *The renormalization transformation* F_{c_1}.

5.7 Analysis of the orbit

Theorems 5.1 and 5.2 show that on hierarchical space-time scale n the block averages behave like a diffusion with drift towards θ (the initial value in (5.4)), with drift strength c_{n+1}, and with diffusion function $F^{(n)}b$ (the n-th renormalized diffusion function), provided N is large enough. Thus, to study the large space-time behavior of our system it remains to study the orbit

$$(F^{(n)}b)_{n \in \mathbb{N}_0} \tag{5.26}$$

and to see whether this orbit converges to an attracting fixed point or an attracting orbit. This is not an easy task, because F_c is a *nonlinear integral transform*, as can be seen from (5.21–5.23). Fortunately, F_c turns out to have certain nice properties, which can be exploited to analyze the asymptotics of the orbit (see the explanation at the end of this section). The following results are due to Baillon et al. (1995, 1997) and show that, for each of the three choices of $\mathcal{S}$, the orbit in (5.26) indeed exists and convergences after appropriate scaling.

Theorem 5.3 $F_c \mathcal{B} \subseteq \mathcal{B}$ *for all* $c \in (0, \infty)$, *for each of the three classes* $\mathcal{B}$ *associated with* $\mathcal{S} = [0, 1], [0, \infty), \mathbb{R}$ *in Section 5.5.*

This result says that the iterates $F^{(n)}$ are well-defined on $\mathcal{B}$, so that indeed we have the right to speak about an orbit. In the following we abbreviate

$$\sigma_n = \sum_{l=1}^{n} \frac{1}{c_l} \qquad (n \in \mathbb{N}). \tag{5.27}$$

- $\mathcal{S} = [0, 1]$:

Theorem 5.4 *Suppose that* $b \in \mathcal{B}$. *Then* $\lim_{n \to \infty} \sigma_n F^{(n)} b = b^*$ *pointwise and uniformly, with* $b^*(x) = x(1 - x)$.

Interpretation: The n-blocks have a diffusion function that is close to $\frac{1}{\sigma_n}b^*$ for large n, irrespective of the diffusion function b for single components (recall Theorem 5.1). The scaling factor σ_n is independent of b. In other words, there is *complete universality* within the class $\mathcal{B}$, with the parabola b^* acting as the global attractor. The diffusion in (5.20) with $b = b^*$ is called the *Fisher-Wright diffusion*.

- $\underline{S = [0, \infty)}$:

Theorem 5.5 *(a) Suppose that $b \in \mathcal{B}$ satisfies $b(x) \sim dx$ $(x \to \infty)$ for some $d > 0$. Then*

$$\lim_{n \to \infty} F^{(n)}b = db^* \qquad \text{pointwise and locally uniformly,} \qquad (5.28)$$

with $b^*(x) = x$.

(b) Suppose that $b \in \mathcal{B}$ satisfies $b(x) \sim x^\alpha L(x)$ $(x \to \infty)$ with $\alpha \in (0, 2)\setminus\{1\}$ and with L slowly varying at infinity. Then there exist constants $0 < K_1 \leq K_2 < \infty$, depending only on α, such that

$$K_1 b^* \leq \liminf_{\sigma_n \to \infty} \frac{\sigma_n}{e_n} F^{(n)}b \leq \limsup_{\sigma_n \to \infty} \frac{\sigma_n}{e_n} F^{(n)}b \leq K_2 b^*$$

$$\text{pointwise and locally uniformly,} \qquad (5.29)$$

with $b^*(x) = x$. Here $(e_n)_{n \in \mathbb{N}}$ is defined by the relation

$$\frac{1}{\sigma_n} = \frac{b(e_n)}{e_n^2} \qquad (n \in \mathbb{N}). \qquad (5.30)$$

Interpretation: The n-blocks have a diffusion function that, modulo a constant, is close to $\frac{e_n}{\sigma_n}b^*$ for large n. Again b^* is the global attractor within the class $\mathcal{B}$, except that this time the scaling factor σ_n/e_n depends on the growth rate of b at infinity through the solution $(e_n)_{n \in \mathbb{N}}$ of (5.30). Although this solution need not be unique (because $x \mapsto b(x)/x^2$ need not be monotone), all solutions are asymptotically equivalent. Thus, there is *partial universality*: within the subclass of $\mathcal{B}$ consisting of all b with the same α and L the scaling is the same. The diffusion in (5.20) with $b = b^*$ is called *Feller's branching diffusion*. Bounds for K_1 and K_2 are known, but they are not sharp. It is conjectured that

$$K_1 = K_2 = (\alpha!)^{\frac{1}{2-\alpha}} 2^{\frac{1-\alpha}{2-\alpha}}. \qquad (5.31)$$

For the special case where $c_l \equiv 1$ and $L \equiv 1$, we find that $\sigma_n = n$ and $e_n \sim n^{1/(2-\alpha)}$ $(n \to \infty)$ and so

$$F^{(n)}b \asymp n^{-\frac{1-\alpha}{2-\alpha}} b^* \qquad (n \to \infty). \qquad (5.32)$$

Thus, for $\alpha \in (1, 2)$ the iterates $F^{(n)}b$ grow with n, while for $\alpha \in (0, 1)$ they shrink with n.

- $\underline{S} = \mathbb{R}$:

Theorem 5.6 *Suppose that $b \in \mathcal{B}$ satisfies $b(x) \to d$ $(|x| \to \infty)$ for some $d > 0$. Then*

$$\lim_{n \to \infty} F^{(n)} b = db^* \qquad \text{pointwise and locally uniformly,} \qquad (5.33)$$

with $b^*(x) = 1$.

Interpretation: The n-blocks have a diffusion function that is close to db^* for large n. This time b^* is the attractor for the subclass of $\mathcal{B}$ consisting of all b that are asymptotically constant, which is the analogue of Theorem 5.5(a). The diffusion in (5.20) with $b = b^*$ is called the *Ornstein-Uhlenbeck diffusion*. There is an analogue of Theorem 5.5(b) as well, but we will not discuss this here. [4]

Let us explain how Theorem 5.4 comes about. The key point is that F_c acts nicely on the parabola $x \mapsto b^*(x) = x(1 - x)$, as follows.

For $b \in \mathcal{B}$ and $c \in (0, \infty)$, let $K^{b,c}$ be the probability kernel on $\mathcal{S} = [0, 1]$ defined by

$$K^{b,c}(y, dx) = \nu^{y,b,c}(dx) \qquad (x, y \in [0, 1]). \qquad (5.34)$$

Then $F_c b = K^{b,c} b$ by (5.21). A direct computation based on the explicit formulas in (5.22–5.23) shows that

$$\int_0^1 x(1 - x) \nu^{y,b,c}(dx) = y(1 - y) - \frac{1}{c} \int_0^1 b(x) \nu^{y,b,c}(dx) \qquad (y \in [0, 1]), \qquad (5.35)$$

in other words,

$$K^{b,c} b^* = b^* - \frac{1}{c} K^{b,c} b. \qquad (5.36)$$

Next, define the iterates of (5.34) as follows:

$$K^{b,(n)} = K^{F^{(n-1)}b, c_n} \circ \cdots \circ K^{b,c_1} \qquad (n \in \mathbb{N}). \qquad (5.37)$$

Then $F^{(n)} b = K^{b,(n)} b$. Applying these iterates to (5.36), we find the relation

$$K^{b,(n)} b^* = b^* - \sigma_n K^{b,(n)} b \qquad (n \in \mathbb{N}). \qquad (5.38)$$

Now, we see from (5.38) that $0 \le \sigma_n K^{b,(n)} b \le b^*$. Since $\lim_{n \to \infty} \sigma_n = \infty$ by (5.14) and (5.27), it follows that $\lim_{n \to \infty} \|K^{b,(n)} b\| = 0$ with $\|\cdot\|$ the supremum norm. Since $b \in \mathcal{B}$ (see Section 5.5), the latter in turn implies that, for every $x \in [0, 1]$, $K^{b,(n)}(x, \cdot)$ concentrates more and more near the boundary of $[0, 1]$ as $n \to \infty$. But also $b^* \in \mathcal{B}$, and hence $\lim_{n \to \infty} \|K^{b,(n)} b^*\| = 0$. Looking back at (5.38), we thus see that

$$\lim_{n \to \infty} \|b^* - \sigma_n K^{b,(n)} b\| = 0, \qquad (5.39)$$

---

[4] The result in Theorem 5.6 is not actually stated in Baillon et al. (1997), but can be easily deduced from the formalism developed there.

which is the claim in Theorem 5.4.

To get Theorems 5.5 and 5.6, the above argument needs to be modified. The point is that F_c acts nicely on parabolas $x \mapsto [\lambda x(\mu - x)]_+$ $(\lambda, \mu > 0)$, but these parabolas are not in $\mathcal{B}$. Still, it is possible to use them as comparison objects. The details are technical and somewhat delicate, which is why Theorems 5.5 and 5.6 are less complete and the scaling depends on the choice of b.

In conclusion, the above three examples show that the program outlined in Section 5.4 can be made explicit for one-dimensional diffusions indexed by the hierarchical lattice in the limit of large dimension. The outcome shows that indeed there is universality, and that the limiting processes correspond to certain *special* types of diffusions.

5.8 Higher-dimensional state spaces

What happens if $\mathcal{S}$ is not one-dimensional but, say, is chosen to be

$$\mathcal{S} \subset \mathbb{R}^d \ (d \geq 2) \text{ convex and compact?} \tag{5.40}$$

This is an important generalization, because now the state $X_i^N(t)$ at site $i \in \Omega_N$ at time t can be used to describe $d+1$ different "types" of individuals (compare with the interpretation in Section 5.3). The typical example is when $\mathcal{S}$ is the d-dimensional simplex:

$$\mathcal{S} = \left\{ x = (x_1, \ldots, x_d) \in \mathbb{R}^d : x_m \geq 0, \sum_{m=1}^{d} x_m \leq 1 \right\}. \tag{5.41}$$

If $X_i(t) = x$, then x_m is the fraction of individuals at site i at time t of type $m = 1, \ldots, d$, while $1 - \sum_{m=1}^{d} x_m$ is the fraction of individuals of type $d + 1$. The class $\mathcal{B}$ is now specified by (compare with Section 5.5):

(i) $b(x) = 0 \ \forall x \in \partial \mathcal{S}$
(ii) $b(x) > 0 \ \forall x \in int(\mathcal{S})$
(iii) b is Lipschitz on $\mathcal{S}$.

The rest of the model is unchanged. [5]

The question we address is: Do Theorems 5.1–5.4 carry over? The answer is: Not immediately because the generalization to (5.40) causes major problems. The reason is twofold:

– Diffusions in higher-dimensional domains with boundaries are presently rather ill understood. For instance, it is not known whether (5.3) has a unique strong solution for all $b \in \mathcal{B}$, nor is it known whether (5.20) has

---

[5] In particular, the diffusion term in (5.3) remains a scalar and is not replaced by a $d \times d$ diffusion matrix. This choice corresponds to an isotropic model. Generalization to anisotropic models is possible. See Swart (2000).

a unique and ergodic equilibrium for all $b \in \mathcal{B}$. [6] These questions are so far settled only for a subclass of $\mathcal{B}$ (see open problem 1 in Section 5.9), but this subclass is too small for our purposes (e.g. it does not contain the b^* in (5.42) below).

- Even in the case of a unique and ergodic equilibrium for a given $b \in \mathcal{B}$, there is no explicit formula for the equilibrium measure $\nu^{y,b,c}$, like in (5.22–5.23) for the one-dimensional case. This means that there is no explicit formula for the renormalization transformation F_c either, like in (5.21) for the one-dimensional case. Consequently, one has to use somewhat indirect techniques to get a handle on the renormalization analysis.

The generalization to (5.40) has been analyzed by den Hollander and Swart (1998). There, Theorems 5.1–5.3 are essentially *assumed* to be true, and the work focusses on the analogue of Theorem 5.4. The following is the main result.

Theorem 5.7 *Suppose that $b \in \mathcal{B}$. Then $\lim_{n \to \infty} \sigma_n F^{(n)} b = b^*$ pointwise and uniformly, with b^* the unique solution of the equation*

$$\begin{cases} \frac{1}{2}\Delta b & = & -1 & \text{on int}(\mathcal{S}) \\ b & = & 0 & \text{on } \partial\mathcal{S}, \end{cases} \qquad (5.42)$$

where Δ is the Laplacian on $\mathbb{R}^d$.

Thus we again get complete universality, this time with b^* as the global attractor. The diffusion in (5.20) with $b = b^*$ is the analogue of the Fisher-Wright diffusion on $[0, 1]$ and depends on the shape of $\mathcal{S}$. The proof of Theorem 5.7 runs via (5.34), (5.37) and (5.38), which carry over.

There are as yet no analogues of Theorem 5.7 for noncompact domains, like $\mathcal{S} = [0, \infty)^d$. The problems alluded to above are even worse for this case, and the scaling with n is not known either. Hopefully some progress will come in the near future.

5.9 Open problems

The above results lead us to the following open problems:

1. Try to prove Theorems 5.1 and 5.2 for the choice in (5.40). This task has been taken up with limited success in den Hollander and Swart (1998), namely, only for a certain subclass of $\mathcal{B}$.

2. Try to extend Theorem 5.7 to $\mathcal{S} = [0, \infty)^d$ and $\mathcal{S} = \mathbb{R}^d$, i.e., try to prove the analogues of Theorem 5.5 and 5.6 for higher dimensions.

---

[6] In den Hollander and Swart (1998) it is shown though that, if (5.3) has a unique weak solution for a given $b \in \mathcal{B}$, then (5.20) has a unique and ergodic equilibrium for this b.

3. Try to replace Ω_N by a different lattice, e.g. $\mathbb{Z}^N$ $(N \in \mathbb{N})$. Do similar results hold in the limit as $N \to \infty$? The answer will certainly depend on the choice of the function $a(\cdot)$ controlling the interaction strength between different components. A different space-time scaling relation $n \mapsto s_n$ will be needed too.

Swart (2000) has proved universality in the *clustering* properties of the system in (5.3) under very general conditions. Namely, for any lattice $\mathcal{L}$ that is a finite or countable Abelian group, for any state space $\mathcal{S}$ that is a convex and compact subset of $\mathbb{R}^d$ $(d \geq 1)$ with a non-empty interior, for any diffusion function $b(\cdot)$ that is continuous on $\mathcal{S}$ with an appropriate behavior near $\partial\mathcal{S}$, and for any interaction function $a(\cdot)$ such that the symmetrized function $a(\cdot) + a(-\cdot)$ is the jump rate of a *recurrent* random walk on $\mathcal{L}$, the result is that the system clusters: it develops large blocks in which the components are almost identical. The limiting distribution of the single components can be identified explicitly and turns out to be independent of $a(\cdot)$ and $b(\cdot)$. In the proof the function $b^*(\cdot)$ solving (5.42) again plays a major role. This analysis considerably enhances our insight into what drives the universality.

5.10 Conclusion

What is described above is the implementation of renormalization ideas from statistical physics to analyze the large space-time behavior of a particular class of interacting diffusions. We have managed to actually renormalize the whole dynamics explicitly. There are very few examples of interacting particle systems for which a picture like the one explained in this paper has been corroborated in full detail, although physicists believe it should be quite general.

5.11 References

J. B. Baillon, Ph. Clément, A. Greven and F. den Hollander, On the attracting orbit of a non-linear transformation arising from renormalization of hierarchically interacting diffusions, Part I: The compact case, *Canad. J. Math.* 47 (1995) 3–27.

J. B. Baillon, Ph. Clément, A. Greven and F. den Hollander, On the attracting orbit of a non-linear transformation arising from renormalization of hierarchically interacting diffusions, Part II: The non-compact case, *J. Funct. Anal.* 146 (1997) 236–298.

D.A. Dawson and A. Greven, Multiple time scale analysis of hierarchically interacting systems, in: *Stochastic Processes*, A Festschrift in Honour of Gopinath Kallianpur (eds. S. Cambanis, J.K. Gosh, R.L. Karandikar and P.K. Sen), Springer, New York, 1993a, pp. 41–50.

D.A. Dawson and A. Greven, Multiple time scale analysis of interacting diffusions, *Probab. Theory Relat. Fields* 95 (1993b) 467–508.

D.A. Dawson and A. Greven, Hierarchical models of interacting diffusions: multiple time scale phenomena, phase transition and pattern of cluster-formation, *Probab. Theory Relat. Fields* 96 (1993c) 435–473.

D.A. Dawson and A. Greven, Multiple space-time scale analysis for interacting branching models, *Electronic J. of Prob.* 1 (1996), paper no. 14, pp. 1–84 (http://math.washington.edu/~ejpecp/).

S.N. Ethier and T.G. Kurtz, *Markov Processes; characterization and convergence*, John Wiley & Sons, New York, 1986.

F. den Hollander and J.M. Swart, Renormalization of hierarchically interacting isotropic diffusions, *J. Stat. Phys.* 93 (1998) 243–291.

S. Sawyer and J. Felsenstein, Isolation by distance in a hierarchically clustered population, *J. Appl. Probab.* 20 (1983) 1–10.

J.M. Swart, Clustering of linearly interacting diffusions and universality of their long-time distribution (2000). To appear in *Probab. Theory Relat. Fields*.

J.M. Swart, Large Space-Time Scale Behavior of Linearly Interacting Diffusions, thesis, University of Nijmegen (1999).

CHAPTER 6

Stein's Method for Epidemic Processes

Gesine Reinert

Cambridge University

This chapter considers a General Stochastic Epidemic with non-Markovian transition behaviour. At time $t = 0$, the population of total size K consists of aK individuals that are infected by a certain disease (and infectious); the remaining bK individuals are susceptible with respect to that disease. An initially susceptible individual i, when infected (call A_i^K its time of infection), stays infectious for a period of length r_i, until it is removed. An initially infected individual i stays infected for a period of length $\hat{r}_i$, until it is removed. Removed individuals can no longer be affected by the disease. A bound to the distance of the empirical measure

$$\xi_K = \frac{1}{K} \sum_{i=1}^{aK} \delta_{(0,\hat{r}_i)} + \frac{1}{K} \sum_{i=1}^{bK} \delta_{(A_i^K, A_i^K + r_i)},$$

describing the average path behaviour, to its mean field limit is established, using Stein's method. The bound, being in fact the first bound available for this mean field approximation for epidemics, gives explicit constants depending on the time length that the epidemic is observed, and on the total population size.

6.1 Introduction

In 1949, Bartlett introduced the General Stochastic Epidemic (GSE). This is a birth-death-process in a closed population where the temporal evolution of one individual depends "uniformly" on those of the others. In a closed population with K individuals, at time $t = 0$ a proportion a of the individuals is infected by a certain disease (and infective); the remaining $bK = (1 - a)K$ individuals are susceptible to that disease. Infectious individuals will get removed after some time, e.g., by lifelong immunity or death, and are then no longer affected by that disease. Thus we have an SIR model, where the abbreviation stands for "susceptible – infected – removed".

In Bartlett (1949), the general stochastic epidemic is defined as a Markov process $(X_K(t))_{t \geq 0}$, where $X_K(t) = (Y_K(t), Z_K(t))$, taking values in the set

$S = \{(r, s) : r \leq (1 - a)K, r + s \leq K, r, s \in \mathbb{N}\}$, with $X_K(0) = (K - a, a)$ and transition probabilities $(\alpha, \beta > 0)$

$$\mathbb{P}[X_K(t + \Delta t) = (r - 1, s + 1)|X_K(t) = (r, s)] = \alpha rs\Delta t + o(\Delta t)$$
$$\mathbb{P}[X_K(t + \Delta t) = (r, s - 1)|X_K(t) = (r, s)] = \beta s\Delta t + o(\Delta t) \quad (6.1)$$
$$\mathbb{P}[X_K(t + \Delta t) = (r, s)|X_K(t) = (r, s)] = 1 - (\alpha r + \beta)s\Delta t + o(\Delta t)$$

for $(r, s), (r, s - 1), (r - 1, s + 1) \in S$. Here, $Y_K(t)$ is interpreted as the number of susceptibles, and $Z_K(t)$ is the number of infectives, respectively, at time t. An infective individual is assumed to be infectious, during its infected period, and no multiple infections are allowed.

Often it is convenient to use the time-scale given by $\tau = \beta t$; the quantity $\rho = \frac{\beta}{\alpha}$, the ratio of removal-rate to infection-rate, is called the *relative removal-rate* and is a critical parameter for the epidemic; see Bailey (1975).

In the following, similar to the construction of Bartlett's GSE by Sellke (1983), we will consider a generalization that was first studied in Reinert (1995); we also apply the term "GSE" to this more general model.

From the viewpoint of an individual, the epidemic process can be described as follows. Let $(l_i, r_i)_{i \in \mathbb{N}}$ be a family of positive i.i.d. random vectors, and let $(\hat{r}_i)_{i \in \mathbb{N}}$ be a family of positive, i.i.d. random variables. Assume that the families $(l_i, r_i)_{i \in \mathbb{N}}$ and $(\hat{r}_i)_{i \in \mathbb{N}}$ are mutually independent. An initially infected individual i stays infectious for a period of length $\hat{r}_i$; then it is removed. (That the $\hat{r}_i$ need not have the same distribution as the r_i reflects the possibility that an infected individual has already been infectious for a certain period before, at time $t = 0$, it is observed.) An initially susceptible individual i, once infected, stays infectious for a period of length r_i, until it is removed. Furthermore, an initially susceptible individual i accumulates exposure to infection with a rate that depends on the evolution of the epidemic; if the total exposure reaches l_i, the individual i becomes infected. The possible dependence between l_i and r_i for each fixed i reflects the fact that both the resistance to infection and the duration of the infection may, for a fixed individual, depend on its physical constitution.

More precisely, an initially susceptible individual i gets infected as soon as a certain functional, depending on the course of the epidemic, exceeds the individual's level l_i of resistance; denote its infection time by A_i^K. If $I_K(t)$ denotes the proportion of infected individuals present in the population at time $t \in \mathbb{R}_+$, then A_i^K is given by

$$A_i^K = \inf \left\{ t \in \mathbb{R}_+ : \int_{(0,t]} \lambda(s, I_K) ds = l_i \right\},$$

for a certain function λ, acting on time as one variable and on a function path as the second variable; this function will be specified later.

Since, for epidemics, the length of the infectious period of an individual is usually very small compared to its life length, we neglect births and removals that are not caused by the disease, as well as any age-dependence of the infectivity or the susceptibility. Thus the model studied is a closed epidemic. Furthermore, the population is idealized to be homogeneously mixing.

From this construction, Bartlett's GSE can be recovered by choosing $\lambda(t, x) = x(t)$, (l_i) being i. i. d. $\exp(\alpha)$, and $(\hat{r}_i), (r_i)$ being i. i. d. $\exp(\beta)$ for some $\alpha > 0, \beta > 0$; for each i, l_i and r_i are independent. The same model is obtained by choosing $\lambda(t, x) = \alpha x(t)$, (l_i) being i. i. d. $\exp(1)$, and $(\hat{r}_i), (r_i)$ being i. i. d. $\exp(\beta)$ (in this sense, Sellke's construction is ambiguous). From now on, for Bartlett's GSE we shall use the parametrization $\lambda(t, x) = \alpha x(t)$, (l_i) being i. i. d. $\exp(1)$, and $(\hat{r}_i), (r_i)$ being i. i. d. $\exp(\beta)$.

For applications see, e.g., Berard et al. (1981), Swinton et al. (1998), to name but a few; often latency periods are included. This model is rather rigid in assuming a Markovian structure. In particular, in fungi infections for instance, the accumulation function should not be modeled as directly proportional to the number of infectives present at any given time, but should rather involve the entire history of the process. Moreover, it is often not reasonable to assume the duration of the infectious period to be exponentially distributed (the memoryless assumption is obviously frequently not appropriate); see, e.g., Keeling and Grenfell (1998). This motivated the above generalization.

In the literature, a generalization in the same direction has been investigated by Wang (1975, 1977), still assuming Markovian transition behaviour, i. e. the law of l_1 being $\mathcal{L}(l_1) = \exp(\alpha)$ for some $\alpha > 0$. Moreover, he assumed that l_i and r_i are independent for each individual, and he chose the special type of accumulation function $\lambda(t, x) = \lambda(x(t))$. For general λ, Solomon (1987) has discussed a related, age-dependent population model that deals only with one class of individuals, with an easier dependence structure. The case of variability among susceptibles, for a Markovian model, has been studied by Picard and Lefevre (1991), for example.

Typically, for epidemics the vector of the proportion of susceptibles, infected and removed individuals is studied. The information in this vector is limited; though. For instance, one might also be interested in the proportion of individuals that were infected before time s and are not removed before time t, that is the infectivity at time t in the population resulting from individuals that were infected before time s. This quantity is of interest if, for example, an immunization campaign was started in the population at time s.

For large populations, stochastic epidemic models are often approximated by their mean field limits, yielding deterministic epidemic models for the dynamic variables. This approximation being warranted only for large populations, bounds on the rate of convergence of the stochastic model towards the deterministic model are needed to make the mean field approximation useful in practice. Yet, so far there are no bounds given in the literature. Here we will derive explicit bounds for any finite time interval. We will show that there is an explicit constant C, depending on the length of the time interval for which we observe the epidemic, such that the distance (in a metric based on smooth test functions) will never be larger than this constant times $K^{-1/2}$. This bound is valid for any finite population.

The quantity we will approximate is not only the vector of the proportion of susceptibles, infected and removed, but rather the whole average path behaviour of the epidemic process, given by the empirical measure

$$\xi_K = \frac{1}{K} \sum_{i=1}^{aK} \delta_{(0,\hat{r}_i)} + \frac{1}{K} \sum_{i=1}^{bK} \delta_{(A_i^K, A_i^K + r_i)}.$$

As not necessarily (and hopefully) not every individual in the population gets infected by the disease, the total mass of ξ_K can be less than 1. Thus ξ_K is a substochastic measure on $[0,\infty)^2$, where the half-open interval $[u,v) \subset [0,\infty)$ is represented by the point $(u,v) \in [0,\infty)^2$. (In general, δ_x shall denote the Dirac measure at the point x). We could think of a point process with values in the positive quadrant $[0,\infty) \times [0,\infty)$, each point marking infection time and removal time for one individual. For a Borel set B, the empirical measure ξ_K then returns for B the number of points of this process that are in B.

From the empirical measure the proportion of infected and the proportion of susceptibles can easily be derived. For instance, if $t \geq 0$,

$$\xi_K([0,t] \times (t,\infty)) = \frac{1}{K} \sum_{i=1}^{aK} 1_{[0,\hat{r}_i)}(t) + \frac{1}{K} \sum_{i=1}^{bK} 1_{[A_i^K, A_i^K + r_i)}(t) =: I_K(t)$$

describes the proportion of infected present at time t. Moreover, we can also investigate quantities like

$$\xi_K([0,s] \times (t,\infty]), \quad t > s,$$

giving the proportion of individuals that were infected before time s and are not removed before time t. Thus we gain new insights concerning the behavior of the epidemic.

As a tool for deriving this result, we will employ Stein's method. In the context of empirical measures it has been derived in Reinert (1994). In Reinert (1995) it has been used to prove convergence of the average path behaviour to its mean field limit, but no bound on the rate of convergence has been given. This void will be filled in the present paper.

The novelties of the approach described here are thus the generality of the considered model (not assuming any Markovian structure), the generality of the quantities that can be approximated, and, perhaps most importantly, the first known bound on the quality of mean field approximations in epidemics.

For didactical purposes, the paper is organized as follows. In Section 6.2

we give a brief review of Stein's method. For measure-valued random elements, some technical background is described, and the main results of Stein's method for mean field limits are recalled. In Section 6.3 we apply Stein's method to the GSE, yielding an explicit bound on the distance of the GSE to its mean field limit. In Section 6.4 we discuss the above results and we point out directions of future research.

6.2 A brief introduction to Stein's method

Stein's method has first been introduced by Stein (1972) for proving normal approximations. This method avoids characteristic functions, but instead relies on a characterizing equation for the normal distribution. The distribution of any random variable would then be approximately normal if it satisfies the characterizing equation approximately. Indeed, the deviation from satisfying the characterizing equation exactly turns out to provide a measure on the distance to the normal distribution.

6.2.1 Stein's method for normal approximations

As this chapter is designed to give a brief introduction to Stein's method, we will first illustrate this method by its original example - normal approximations. First described in Stein (1972), in more detail it can be found in Stein (1986); improved bounds are in Baldi et al. (1989). It can be sketched as follows.

1. A random variable Z has standard normal distribution, that is, $\mathcal{L}(Z) = \mathcal{N}(0,1)$, if and only if for all smooth f,

$$\mathbb{E}\{f'(Z) - Zf(Z)\} = 0.$$

2. Let $\mathcal{L}(Z) = \mathcal{N}(0,1)$. For any smooth h there is a function $f = f_h$ solving

$$h(x) - \mathbb{E}h(Z) = f'(x) - xf(x) \quad (\textit{Stein equation})$$

   such that

$$\|f\| \ \leq \ \sqrt{\frac{\pi}{2}}\|h - \mathbb{E}h(Z)\|$$
$$\|f'\| \ \leq \ (\sup h - \inf h)$$
$$\|f^{(2)}\| \ \leq \ 2\|h'\|.$$

   Here, $\| f \| = \sup_{x \in \mathbb{R}} |f(x)|$ denotes the supremum norm.

3. So for any random variable W, any smooth h, substituting W for x in the Stein equation and taking expectations on both sides gives

$$\mathbb{E}h(W) - \mathbb{E}h(Z) = \mathbb{E}f'(W) - \mathbb{E}Wf(W). \tag{6.2}$$

Now if $\mathcal{H}$ is a convergence-determining class for weak convergence such that, for all $h \in \mathcal{H}$ we have that $f_h \in \mathcal{H}$, then taking absolute values in Equation (6.2) gives

$$\sup_{h \in \mathcal{H}} |\mathbb{E}h(W) - \mathbb{E}h(Z)| = \sup_{f \in \mathcal{H}} |\mathbb{E}f'(W) - \mathbb{E}W f(W)|,$$

and if the r.h.s., for some quantity $W = W_n$, tends to zero as $n \to \infty$, then we obtain a central limit theorem. In particular, we can choose $\mathcal{H}$ as the space of all continuous, bounded functions with piecewise continuous, bounded first derivatives. Similar results for nonsmooth test functions have been obtained by Bolthausen (1984), by Götze (1991), and by Rinott and Rotar (1996).

To see why this approach might be useful at all, consider a very classical example. Let $X, X_1, \ldots, X_n$ i.i.d. random variables with $\mathbb{E}X = 0$ and $\mathrm{Var}X = \frac{1}{n}$. Then

$$W = \sum_{i=1}^{n} X_i$$

has mean zero and variance 1. Put

$$W_i = W - X_i = \sum_{j \neq i} X_j.$$

To evaluate the right-hand side of Equation (6.2) we first calculate

$$
\begin{aligned}
\mathbb{E}W f(W) &= \sum_{i=1}^{n} \mathbb{E}X_i f(W) \\
&= \sum_{i=1}^{n} \mathbb{E}X_i f(W_i) + \sum_{i=1}^{n} \mathbb{E}X_i^2 f'(W_i) + R \\
&= \frac{1}{n} \sum_{i=1}^{n} \mathbb{E}f'(W_i) + R,
\end{aligned}
$$

where we used Taylor expansion, and for the Taylor remainder term R we have

$$|R| \leq \|f''\| \sum_{i=1}^{n} \mathbb{E}|X_i|^3.$$

So

$$\mathbb{E}f'(W) - \mathbb{E}W f(W) = \frac{1}{n} \sum_{i=1}^{n} \mathbb{E}\{f'(W) - f'(W_i)\} + R.$$

Applying Taylor expansion again we obtain the following result (see, e.g. Stein (1986)).

Theorem 6.1 *Let* $\mathcal{L}(Z) = \mathcal{N}(0,1)$. *For any smooth* h

$$|\mathbb{E}h(W) - \mathbb{E}h(Z)| \leq \|h'\| \left(\frac{2}{\sqrt{n}} + \sum_{i=1}^{n} \mathbb{E}|X_i^3| \right).$$

Note that the bound on the r.h.s. does not involve any asymptotic statement; rather, it is valid for any number n. Thus, even if convergence might not hold, a bound on the distance is still obtainable.

Stein's method has been particularly useful for proving results for sums of dependent observations. Consider, for example, the case that $X, X_1, \ldots, X_n$ are random variables with $\mathbb{E}X = 0$ and $\text{Var}X = \frac{1}{n}$, such that for each X_i there is a set S_i such that X_i is independent of $\sigma(X_j, j \notin S_i)$. (A special case would be m-dependent random variables.) For simplicity assume that $|S_i| = \gamma$ for some γ, and that $X_i \leq \frac{C}{\sqrt{n}}$ for some constant C that does not depend on i. Again put

$$W = \sum_{i=1}^{n} X_i.$$

If γ is small, then the summands are approximately independent, so that a normal approximation should hold. Indeed, we will prove the following result.

Theorem 6.2 *Let* $X, X_1, \ldots, X_n$ *be random variables with* $\mathbb{E}X = 0$ *and* $\text{Var}X = \frac{1}{n}$, *such that for each* X_i *there is a set* S_i *such that* X_i *is independent of* $\sigma(X_j, j \notin S_i)$. *Assume that* $|S_i| = \gamma$ *for some* γ. *Assume that* $X_i \leq \frac{C}{\sqrt{n}}$ *for some constant* C *that does not depend on* i. *Let* $\mathcal{L}(Z) = \mathcal{N}(0,1)$. *For all continuous, bounded functions* h *with piecewise continuous, bounded first derivatives,*

$$|\mathbb{E}h(W) - \mathbb{E}h(Z)|$$
$$\leq \|h'\| \frac{10\gamma^2 C^3}{\sqrt{n}} + (\sup h - \inf h) \sum_{i=1}^{n} \sum_{j \in S_i, j \neq i} \mathbb{E}|X_i X_j|.$$

Note that Theorem 6.2 gives an explicit bound in terms of neighborhood size and number of observations, as well as correlations. If the covariance term is large, one would rather approximate the sum with a normal distribution having as variance $\text{Var}(\sum_{i=1}^{n} X_i)$; then the second error term in Theorem 6.2 vanishes.

Proof of Theorem 6.2
Put

$$W_i = \sum_{j \notin S_i} X_j \quad \text{and} \quad W_{i,j} = \sum_{k \notin S_i \cup S_j} X_k.$$

Then W_i is independent of X_i, and $W_{i,j}$ is independent of X_i and X_j. Similarly as above, we expand the right-hand side of Equation (6.2). We have

$$
\begin{aligned}
\mathbb{E}W f(W) &= \sum_{i=1}^{n} \mathbb{E}X_i f(W) \\
&= \sum_{i=1}^{n} \mathbb{E}X_i f(W_i) + \sum_{i=1}^{n} \sum_{j \in S_i} \mathbb{E}X_i X_j f'(W_i) + R_1 \\
&= \sum_{i=1}^{n} \sum_{j \in S_i} \mathbb{E}X_i X_j f'(W_i) + R,
\end{aligned}
$$

where

$$
\begin{aligned}
|R_1| &\leq \|f''\| \sum_{i=1}^{n} \sum_{j \in S_i} \sum_{k \in S_i} \mathbb{E}|X_i X_j X_k| \\
&\leq \|f''\| \frac{\gamma^2 C^3}{\sqrt{n}}.
\end{aligned}
$$

Moreover,

$$
\sum_{i=1}^{n} \sum_{j \in S_i} \mathbb{E}X_i X_j f'(W_i) = \sum_{i=1}^{n} \sum_{j \in S_i} \mathbb{E}X_i X_j f'(W_{i,j}) + R_2,
$$

where

$$
\begin{aligned}
|R_2| &\leq \|f''\| \sum_{i=1}^{n} \sum_{j \in S_i} \sum_{k \in S_i \cup S_j} \mathbb{E}|X_i X_j| \mathbb{E}|X_k| \\
&\leq \|f''\| \frac{2\gamma^2 C^3}{\sqrt{n}}.
\end{aligned}
$$

Using Taylor expansion again, we obtain

$$
\begin{aligned}
\sum_{i=1}^{n} \sum_{j \in S_i} \mathbb{E}X_i X_j f'(W_{i,j}) &= \sum_{i=1}^{n} \sum_{j \in S_i} \mathbb{E}X_i X_j \mathbb{E}f'(W_{i,j}) \\
&= \sum_{i=1}^{n} \sum_{j \in S_i} \mathbb{E}X_i X_j \mathbb{E}f'(W) + R_3,
\end{aligned}
$$

with

$$
|R_3| \leq \|f''\| \frac{2\gamma^2 C^3}{\sqrt{n}};
$$

thus,

$$
\mathbb{E}W f(W) = Ef'(W) + \sum_{i=1}^{n} \sum_{j \in S_i, j \neq i} \mathbb{E}X_i X_j \mathbb{E}f'(W) + R_1 + R_2 + R_3.
$$

This yields the result.   □

A more detailed survey on Stein's method for normal approximations is in Reinert (1998a).

6.2.2 Stein's method in general

Stein's method in general can briefly be described as follows. Find a good characterization (that is, an operator $\mathcal{A}$) of the target distribution μ that is of the type

$$\mathcal{L}(X) = \mu \iff \mathbb{E}\mathcal{A}f(X) = 0, \text{ for all smooth functions } f,$$

where X is a random element, $\mathcal{L}(X)$ denotes the law of X and $\mathcal{A}$ is an operator associated with the distribution μ. Then assume X to have distribution μ, and consider the *Stein equation*

$$h(x) - \mathbb{E}h(X) = \mathcal{A}f(x), \quad x \in \mathbb{R}. \tag{6.3}$$

For every smooth h, find a corresponding solution f of this equation. Then, for any random element W, we obtain

$$\mathbb{E}h(W) - \mathbb{E}h(X) = \mathbb{E}\mathcal{A}f(W).$$

Hence, to estimate the proximity of W and X, it is sufficient to estimate $\mathbb{E}\mathcal{A}f(W)$ for all possible solutions f of (6.3). Moreover, bounding $\mathbb{E}\mathcal{A}f(W)$ for all smooth f automatically yields a bound on the distance of $\mathcal{L}(W)$ to μ in a smooth metric, regardless of asymptotics.

For μ being the standard normal distribution, the corresponding operator is $\mathcal{A}f(x) = f'(x) - xf(x)$. Of course the operator could also be defined as a second-order operator, namely $\mathcal{A}f(x) = f''(x) - xf'(x)$, the Ornstein-Uhlenbeck generator.

Barbour (1990) suggests employing as operator $\mathcal{A}$ in Equation (6.3) the generator of a Markov process, as this provides a way to look for solutions of (6.3). This is what in the following will be called the generator method. Suppose we can find a Markov process $(X(t))_{t \geq 0}$ with generator $\mathcal{A}$ and unique stationary distribution μ, such that $\mathcal{L}(X(t)) \overset{w}{\Rightarrow} \mu$ $(t \to \infty)$; here, $\overset{w}{\Rightarrow}$ denotes weak convergence. Then, if a random variable X has distribution μ,

$$\mathbb{E}\mathcal{A}f(X) = 0$$

for all f such that $\mathcal{A}f$ is defined. Now a method for solving Equation (6.3) is provided by Proposition 1.5 of Ethier and Kurtz (1986, p. 9; for the argument, see Barbour (1990)). Let $(T_t)_{t \geq 0}$ be the transition semigroup of the Markov process $(X(t))_{t \geq 0}$. Then

$$T_t h - h = \mathcal{A}\left(\int_0^t T_u h\, du \right).$$

As $(T_t)_{t\geq 0}$ is a strongly continuous contraction semigroup, $\mathcal{A}$ is closed (Ethier and Kurtz (1986), Corollary 1.6), and we could formally take limits:

$$h(x) - \mathbb{E}h(X) = -\mathcal{A}\left(\int_0^\infty T_u h\, du\right).$$

Thus $f = -\int_0^\infty T_u h\, du$ would be a solution of (6.3), if this expression exists and if $f \in \mathcal{D}(\mathcal{A})$. This will in general be the case only for a certain class of functions h. However, the latter conditions can usually be checked.

Stein's method has been generalized to many other distributions, foremost the Poisson distribution (see Chen (1975), Arratia et al. (1989), Barbour et al. (1992b), Aldous (1989), to cite but a few). Other distributions include the uniform distribution (Diaconis (1989)), the binomial distribution (Ehm (1991)), the compound Poisson distribution (Barbour et al. (1992a), Barbour and Utev (1998), Roos (1993)), the multinomial distribution (Loh (1992)), the gamma distribution (Luk (1994); for the χ^2 distribution see also Mann (1997)), the geometric distribution (Peköz (1995)) and, more generally, Pearson curves (Diaconis and Zabell (1991), Loh (1997)).

The most obvious advantage of Stein's method is that it yields immediate bounds. Moreover, in many situations where dependence comes into play the application is straightforward; many examples are of combinatorial nature. An early success of Stein's method is the work by Bolthausen (1984) for a combinatorial central limit theorem; he was the first to obtain the correct order for this approximation. In examples from random graph theory, where the method of moments used to be the most popular technique, Stein's method allowed not only to provide rates of convergence for the first time, but also to weaken conditions; see, for instance, Barbour et al. (1989). Another advantage of Stein's method is that it can also be used to derive lower bounds for the approximations; Hall and Barbour (1984) applied it to give lower bounds for the rate of convergence in the central limit theorem for independent random variables.

Unfortunately such a straightforward application of Stein's method may not yield the correct order for the rate of convergence; for an example, see, e.g. Reinert (1998a).

6.2.3 Stein's method for the weak law of large numbers

Here we will be interested in applying Stein's method to point mass in measure space, as developed in Reinert (1994), resulting in weak laws of large numbers. Point mass can be seen as an extreme case of the normal distribution with zero variance. Hence we put

$$\mathcal{A}f(x) = -xf'(x), \quad x \in \mathbb{R}.$$

Note that $\mathcal{A}$ is the generator of the deterministic Markov process $(Y(t))_{t\geq 0}$ given by

$$\mathbb{P}[Y(t) = xe^{-t} \mid Y(0) = x] = 1, \quad x \in \mathbb{R}.$$

The corresponding transition semigroup is

$$T_t h(x) = h(xe^{-t}),$$

and the unique stationary distribution is δ_0.

According to the general Equation (6.3), the Stein equation in this context is

$$h(x) - h(0) = -xf'(x), \quad x \in \mathbb{R}. \tag{6.4}$$

Let $C_b^2(\mathbb{R})$ be the space of all bounded, twice continuously differentiable real-valued functions on $\mathbb{R}$ with bounded first and second derivatives, and let $D_b^2(\mathbb{R})$ be the space of all twice continuously differentiable functions $f : \mathbb{R} \to \mathbb{R}$ with bounded first and second derivatives. Using the semigroup approach the following proposition is easy to derive.

Proposition 6.3 *For any $h \in C_b^2(\mathbb{R})$, there is a function $f = \phi(h) \in D_b^2(\mathbb{R})$ that solves the Stein equation (6.4) for h. Furthermore, for the derivatives, $\|f'\| \leq \|h'\|$, and $\|f''\| \leq \|h''\|$.*

Now we have all the ingredients to derive weak laws of large numbers.

Theorem 6.4 *Let $(X_i)_{i\in\mathbb{N}}$ be a family of random variables on $\mathbb{R}$, defined on the same probability space, with finite variances. Put*

$$Y_n = \frac{1}{n} \sum_{i=1}^{n} (X_i - \mathbb{E}X_i).$$

Then, for all $h \in C_b^2(\mathbb{R})$

$$|\mathbb{E}h(Y_n) - h(0)| \leq \|h''\| \, Var\Big(\frac{1}{n} \sum_{i=1}^{n} X_i\Big).$$

The proof of Theorem 6.4 follows easily using Taylor expansion (indeed, it is not even necessary to employ Stein's method).

As $C_b^2(\mathbb{R})$ is convergence-determining for weak convergence of the laws of real-valued random variables, a weak law of large numbers follows from Theorem 6.4 provided that

$$Var\Big(\frac{1}{n} \sum_{i=1}^{n} X_i\Big) \to 0 \quad (n \to \infty).$$

Moreover, we obtain an explicit bound on the distance to point mass.

6.2.4 Stein's method for the weak law in measure space

As the quantity ξ_K we want to approximate is a random element taking values in the space of substochastic measures, some technical details on convergence in this space are needed (see, e.g., Reinert (1994), (1998b)). Let E be a locally compact Hausdorff space with a countable basis (for instance, $E = \mathbb{R}^2$), let $\mathcal{E} = \mathcal{B}(E)$ be the Borel-σ-field of E, and let $M^b(E)$ the space of all bounded Radon measures on E, equipped with the vague topology. Let $C_c(E)$ be the space of real-valued continuous functions on E with support contained in a compact set. Abbreviate the integral

$$\langle \mu, \phi \rangle = \int_E \phi \, d\mu \,.$$

Convergence in the vague topology is defined as

$$\nu_n \overset{v}{\Rightarrow} \nu \iff \text{for all } f \in C_c(E) : \langle \nu_n, f \rangle \to \langle \nu, f \rangle \quad (n \to \infty).$$

For probability measures this differs from weak convergence in that the class of test functions is not $C_b(E)$, the class of bounded continuous functions on E. Thus, for example, if $E = \mathbb{R}$ and $\nu_n = \delta_n$, then $\nu_n \overset{v}{\Rightarrow} 0$, but $(\nu_n)_n$ does not converge weakly. Here, 0 denotes the measure that assigns measure 0 to any set. In what follows, weak convergence is denoted by $\overset{w}{\Rightarrow}$.

For $\mu \in M^b(E)$, set $\|\mu\| = \sup_{A \in \mathcal{E}} |\mu(A)|$. Let

$$M_1(E) = \{\mu \in M^b(E) : \mu \text{ positive}, \|\mu\| \leq 1\}$$

be the space of all positive Radon measures with total mass smaller or equal to 1 (the space of substochastic measures). As E has a countable basis, $M_1(E)$ is Polish with respect to the vague topology. Moreover, $M_1(E)$ is vaguely compact with a countable basis. Thus the considerations about vague convergence are valid for both $E = \mathbb{R}^2$ and $E = M_1(\mathbb{R}^2)$.

The next ingredient needed is a convergence-determining class of functions. Put

$$
\begin{aligned}
\mathcal{F} \quad := \quad & \{F \in C_b(M_1(\mathbb{R}^2)) : F \text{ has the form} \\
& F(\mu) = f(\langle \mu, \phi_1 \rangle, \ldots, \langle \mu, \phi_m \rangle) \qquad\qquad (6.5) \\
& \text{for an } m \in \mathbb{N}, f \in C_b^\infty(\mathbb{R}^m) \\
& \text{with } \|f'\| \leq 1, \|f''\| \leq 1, \|f'''\| \leq 1, \text{ and for} \\
& \phi_i \in C_b^\infty(\mathbb{R}^2) \text{ with } \|\phi_i\| \leq 1, \|\phi_i'\| \leq 1, i = 1, \ldots, m\}.
\end{aligned}
$$

Here the superscript $'$ indicates the total derivative, and the norm $\|\cdot\|$ indicates the sum of the supremum norms of the components, so that, for $f \in C_b^\infty(\mathbb{R}^m)$, the space of infinitely often differentiable continuous functions from $\mathbb{R}^m$ to $\mathbb{R}$ with bounded derivatives, $\|f'\| = \sum_{i=1}^m \|f_{(j)}\|$, where $f_{(j)}$ is the partial derivative of f in direction x_j.

This construction is similar to the algebra of polynomials used by Dawson (1993). In Reinert (1994) it is shown that this class of functions is convergence-determining for vague convergence. Thus it is a suitable class for Stein's method. One could even introduce a Zolotarev-type metric using $\mathcal{F}$; see Reinert (1998b) for details.

Now we have all the ingredients needed to set up Stein's method on the space of substochastic measures. Following Reinert (1994), the corresponding generator for the weak law of large numbers for random measures with target measure δ_μ, for some $\mu \in M_1(\mathbb{R}^2)$, is, for $F \in \mathcal{F}$ of the form (6.5)

$$\mathcal{A}F(\nu) = \sum_{i=1}^{m} f_{(i)}(\langle \nu, \phi_j \rangle, j = 1, \ldots, m)\langle \nu - \mu, \phi_i \rangle. \tag{6.6}$$

(This can be seen as a Gâteaux differential operator.) Moreover, the Stein equation has smooth solutions; we have

Proposition 6.5 *For any function $H \in \mathcal{F}$, there is a function $\psi(H) \in \mathcal{F}$ that solves the Stein equation with the operator (6.6) for H. If*

$$H(\nu) = h(\langle \nu, \phi_1 \rangle, \ldots, \langle \nu, \phi_m \rangle),$$

then

$$\psi(H)(\nu) = f(\langle \nu, \phi_1 \rangle, \ldots, \langle \nu, \phi_m \rangle)$$

for a function $f \in C_b^\infty(\mathbb{R}^m)$ with $\|f^{(k)}\| \leq \|h^{(k)}\|$ for all $k \in \mathbb{N}$.

Stein's method then immediately yields the following theorem.

Theorem 6.6 *Let ξ_n be a random element with values in $M_1(\mathbb{R}^2)$, let $\mu \in M_1(\mathbb{R}^2)$, and let $\mathcal{A}$ be as in (6.6). Then*

$$\sup\{|\mathbb{E}H(\xi_n) - H(\mu)| : H \in \mathcal{F}\} \quad \leq \quad \sup\{|\mathbb{E}\mathcal{A}F(\xi_n)| : F \in \mathcal{F}\}.$$

Many examples on how to apply this theorem are given in Reinert (1994). Here we will apply it to the GSE as constructed above.

6.3 The distance of the GSE to its mean field limit

6.3.1 Assumptions

As described in the introduction, let $(l_i, r_i)_{i \in \mathbb{N}}$ be a family of positive i.i.d. random vectors, let Ψ be the common distribution function of the $(l_i)_{i \in \mathbb{N}}$, let Φ be the common distribution function of the $(r_i)_{i \in \mathbb{N}}$, and let $(\hat{r}_i)_{i \in \mathbb{N}}$ be a family of positive, i.i.d. random variables with distribution function $\hat{\Phi}$ and law $\hat{\mu}$. Assume the $(l_i, r_i)_{i \in \mathbb{N}}$, $(\hat{r}_i)_{i \in \mathbb{N}}$ to be mutually independent (whereas, for each fixed i, l_i and r_i may be dependent). (In Reinert (1995), the $\hat{r}_i$ were allowed to have different distributions; however, although it would not be difficult to incorporate this inhomogeneity, for the sake of presentation it has been omitted.) Let $D_+ = \{x : [0, \infty) \to [-1, 1]$ right continuous with left-hand limits$\}$, and let $\lambda : \mathbb{R}_+ \times D_+ \to \mathbb{R}_+$ be the "accumulation" function. We use the notation $1_C(t)$ to denote the indicator function on the set C; the notation $I[t \in C]$ refers to the indicator of a set, not considered as a function. Then, for an initially susceptible individual i, its infection time A_i^K is given by

$$A_i^K = \inf\left\{t \in \mathbb{R}_+ : \int_{(0,t]} \lambda(s, I_K)ds = l_i\right\}, \tag{6.7}$$

with

$$I_K(t) = \frac{1}{K}\sum_{j=1}^{aK} 1_{[0,\hat{r}_j)}(t) + \frac{1}{K}\sum_{j=1}^{bK} 1_{[A_j^K, A_j^K + r_j)}(t)$$

being the proportion of infected individuals present in the population at time $t \in \mathbb{R}_+$. This gives a recursive definition of the A_i^K's: If $l_{(j)}$ is the jth order statistics of $l_1, ..., l_{bK}$, corresponding to the individual i_j, say, then

$$A_{i_j}^K = \inf\left\{t > 0 :\right.$$

$$\left. \int_{(0,t]} \lambda\left(s, \frac{1}{K}\sum_{m=1}^{aK} 1_{[0,\hat{r}_m)} + \frac{1}{K}\sum_{k=1}^{j-1} 1_{[A_{i_k}^K, A_{i_k}^K + r_{i_k})}\right) ds = l_{(j)}\right\}.$$

This completes the description of the model. Furthermore, we make some technical assumptions.

- The function $\lambda : \mathbb{R}_+ \times D_+ \to \mathbb{R}_+$ satisfies for all $t \in \mathbb{R}_+$, $x, y \in D_+$

  1. it is non-anticipating: $\lambda(t, x) = \lambda(t, x_t)$, where, for $t, u \in \mathbb{R}_+$, $x \in D_+$, $x_t(u) = x(t \wedge u)$;
  2. it is Lipschitz: there is a constant $\alpha > 0$ such that, for all t,

     $$|\lambda(t, x) - \lambda(t, y)| \le \alpha \sup_{0 \le s \le t} |x(s) - y(s)|; \tag{6.8}$$

  3. it is bounded: there is a constant $\gamma > 0$ such that, for all t,

     $$\sup_{0 \le s \le t} \lambda(s, x) \le \gamma;$$

  4. we have that, for all t,

     $$\lambda(t, x) = 0 \iff x(t) = 0.$$

- There is a constant $\beta > 0$ such that, for each $x \in \mathbb{R}_+$, the conditional distribution function $\Psi_x(t) := \mathbb{P}[l_1 \le t | r_1 = x]$ has a density $\psi_x(t)$ that is uniformly bounded from above by β;

  $$\psi_x(t) \le \beta \text{ for all } x \in \mathbb{R}_+, t \in \mathbb{R}_+. \tag{6.9}$$

  We assume that Ψ has a density ψ. (In Reinert (1995) it was only assumed that the conditional distribution function is Lipschitz continuous; however, imposing the existence of a density is more convenient.)

- We assume that $\hat{\Phi}(0) = 0$ and $\Phi(0) = 0$, so that infected individuals do not immediately get removed.

For Bartlett's GSE defined in Equations (6.1), the above assumptions are satisfied. Choosing $\lambda(t,x) = \alpha x(t)$, (l_i) being i.i.d. exp(1), and $(\hat{r}_i), (r_i)$ being i.i.d. exp(ρ), where for each i, l_i and r_i are independent, we obtain

$$|\lambda(t,x) - \lambda(t,y)| \leq \alpha \sup_{0 \leq s \leq t} |x(s) - y(s)|.$$

Moreover, λ is bounded by $\gamma = 1$, and $\mathbb{P}[l_1 \leq t | r_1 = x] = 1 - e^{-t}$ has density that is uniformly bounded by 1.

Note that we think of x as resembling the proportion of infected individuals. Thus the Condition 4. on λ guarantees that, once there are no infectives left, the epidemic stops. In Reinert (1995) this assumption was not made, which resulted in lengthy calculations.

6.3.2 Heuristics

To understand the argument for the limiting distribution, here is a heuristic explanation. Put

$$F_K(t) = \int_0^t \lambda(s, I_K)ds, \tag{6.10}$$

then

$$A_i^K = F_K^{-1}(l_i).$$

Moreover, the proportion $I_K(t)$ of infectives present at time t depends itself on F_K; we have

$$I_K(t) = \frac{1}{K}\sum_{i=1}^{aK} \mathbf{1}(\hat{r}_i > t) + \frac{1}{K}\sum_{j=1}^{bK} \mathbf{1}(F_K^{-1}(l_j) \leq t < F_K^{-1}(l_j) + r_j).$$

For $f \in D(\mathbb{R}_+)$, the space of right-continuous functions from $\mathbb{R}_+$ to $\mathbb{R}$ with left-hand limits, and for $t \in \mathbb{R}_+$, we define operators

$$\mathcal{Z}_K f(t) = \frac{1}{K}\sum_{i=1}^{aK} \mathbf{1}(\hat{r}_i > t) + \frac{1}{K}\sum_{j=1}^{bK} \mathbf{1}(f(t - r_j) < l_j \leq f(t)) \tag{6.11}$$

$$L_K f(t) = \int_{(0,t]} \lambda(s, \mathcal{Z}_K f)ds. \tag{6.12}$$

Then $F_K = L_K F_K$; thus F_K can be described as a fixed point of an operator. Moreover, for the operator $\mathcal{Z}_K$ the weak law of large numbers suggests as limiting operator, for $f \in C(\mathbb{R}_+), t \in \mathbb{R}_+$,

$$\mathcal{Z}f(t) = a(1 - \hat{\Phi}(t)) + b\mathbb{P}(f(t - r_1) < l_1 \leq f(t)). \tag{6.13}$$

Accordingly we define

$$Lf(t) = \int_{(0,t]} \lambda(s, \mathcal{Z}f)ds. \tag{6.14}$$

Using the Contraction Theorem it can be shown (see Reinert (1995)) that on each finite time interval the equation $Lf = f$ has a unique solution G. It thus heuristically follows that

$$F_K \approx G.$$

As G is deterministic, the desired limiting distribution of the process can now easily be derived, and is given in the next section.

6.3.3 Previous results

Under the above assumptions, in Reinert (1995) the following results were obtained. For the required measure μ, observe that during the course of the epidemic not necessarily (hopefully) every susceptible will get infected; $A_i^K = \infty$ for some i is possible. Therefore, if such a μ exists, it will in general not be a probability measure but a positive measure with total mass ≤ 1. Furthermore, as the existence of $\mathbb{E}r_i$ or $\mathbb{E}\hat{r}_i$, $i \in \mathbb{N}$, is not assumed, we restrict the observations to finite intervals $[0, T] \times [0, T]$ for a $T \in \mathbb{R}_+$ arbitrary, fixed. This leads to some notation. For $T \in \mathbb{R}_+$, put $[0, T]^2 = [0, T] \times [0, T]$, and $\mathcal{B}_T = \mathcal{B}([0, T]^2)$. Let $\nu \in M_1(\mathbb{R}_+^2)$, then

$$\nu^T = \nu|_{\mathcal{B}_T}$$

is the restriction of ν on $\mathcal{B}_T$ (hence, $\nu^T \in M_1([0, T]^2)$). For $A \in \mathcal{B}(\mathbb{R}^2)$, put

$$\nu^T(A) = \nu(A \cap [0, T]^2);$$

this defines ν^T also on $\mathcal{B}(\mathbb{R}^2)$. If in addition X is a random element with $\mathcal{L}(X) = \nu$, then, for all $T \in \mathbb{R}_+, f \in L_1(\nu), A \in \mathcal{B}(\mathbb{R}^2)$, the corresponding restrictions are

$$\mathbb{E}^T f(X) = \int f(x)\nu^T(dx)$$

$$\mathbb{P}^T[X \in A] = \int 1_A(x)\nu^T(dx)$$

$$\mathcal{L}^T f(X) = \mathcal{L}(f(X))|_{\mathcal{B}_T}.$$

Moreover, for any $T > 0$ we use the notation

$$\| f \|_T = \sup_{t \in [0,T]} |f(t)|.$$

Theorem 6.7 *For $T \in \mathbb{R}_+$, the equation*

$$f(t) = \int_{(0,t]} \lambda(s, \mathcal{Z}f)ds, \quad 0 \leq t \leq T, \tag{6.15}$$

has a unique solution G_T. This solution can be obtained by an iteration procedure: Choose an arbitrary $f_0 \in C([0, T])$, put $f_1 = Lf_0$, $f_n = Lf_{n-1}$

for $n \in \mathbb{N}$. Then,

$$\| f_n(t) - G_T(t) \|_T$$

$$\leq \frac{\left(\frac{b}{2}\right)^n}{1 - \frac{b}{2}} \, \frac{(1 + 4\alpha\beta T(n+1))^{\frac{I}{\eta}+2} - 1}{4\alpha\beta T(n+1)} \, \| f_0 - L f_0 \|_T,$$

where

$$\eta = \sup \left\{ t \leq T : \int_{(0,T]} (1 + \Phi(s))\, ds \leq \frac{1}{2\alpha\beta} \right\}.$$

Theorem 6.8 *For $T \in \mathbb{R}_+$, let G_T be the solution of (6.15) and $\tilde{\mu}^T \in M_1(\mathbb{R}_+^2)$ be given for $r, s \in (0, T]$ by*

$$\tilde{\mu}^T([0,r] \times [0,s]) = \int_{(0,(s-r)\vee 0]} \Psi_x(G_T(r)) d\Phi(x)$$

$$+ \int_{((s-r)\vee 0, s]} \Psi_x(G_T(s-x)) d\Phi(x).$$

Put

$$\mu^T = a(\delta_0 \times \hat{\mu})^T + b\tilde{\mu}^T.$$

Then

$$\frac{1}{K} \sum_{i=1}^{aK} \mathcal{L}^T((0, \hat{r}_i)) + \frac{1}{K} \sum_{i=1}^{bK} \mathcal{L}^T((A_i^K, A_i^K + r_i)) \overset{v}{\Rightarrow} \mu^T \quad (K \to \infty).$$

Note that it might be more intuitive to think of $\tilde{\mu}^T([0,r] \times [0,s])$ as

$$\tilde{\mu}^T([0,r] \times [0,s]) = \mathbb{P}^T[l_1 \leq G_T(r), l_1 \leq G_T(s - r_1)].$$

In practice, Theorem 6.7 thus provides a numerical iteration procedure for finding the limiting function G_T; from Theorem 6.8 it follows that this function suffices to find the deterministic approximation.

Theorem 6.9 *Let μ^T be as in Theorem 6.8. Then, for all $T \in \mathbb{R}_+$,*

$$\mathcal{L}(\xi_K^T) \overset{w}{\Rightarrow} \delta_{\mu^T} \quad (K \to \infty).$$

However, in Reinert (1995), no bound on the rate of convergence was given. As the proof employed the Glivenko-Cantelli theorem and thus almost-sure convergence results, below we will give a different proof of Theorem 6.9. First note, though, that Theorem 6.9 gives an approximation of the GSE by its mean field limit. To make this more obvious, in the next subsection we consider the proportion of infected individuals and the proportion of susceptible individuals.

The proportion of infected individuals and the proportion of susceptible individuals

As mentioned in the introduction, the proportion of infected individuals and the proportion of susceptible individuals can be reconstructed via ξ_K. For $t > 0$,

$$\xi_K([0,t] \times (t,\infty)) =: I_K(t)$$

gives the proportion of infected present at t,

$$\xi_K((t,\infty] \times [0,\infty]) = \frac{1}{K} \sum_{i=1}^{bK} I[A_i^K > t] =: S_K(t)$$

gives the proportion of susceptibles present at time t, and

$$\xi_K([0,t] \times [0,t]) =: R_K(t)$$

is the proportion of removed at time t. We obtain as weak limits for all $t \in \mathbb{R}_+$ with $\hat{\mu}(\{t\}) = 0$ and for all $T > t$ with $\mathbb{P}[r_1 = T - t] = 0$

$$
\begin{aligned}
R(t) &= \mathbb{P}\text{-}\lim_{K\to\infty} R_K(t) \\
&= a\hat{\Phi}(t) + b \int_{(0,t]} \Psi_x(G_T(t-x)) d\Phi(x) \\
S(t) &= \mathbb{P}\text{-}\lim_{K\to\infty} S_K(t) \\
&= b(1 - \Psi(G_T(t))) \\
I(t) &= \mathbb{P}\text{-}\lim_{K\to\infty} I_K(t) \\
&= a(1 - \hat{\Phi}(t)) + b\left(\Psi(G_T(t)) - \int_{(0,t]} \Psi_x(G_T(t-x)) d\Phi(x)\right).
\end{aligned}
$$

Moreover, additional information about the epidemic is provided. Suppose, for example, as mentioned in the introduction, that an epidemic is known to be taking place in a region, and that after some time t_0 every remaining susceptible in that region is immunized. Thus there are no new cases, though infectives may still be present. To decide at what time the region, which was probably put under quarantine, can be opened to the public again, we are interested in estimating the remaining infectivity in the population at times $s > t_0$. This is given by $\xi_K([0,t_0] \times (s,\infty))$, which is the proportion of individuals that were infected before the time t_0 and are still present in the population at time s. For large K and T,

$$\xi_K([0,t_0] \times (s,\infty))$$
$$\approx a(1 - \hat{\Phi}(s)) + b\left(\Psi(G_T(t_0)) - \int_{(0,s]} \Psi_x(G_T(s-x)) d\Phi(x)\right).$$

Thus, as soon as this expression is smaller than a certain critical level, we can abandon the isolation.

The above expressions for the limiting quantities can be translated into a system of deterministic differential equations. In the next subsection we will see that this system coincides with results previously obtained by Wang (1975, 1977) in the special case that he considers. Yet, without an explicit bound on the distance to the limit these approximations remain naive.

Example: A Markovian epidemic

The above asymptotic results can be compared with those obtained by Wang (1975), which seem to be the most general ones for the GSE known so far. As is made more explicit in Wang (1977), Wang (1975) considers a population of total size $N = K$ and assumed, in our notation,

1. $\mathbb{P}$ [a particular susceptible individual becomes infected during the time interval $[t, t + \delta t]] = \lambda(I_N(t))\Delta t + o(\Delta t)$; for a function λ that is positive, bounded and Lipschitz on $[0, 1]$, and $\lambda(0) = 0$.

2. $\mathbb{P}$[a particular infected individual stays infected for at least a period of length $t] = F(t)$; for a function F with $F(0) = 1$ and $F(t) \searrow 0$ as $t \to \infty$.

3. At time 0 there are $NI_N(0) = x(N)$ infected individuals $s_1, \cdots, s_{x(N)}$ present, where s_i represents also the total time that the ith individual has been infected up to time 0. There is assumed to exist a positive density $q \in L_1(\mathbb{R}_+)$ such that for all $s \in \mathbb{R}_+$

$$\lim_{N \to \infty} \frac{1}{N} \sum_{i=1}^{x(N)} 1_{[0,s]}(s_i) = \int_0^s q(u)du.$$

With

$$g(s,t) = \frac{F(t+s)}{F(s)} I[F(s) > 0],$$

$$\gamma(t) = \int_0^\infty g(s,t)q(s)ds,$$

Wang proves that $(I_N(t), I_N(t) + R_N(t))$ converges to the unique positive solution $(P(t), B(t))$ of the system

$$P(t) = \gamma(t) + \int_{(0,t]} \lambda(P(u))(1 - B(u))F(t - u)du$$

$$B(t) = P(0) + \int_{(0,t]} \lambda(P(u))(1 - B(u))du, \qquad (6.16)$$

in the sense that for every $\epsilon > 0$,

$$\lim_{N \to \infty} \mathbb{P}[\sup_{u \in [0,T]} |I_N(u) - P(u)| + |I_N(u) + R_N(u) - B(u)| > \epsilon] = 0.$$

Now consider in our model the special case that the (l_i) have an exp(1)-distribution, that l_i and r_i are independent for each i, that there are $s_i \in \mathbb{R}_+, i \in \mathbb{N}$ such that, with $\Phi(t) = 1 - F(t)$,

$$\mathbb{P}[\hat{r}_i > t] = \frac{1 - \Phi(t + s_i)}{1 - \Phi(s_i)}, \quad i \in \mathbb{N},$$

that $\lambda(t, x) = \lambda(x(t))$, and that there is a positive density $q \in L_1(\mathbb{R}_+)$ with

$$\lim_{K \to \infty} \frac{1}{K} \sum_{i=1}^{aK} 1_{[0,s]}(s_i) = \int_0^s q(u) du.$$

Under these assumptions, we recover Wang's model, as well as his characterization of the deterministic limit; a derivation can be found in Reinert (1995). However, our formulation covers a much wider class of models, and gives much more detailed information about the process.

In Bartlett's GSE, we moreover have $\lambda(t, x) = \alpha x(t)$, and that $s_i = 0$ for all i, so that $\gamma(t) = 0$ for all t. The above deterministic system then leads to the classical deterministic approximation; see, for example, Bailey (1975) or Isham (1991).

6.3.4 A bound on the distance to the mean field limit

In this subsection we will prove a refinement of Theorem 6.9 that will provide us with an explicit bound on the distance in the mean field approximation.

Theorem 6.10 Let μ^T be as in Theorem 6.8. Then, for all $T \in \mathbb{R}_+$, and for all $H \in \mathcal{F}$,

$$|\mathbb{E}^T H(\xi_K) - H(\mu)|$$

$$\leq \frac{\sqrt{a} + \sqrt{b}}{\sqrt{K}} + \alpha b \beta T (T + 2) \exp(b\lceil 2\alpha\beta T\rceil) \left\{ (1 + b)\sqrt{\frac{1}{K}} + \frac{2}{K} \right\},$$

where α and β are as in (6.8) and (6.9), and $\lceil x \rceil$ is the smallest integer larger than x.

Remarks. Note that, in Theorem 6.10, the constants α and β always occur as a combination. This relates to the ambiguity in Sellke's construction, referred to before: we could choose $l_i \sim \exp(\beta)$ and $\lambda(t, x) = \alpha x(t)$, or, equally, choose $l_i \sim \exp(1)$ and $\lambda(t, x) = \alpha\beta x(t)$, for example. Furthermore, the bound does not depend on the distribution of the infectious period. This is due to Lemma 6.11, a uniform convergence result. Of course the distribution of the infectious period is reflected in μ.

Theorem 6.10 gives the result in terms of expectations. Using Markov's inequality, we also obtain information for the distribution function. Namely,

$$P\left(|\mathbb{E}^T H(\xi_K) - H(\mu)| > \frac{\log K}{\sqrt{K}}\right)$$

$$\leq \frac{1}{\log K}\left\{\sqrt{a} + \sqrt{b} + \alpha b \beta T(T+2)\exp(b\lceil 2\alpha\beta T\rceil)\left\{(1+b) + \frac{2}{\sqrt{K}}\right\}\right\}.$$

As the proof of Theorem 6.10 employs uniform convergence for the empirical distribution function, we first prove the following bound.

Lemma 6.11 *Let $X_1, \ldots, X_n$ be i. i. d. real-valued random variables from a distribution with distribution function F, and let F_n denote the empirical distribution function*

$$F_n(t) = \frac{1}{n}\sum_{i=1}^{n}\mathbf{1}(X_i \leq t).$$

Then

$$\mathbb{E}\sup_{x\in\mathbb{R}}|F_n(x) - F(x)| \leq \frac{1}{\sqrt{n}}.$$

Proof of Lemma 6.11
To prove this lemma, we employ the following bound from Massart (1990). For all $\epsilon > 0$,

$$\mathbb{P}(\sup_{x\in\mathbb{R}}|F_n(x) - F(x)| > \epsilon) \leq 2e^{-2n\epsilon^2}.$$

The above bound is trivial for $\epsilon \leq \sqrt{\frac{\ln 2}{2n}}$. Thus we have

$$\mathbb{E}\sup_{x\in\mathbb{R}}|F_n(x) - F(x)| = \int_0^\infty \mathbb{P}(\sup_{x\in\mathbb{R}}|F_n(x) - F(x)| > \epsilon)d\epsilon$$

$$\leq \sqrt{\frac{\ln 2}{2n}} + \int_{\sqrt{\frac{\ln 2}{2n}}}^1 2e^{-2n\epsilon^2}d\epsilon.$$

Using a change of variable we obtain

$$\mathbb{E}\sup_{x\in\mathbb{R}}|F_n(x) - F(x)| < \sqrt{\frac{\ln 2}{n}} + \sqrt{\frac{2\pi}{n}}(1 - \Phi_{\mathcal{N}(0,1)}(\sqrt{2\ln 2}))$$

$$< \frac{1}{\sqrt{n}},$$

where $\Phi_{\mathcal{N}(0,1)}$ denotes the standard normal distribution function. □

Secondly, we shall need a bound on the expected uniform distance between the stochastic function F_K and its deterministic approximation G_T. More precisely, we consider the distance between a slight perturbation of F_K and G_T, caused by omitting individual 1 from the population. To make this precise, put

$$\mathcal{H}_T = D([0, T]),$$

the space of right-continuous functions with left-hand limits from $[0, T]$ to $\mathbb{R}$. Similarly to (6.10), (6.11), and (6.12), define for $h \in \mathcal{H}_T$ the operators

$$\mathcal{Z}_{K,1}h(t) \;=\; \frac{1}{K}\sum_{i=1}^{aK}\mathbf{1}(\hat{r}_i > t) + \frac{1}{K}\sum_{j=2}^{bK}\mathbf{1}(h(t - r_j) < l_j \le h(t))$$

$$L_{K,1}h(t) \;=\; \int_{(0,t]} \lambda(s, \mathcal{Z}_{K,1}h)\,ds.$$

and let $F_{K,1}$ be the unique fixed point of $L_{K,1}h = h$.

Lemma 6.12 *Let $F_{K,1}$ be the unique solution of the equation $L_{K,1}h = h$ on $[0, T]$, and let G_T be the unique solution of Equation (6.15) on $[0, T]$. Then*

$$\mathbb{E}\,\|\,F_{K,1} - G_T\,\|_T \;\le\; \alpha T \exp(b\lceil 2\alpha\beta T\rceil)\left\{(1 + b)\sqrt{\frac{1}{K}} + \frac{2}{K}\right\},$$

where α, β and γ are as in (6.8) and (6.9), and $\lceil x \rceil$ is the smallest integer larger than x.

Proof of Lemma 6.12

To bound this quantity, we proceed by an inductive argument, similarly to the proof of Theorem 6.7 in Reinert (1995). For any t we have

$$\begin{aligned}
F_{K,1}(t) - G_T(t) &= L_{K,1}F_{K,1}(t) - LG_T(t) \\
&= (L_{K,1}F_{K,1}(t) - LF_{K,1}(t)) + (LF_{K,1}(t) - LG_T(t)),
\end{aligned}$$

so that, for all $t \le T$,

$$\mathbb{E}\,\|\,F_{K,1} - G_T\,\|_t \;\le\; \mathbb{E}\sup_{h\in\mathcal{H}_t}\|\,L_{K,1}h - Lh\,\|_T + \mathbb{E}\,\|\,LF_{K,1} - LG_T\,\|_t.$$

First we bound $\mathbb{E}\sup_{h\in\mathcal{H}_t}\|\,L_{K,1}h - Lh\,\|_T$.

Bound on $\mathbb{E}\sup_{h\in\mathcal{H}_t}\|\,L_{K,1}h - Lh\,\|_T$

For $h \in \mathcal{H}_T$ we have, due to the Lipschitz property of λ, that

$$\begin{aligned}
\|\,L_{K,1}h - Lh\,\|_T &\le \alpha\int_0^T \sup_{s\le x}|\mathcal{Z}_{K,1}h(s) - \mathcal{Z}h(s)|\,dx \\
&\le \alpha T\left(aR_2 + 2bR_3 + \frac{2}{K}\right),
\end{aligned}$$

where

$$R_2 = \sup_s \left| \frac{1}{aK} \sum_{i=1}^{aK} \mathbf{1}(\hat{r}_i \le s) - \hat{\Phi}(s) \right|$$

and

$$R_3 = \sup_s \left| \frac{1}{bK-1} \sum_{i=2}^{bK} \mathbf{1}(l_i \le s) - \Psi(s) \right|.$$

From Lemma 6.11 we have that for both $u = 1$ and $u = 2$, $\mathbb{E}R_u \le \sqrt{\frac{1}{K}}$. As $a + b = 1$, this yields

$$\mathbb{E} \sup_{h \in \mathcal{H}_t} \| L_{K,1}h - Lh \|_T \le aT \left\{ (1+b)\sqrt{\frac{1}{K}} + \frac{2}{K} \right\}$$

$$= : S(K). \tag{6.17}$$

Bound on $\mathbb{E} \| LF_{K,1} - LG_T \|_t$

Now we bound $\mathbb{E} \| LF_{K,1} - LG_T \|_t$. We have

$$|LF_{K,1}(t) - LG_T(t)|$$

$$\le \ ab \int_0^t \sup_{s \le x} \Big| \Psi(F_{K,1}(s)) - \Psi(G_T(s))$$

$$+ \int_0^s \{ \Psi_u(F_{K,1}(s-u)) - \Psi_u(G_T(s-u)) \} \mathbb{P}(r_1 \in du) \Big| dx$$

$$\le \ ab\beta \int_0^t \| F_{K,1} - G_T \|_x (1 + \Phi(x)) dx.$$

Thus we obtain

$$\mathbb{E} \| F_{K,1} - G_T \|_t$$

$$\le \ S(K) + ab\beta \int_0^t \mathbb{E} \| F_{K,1} - G_T \|_x (1 + \Phi(x)) dx. \tag{6.18}$$

Fix $c \ge b$ arbitrary, and define

$$\eta = \frac{1}{2ca\beta}. \tag{6.19}$$

Then we have that

$$2\eta a\beta b \le \frac{b}{c}.$$

Hence, as $\Phi(s) \le 1$ always,

$$\mathbb{E} \| F_{K,1} - G_T \|_\eta \le S(K) + \frac{b}{c}\mathbb{E} \| F_{K,1} - G_T \|_\eta;$$

yielding

$$
\begin{aligned}
\mathbb{E} \parallel F_{K,1} - G_T \parallel_\eta \; &\leq \; \frac{1}{1 - \frac{b}{c}} S(K) \\
&= \; \frac{c}{c-b} S(K).
\end{aligned}
$$

We now prove by induction that, for any $k \in \mathbb{N}$,

$$
\mathbb{E} \parallel F_{K,1} - G_T \parallel_{k\eta} \; \leq \; \left(\frac{c}{c-b} \right)^k S(K). \tag{6.20}
$$

The case $k = 1$ has already been proven above. Suppose (6.20) is true for k. Then, from (6.18),

$$
\begin{aligned}
\mathbb{E} &\parallel F_{K,1} - G_T \parallel_{(k+1)\eta} \\
&\leq \; S(K) + \alpha b \beta \int_0^{(k+1)\eta} \mathbb{E} \parallel F_{K,1} - G_T \parallel_x (1 + \Phi(x)) dx \\
&\leq \; S(K) + \alpha b \beta \sum_{l=1}^{k} \int_{(l-1)\eta}^{l\eta} \mathbb{E} \parallel F_{K,1} - G_T \parallel_x (1 + \Phi(x)) dx \\
&\quad + \alpha b \beta \int_{k\eta}^{(k+1)\eta} \mathbb{E} \parallel F_{K,1} - G_T \parallel_x (1 + \Phi(x)) dx \\
&\leq \; S(K) + \alpha b \beta \eta \sum_{l=1}^{k} \left(\frac{c}{c-b} \right)^l S(K) \\
&\quad + \alpha b \beta \int_{k\eta}^{(k+1)\eta} \mathbb{E} \parallel F_{K,1} - G_T \parallel_x (1 + \Phi(x)) dx,
\end{aligned}
$$

where we used the induction assumption. From the definition (6.19) of η we now obtain

$$
\begin{aligned}
\mathbb{E} &\parallel F_{K,1} - G_T \parallel_{(k+1)\eta} \\
&\leq \; S(K) + \frac{b}{c} \sum_{l=1}^{k} \left(\frac{c}{c-b} \right)^l S(K) + \frac{b}{c} \mathbb{E} \parallel F_{K,1} - G_T \parallel_{(k+1)\eta};
\end{aligned}
$$

thus

$$
\begin{aligned}
\mathbb{E} \parallel F_{K,1} - G_T \parallel_{(k+1)\eta} \; &\leq \; \frac{1}{1 - \frac{b}{c}} S(K) \left(1 + \frac{b}{c} \sum_{l=1}^{k} \left(\frac{c}{c-b} \right)^l \right) \\
&= \; \left(\frac{c}{c-b} \right)^{k+1} S(K).
\end{aligned}
$$

This proves (6.20). Denoting by $\lceil \frac{T}{\eta} \rceil$ the smallest integer at least as large as $\frac{T}{\eta}$, we have hence shown that, with (6.19),

$$
\mathbb{E} \parallel F_{K,1} - G_T \parallel_T \quad \leq \quad \left(\frac{c}{c-b}\right)^{\lceil \frac{T}{\eta} \rceil} S(K)
$$

$$
= \quad \exp(\lceil 2c\alpha\beta T \rceil (\ln(c) - \ln(c-b)))S(K).
$$

As $c > b$ was arbitrarily chosen, we may take the limit $c \to \infty$ and obtain

$$
\mathbb{E} \parallel F_{K,1} - G_T \parallel_T \quad \leq \quad \exp(b\lceil 2\alpha\beta T \rceil)S(K)
$$

$$
= \quad \alpha T \exp(b\lceil 2\alpha\beta T \rceil) \left\{ (1+b)\sqrt{\frac{1}{K}} + \frac{2}{K} \right\},
$$

using (6.17). This completes the proof. $\qquad\square$

Now we have all the ingredients to prove Theorem 6.10. The first part of the proof uses the Cauchy-Schwarz inequality, applied to sums of functions of independent observations, whereas the second part of the proof disentangles the dependence between the observations.

Proof of Theorem 6.10
From Theorem 6.6, it suffices to bound, for all $m \in \mathbb{N}, f \in C_b^\infty(\mathbb{R}^m)$, and for all $\phi_1, \ldots, \phi_m \in C_b^\infty([0,T]^2)$ satisfying (6.5)

$$
\sum_{j=1}^m \mathbb{E} f_{(j)}(\langle \xi_K^T, \phi_k \rangle, k = 1, \ldots, m)\langle \mu^T - \xi_K^T, \phi_j \rangle.
$$

We abbreviate

$$
\zeta_K = \frac{1}{bK} \sum_{i=1}^{bK} \delta_{(A_i^K, A_i^K + r_i)},
$$

so that

$$
\zeta_K^T = \frac{1}{bK} \sum_{i=1}^{bK} \delta_{(A_i^K, A_i^K + r_i)} \mathbf{1}(A_i^K + r_i \leq T).
$$

Then we have

$$
\left| \sum_{j=1}^m \mathbb{E} f_{(j)}(\langle \xi_K^T, \phi_k \rangle, k = 1, \ldots, m)\langle \mu^T - \xi_K^T, \phi_j \rangle \right|
$$

$$
= \left| a \sum_{j=1}^m \mathbb{E} f_{(j)}(\langle \xi_K^T, \phi_k \rangle, k = 1, \ldots, m)\langle (\delta_0 \times \hat{\mu})^T - \frac{1}{aK} \sum_{i=1}^{aK} \delta_{(0,\hat{r}_i)}^T, \phi_j \rangle \right.
$$

$$
\left. + b \sum_{j=1}^m \mathbb{E} f_{(j)}(\langle \xi_K^T, \phi_k \rangle, k = 1, \ldots, m)\langle \tilde{\mu}^T - \zeta_K^T, \phi_j \rangle \right|
$$

$$\leq \ a\sum_{j=1}^{m} \| f_{(j)} \| \, \mathbb{E}\left|\frac{1}{aK}\sum_{i=1}^{aK}(\phi_j(0,\hat{r}_i)\mathbf{1}(\hat{r}_i \leq T) - \mathbb{E}\phi_j(0,\hat{r}_i)\mathbf{1}(\hat{r}_i \leq T))\right|$$

$$+b\sum_{j=1}^{m} \| f_{(j)} \| \, \mathbb{E}\left|\frac{1}{bK}\sum_{i=1}^{bK}(\phi_j(A_i^K, A_i^K + r_i)\mathbf{1}(A_i^K + r_i \leq T)\right.$$

$$\left.-\mathbb{E}\phi_j(G_T^{-1}(l_i), G_T^{-1}(l_i) + r_i)\mathbf{1}(G_T^{-1}(l_i) + r_i \leq T)\right|.$$

For the first summand, using the Cauchy-Schwarz inequality we obtain

$$a\sum_{j=1}^{m} \| f_{(j)} \| \, \mathbb{E}\left|\frac{1}{aK}\sum_{i=1}^{aK}(\phi_j(0,\hat{r}_i))\mathbf{1}(\hat{r}_i \leq T) - \mathbb{E}\phi_j(0,\hat{r}_i))\mathbf{1}(\hat{r}_i \leq T))\right|$$

$$\leq \ a\sum_{j=1}^{m} \| f_{(j)} \| \left(\mathrm{Var}\left(\frac{1}{aK}\sum_{i=1}^{aK}\phi_j(0,\hat{r}_i)\mathbf{1}(\hat{r}_i \leq T)\right)\right)^{\frac{1}{2}}$$

$$\leq \ \frac{\sqrt{a}}{\sqrt{K}},$$

where we used the boundedness assumptions from (6.5) in the last step. For the second summand,

$$b\sum_{j=1}^{m} \| f_{(j)} \| \, \mathbb{E}\left|\frac{1}{bK}\sum_{i=1}^{bK}(\phi_j(A_i^K, A_i^K + r_i)\mathbf{1}(A_i^K + r_i \leq T)\right.$$

$$\left.-\mathbb{E}\phi_j(G_T^{-1}(l_i), G_T^{-1}(l_i) + r_i)\mathbf{1}(G_T^{-1}(l_i) + r_i \leq T))\right|$$

$$\leq \ b\sum_{j=1}^{m} \| f_{(j)} \mathbb{E}\left|\frac{1}{bK}\sum_{i=1}^{bK}(\phi_j(G_T^{-1}(l_i), G_T^{-1}(l_i) + r_i)\mathbf{1}(G_T^{-1}(l_i) + r_i \leq T)\right.$$

$$\left.-\mathbb{E}\phi_j(G_T^{-1}(l_i), G_T^{-1}(l_i) + r_i)\mathbf{1}(G_T^{-1}(l_i) + r_i \leq T))\right|$$

$$+b\sum_{j=1}^{m} \| f_{(j)} \mathbb{E}\left|\frac{1}{bK}\sum_{i=1}^{bK}(\phi_j(A_i^K, A_i^K + r_i)\mathbf{1}(A_i^K + r_i \leq T)\right.$$

$$\left.-\phi_j(G_T^{-1}(l_i), G_T^{-1}(l_i) + r_i)\mathbf{1}(G_T^{-1}(l_i) + r_i \leq T)\right|.$$

Similarly as above, we obtain

$$b\mathbb{E}\left|\frac{1}{bK}\sum_{i=1}^{bK}(\phi(G_T^{-1}(l_i), G_T^{-1}(l_i) + r_i)\mathbf{1}(G_T^{-1}(l_i) + r_i \leq T)\right.$$

$$\left.-\mathbb{E}\phi(G_T^{-1}(l_i), G_T^{-1}(l_i) + r_i)\mathbf{1}(G_T^{-1}(l_i) + r_i \leq T))\right|$$

$$\leq \ \frac{\sqrt{b}}{\sqrt{K}}.$$

Thus

$$\left| \sum_{j=1}^{m} \mathbb{E} f_{(j)}(\langle \xi_K^T, \phi_k \rangle, k = 1, \ldots, m) \langle \mu^T - \xi_K^T, \phi_j \rangle \right|$$

$$\leq \quad \frac{\sqrt{a} + \sqrt{b}}{\sqrt{K}} + b \mathbb{E} \left| \frac{1}{bK} \sum_{i=1}^{bK} (\phi(A_i^K, A_i^K + r_i) \mathbf{1}(A_i^K + r_i \leq T) \right.$$

$$\left. - \phi(G_T^{-1}(l_i), G_T^{-1}(l_i) + r_i) \mathbf{1}(G_T^{-1}(l_i) + r_i \leq T) \right|.$$

Bounding

$$\mathbb{E} \left| \frac{1}{bK} \sum_{i=1}^{bK} (\phi(A_i^K, A_i^K + r_i) \mathbf{1}(A_i^K + r_i \leq T) \right.$$

$$\left. - \phi(G_T^{-1}(l_i), G_T^{-1}(l_i) + r_i) \mathbf{1}(G_T^{-1}(l_i) + r_i \leq T) \right|$$

is more complicated, partly because the derivative of G_T^{-1} is not necessarily bounded. Instead, similarly as in Reinert (1995) we mimic differentiation, making use of the additional stochasticity introduced by the random point l_1. First observe that, as λ is non-anticipating, if we omit individual 1 from the population, the course of the epidemic up to time $F_K^{-1}(l_1)$ is not affected. Let $F_{K,1}$ be as in Lemma 6.12. Then we have $F_K^{-1}(l_1) = F_{K,1}^{-1}(l_1)$ by construction. Thus

$$\mathbb{E} \left| \frac{1}{bK} \sum_{i=1}^{bK} (\phi(A_i^K, A_i^K + r_i) \mathbf{1}(A_i^K + r_i \leq T) \right.$$

$$\left. - \phi(G_T^{-1}(l_i), G_T^{-1}(l_i) + r_i) \mathbf{1}(G_T^{-1}(l_i) + r_i \leq T)) \right|$$

$$\leq \quad \mathbb{E} \left| \phi(A_1^K, A_1^K + r_1) \mathbf{1}(A_1^K + r_1 \leq T) \right.$$

$$\left. - \phi(G_T^{-1}(l_1), G_T^{-1}(l_1) + r_1) \mathbf{1}(G_T^{-1}(l_1) + r_1 \leq T) \right|.$$

Expanding the indicators, we have

$$\phi(A_1^K, A_1^K + r_1) \mathbf{1}(A_1^K + r_1 \leq T)$$
$$- \phi(G_T^{-1}(l_1), G_T^{-1}(l_1) + r_1) \mathbf{1}(G_T^{-1}(l_1) + r_1 \leq T)$$

$$= \quad \phi(F_{K,1}^{-1}(l_1), F_{K,1}^{-1}(l_1) + r_1) \mathbf{1}(F_{K,1}^{-1}(l_1) + r_1 \leq T)$$
$$- \phi(G_T^{-1}(l_1), G_T^{-1}(l_1) + r_1) \mathbf{1}(G_T^{-1}(l_1) + r_1 \leq T) t$$

$$= \quad \phi(F_{K,1}^{-1}(l_1), F_{K,1}^{-1}(l_1) + r_1) \mathbf{1}(F_{K,1}^{-1}(l_1) + r_1 \leq T, G_T^{-1}(l_1) + r_1 \leq T)$$
$$+ \phi(F_{K,1}^{-1}(l_1), F_{K,1}^{-1}(l_1) + r_1) \mathbf{1}(F_{K,1}^{-1}(l_1) + r_1 \leq T, G_T^{-1}(l_1) + r_1 > T)$$
$$- \phi(G_T^{-1}(l_1), G_T^{-1}(l_1) + r_1) \mathbf{1}(F_{K,1}^{-1}(l_1) + r_1 \leq T, G_T^{-1}(l_1) + r_1 \leq T)$$
$$- \phi(G_T^{-1}(l_1), G_T^{-1}(l_1) + r_1) \mathbf{1}(F_{K,1}^{-1}(l_1) + r_1 > T, G_T^{-1}(l_1) + r_1 \leq T).$$

(For simplicity of notation we omit the superscript T for $F_{K,1}$ and for the expectation.) Using Taylor expansion and (6.5) we thus may bound

$$
\mathbb{E}\left| \frac{1}{bK} \sum_{i=1}^{bK} (\phi(A_i^K, A_i^K + r_i)\mathbf{1}(A_i^K + r_i \le T) \right.
$$

$$
\left. - \phi(G_T^{-1}(l_i), G_T^{-1}(l_i) + r_i)\mathbf{1}(G_T^{-1}(l_i) + r_i \le T)) \right|
$$

$$
\le \quad \mathbb{E}\left| ((F_{K,1})^{-1}(l_1) - G_T^{-1}(l_1)) \right.
$$

$$
\left. \mathbf{1}(F_{K,1}^{-1}(l_1) + r_1 \le T, G_T^{-1}(l_1) + r_1 \le T) \right|
$$

$$
+ \mathbb{P}(F_{K,1}^{-1}(l_1) + r_1 > T, G_T^{-1}(l_1) + r_1 \le T)
$$

$$
+ \mathbb{P}(F_{K,1}^{-1}(l_1) + r_1 \le T, G_T^{-1}(l_1) + r_1 > T).
$$

Firstly,

$$
\mathbb{P}(F_{K,1}^{-1}(l_1) + r_1 > T, G_T^{-1}(l_1) + r_1 \le T) \quad \le \quad \beta \mathbb{E} \parallel F_{K,1} - G_T \parallel_T .
$$

Thus, by symmetry,

$$
\mathbb{E}\left| ((F_{K,1})^{-1}(l_1) - G_T^{-1}(l_1))\mathbf{1}(F_{K,1}^{-1}(l_1) + r_1 \le T, G_T^{-1}(l_1) + r_1 \le T) \right|
$$

$$
+ \quad \mathbb{P}(F_{K,1}^{-1}(l_1) + r_1 > T, G_T^{-1}(l_1) + r_1 \le T)
$$

$$
+ \quad \mathbb{P}(F_{K,1}^{-1}(l_1) + r_1 \le T, G_T^{-1}(l_1) + r_1 > T)
$$

$$
\le \quad \mathbb{E}\left| (F_{K,1})^{-1}(l_1) - G_T^{-1}(l_1))\mathbf{1}(F_{K,1}^{-1}(l_1) + r_1 \le T, G_T^{-1}(l_1) + r_1 \le T) \right|
$$

$$
+ 2\beta \mathbb{E} \parallel F_{K,1} - G_T \parallel_T .
$$

Using the existence of the density ψ, for the first term we have

$$
\mathbb{E}\left| ((F_{K,1})^{-1}(l_1) - G_T^{-1}(l_1))\mathbf{1}(F_{K,1}^{-1}(l_1) + r_1 \le T, G_T^{-1}(l_1) + r_1 \le T) \right|
$$

$$
\le \quad \mathbb{E}\left| (F_{K,1})^{-1}(l_1) - G_T^{-1}(l_1) \right| \mathbf{1}(F_{K,1}^{-1}(l_1) \le T, G_T^{-1}(l_1) \le T)
$$

$$
= \quad \mathbb{E}\int_0^{F_{K,1}(T) \wedge G_T(T)} \left| (F_{K,1})^{-1}(x) - G_T^{-1}(x) \right|
$$

$$
\mathbf{1}(F_{K,1}^{-1}(x) \le T, G_T^{-1}(x) \le T)\psi(x)dx
$$

$$
= \quad \mathbb{E}\int_0^{F_{K,1}(T) \wedge G_T(T)} \left| G_T^{-1}(G_T((F_{K,1})^{-1}(x))) - G_T^{-1}(x) \right|
$$

$$
\mathbf{1}(F_{K,1}^{-1}(x) \le T, G_T^{-1}(x) \le T)\psi(x)dx.
$$

From Condition 4. on λ we have that $F_{K,1}$ and G_T are strictly increasing until they stay constant forever after. Thus the derivatives $(G_T^{-1})'$ and $(F_{K,1}^{-1})'$ exist for all t for which G_T and $F_{K,1}$ are not constant; in particular they exist for $t < F_{K,1}^{-1}(x)$ when $F_{K,1}^{-1}(x) \le T$, and for $t < G_T^{-1}(x)$ when

$G_T^{-1}(x) \leq T$. Thus we may integrate as follows.

$$\mathbb{E}\int_0^{F_{K,1}(T)\wedge G_T(T)} \left|G_T^{-1}(G_T((F_{K,1})^{-1}(x))) - G_T^{-1}(x)\right|$$

$$\mathbf{1}(F_{K,1}^{-1}(x) \leq T, G_T^{-1}(x) \leq T)\psi(x)dx$$

$$= \mathbb{E}\int_0^{F_{K,1}(T)\wedge G_T(T)} \left|\int_x^{G_T(F_{K,1})^{-1}(x))} (G_T^{-1})'(y)dy\right| \psi(x)dx$$

$$= \mathbb{E}\int_0^{F_{K,1}(T)\wedge G_T(T)} \int_x^{G_T((F_{K,1})^{-1}(x))} (G_T^{-1})'(y)dy$$

$$\mathbf{1}(x < G_T((F_{K,1})^{-1}(x)))\psi(x)dx$$

$$+\mathbb{E}\int_0^{F_{K,1}(T)\wedge G_T(T)} \int_{G_T((F_{K,1})^{-1}(x))}^x (G_T^{-1})'(y)dy$$

$$\mathbf{1}(x > G_T((F_{K,1})^{-1}(x)))\psi(x)dx.$$

As all integrals are finite, we may interchange the order of integration and obtain for the above expression

$$\mathbb{E}\int_0^{G_T(F_{K,1}^{-1}(F_{K,1}(T)\wedge G_T(T)))} \int_{F_{K,1}(G_T^{-1}(y))}^{G_T(G_T^{-1}(y))} \mathbf{1}(x < G_T((F_{K,1})^{-1}(x)))$$

$$\psi(x)dx(G_T^{-1})'(y)dy$$

$$+ \mathbb{E}\int_0^{G_T(F_{K,1}^{-1}(F_{K,1}(T)\wedge G_T(T)))} \int_{G_T(G_T^{-1}(y))}^{F_{K,1}(G_T^{-1}(y))} \mathbf{1}(x > G_T((F_{K,1})^{-1}(x)))$$

$$\psi(x)dx(G_T^{-1})'(y)dy$$

$$\leq \beta T\mathbb{E}\| F_{K,1} - G_T \|_T .$$

Thus we have derived that

$$b\mathbb{E}\left|\frac{1}{bK}\sum_{i=1}^{bK}(\phi(A_i^K, A_i^K + r_i)\mathbf{1}(A_i^K + r_i \leq T)\right.$$

$$\left. -\phi(G_T^{-1}(l_i), G_T^{-1}(l_i) + r_i)\mathbf{1}(G_T^{-1}(l_i) + r_i \leq T)\right|$$

$$\leq b\beta(T+2)\mathbb{E}\| F_{K,1} - G_T \|_T .$$

Employing Lemma 6.12 we obtain the assertion.   □

In practice, the above bound may become rather large. If K is very large, and $b \approx 1$, then the bound is less than 1 only if

$$K > 4(\alpha\beta)^2 T^4 e^{4\alpha\beta T},$$

which often would be valid only for the initial stages of an epidemic, or if $\alpha\beta$

is tiny. In the latter case, most of the susceptibles become infected almost instantly, so that the epidemic process behaves nearly as a simple death process. In this case, the deterministic approximation should be good, as uniform approximations could be in place, see Lemma 6.11.

The fact that the deterministic approximation is not as good as one would expect when inspecting epidemic curves has already been noticed by Metz (1978). He shows heuristically that epidemic curves all have essentially the same form, differing only by a random transformation. The offset caused by this random transformation might be the major source of error in the deterministic approximation. In the following subsection, we will make this heuristically precise, for a special case.

6.3.5 A special case: $\lambda(t, x) = \alpha x(t)$

In the case that $\lambda(t, x) = \alpha x(t)$, much more can be said. Here we consider this special case, and we furthermore assume that l_i and r_i are independent, and that r_i possesses a density ϕ. (Note that this case includes Bartlett's GSE as a special case.) Much of the uncertainty in approximating an epidemic lies in the initial stages. Initially, only few individuals would typically be infected, so it might take a long time until the epidemic takes off. If, instead, we use a deterministic approximation only after the epidemic has acquired a substantial size, the approximation would be much improved.

With the notation above, choose a "threshold" value d and let t_0 be such that $G_T(t_0) = d$. Assume that, for some $\rho > 0$, $I(t_0 + t) \geq \rho$ and $I_K(\tau + t) \geq \rho$ for all $t \leq T$. Define

$$\tau = \inf\{t : F_K(t) = d\}.$$

Put, for all $t > 0$,

$$G_d(t) = G_T(t_0 + t)$$

and

$$F_d(t) = F_K(\tau + t).$$

Thus we only approximate after the total infectivity in the population has reached the level d. Our new empirical measure is

$$\xi^d = \frac{1}{bK} \sum_{i=1}^{bK} \delta_{(F_d^{-1}(l_i), F_d^{-1}(l_i) + r_i)},$$

considered on $\mathbb{R}_+ \times \mathbb{R}_+$, and the approximating measure $\tilde{\mu}^d$ is given by

$$\tilde{\mu}^d([0, r] \times [0, s]) = \mathbb{P}[l_1 \leq G_d(r), l_1 \leq G_d(s - r_1)]. \qquad (6.21)$$

Furthermore assume that

$$\alpha b \psi(d) \mathbb{E}^T(r_1) < 1. \qquad (6.22)$$

We obtain the following proposition.

Proposition 6.13 *Suppose that* $\lambda(t, x) = \alpha x(t)$. *Assume that* ψ *decreases monotonically on* $[t_0, T]$. *Let* $\tilde{\mu}^d$ *be given in (6.21), and let* d *and* T *be such that (6.22) holds. Then, for all* $T \in \mathbb{R}_+$, *and for all* $H \in \mathcal{F}$,

$$|\mathbb{E}^T H(\xi^d) - H(\tilde{\mu}^d)|$$

$$\leq \frac{1}{\sqrt{bK}} + \frac{\alpha bT}{\rho - \rho\alpha b\psi(d)\mathbb{E}^T(r_1)} \left\{(1+b)\sqrt{\frac{1}{K}} + \frac{2}{K}\right\}.$$

Remark. Note that the bound is now only linear in T, instead of exponential in T. Moreover, in the next subsection some plots will illustrate that the assumptions can be fulfilled in reasonable cases.

The proof of Proposition 6.13 is rather similar to the proof of Theorem 6.10. Due to the additional assumptions, though, the contraction argument is much simplified.

Proof of Proposition 6.13 From Theorem 6.6, it suffices to bound, for

all $m \in \mathbb{N}, f \in C_b^\infty(\mathbb{R}^m)$, and for all $\phi_1, \ldots, \phi_m \in C_b^\infty([0,T]^2)$ satisfying (6.5)

$$\sum_{j=1}^{m} \mathbb{E}^T f_{(j)}(\langle \xi^d, \phi_k \rangle, k = 1, \ldots, m) \langle \tilde{\mu}^d - \xi^d, \phi_j \rangle.$$

We have

$$\left| \sum_{j=1}^{m} \mathbb{E}^T f_{(j)}(\langle \xi^d, \phi_k \rangle, k = 1, \ldots, m) \langle \tilde{\mu}^d - \xi^d, \phi_j \rangle \right|$$

$$\leq \sum_{j=1}^{m} \| f_{(j)} \| \mathbb{E}^T \left| \frac{1}{bK} \sum_{i=1}^{bK} (\phi_j(F_d^{-1}(l_i), F_d^{-1}(l_i) + r_i) \right.$$

$$\left. - \mathbb{E}\phi_j(G_d^{-1}(l_i), G_d^{-1}(l_i) + r_i) \right|$$

$$\leq \mathbb{E}^T \left| \frac{1}{bK} \sum_{i=1}^{bK} (\phi(G_d^{-1}(l_i), G_d^{-1}(l_i) + r_i) \right.$$

$$\left. - \mathbb{E}^T \phi(G_d^{-1}(l_i), G_d^{-1}(l_i) + r_i)) \right|$$

$$+ \mathbb{E}^T \left| \frac{1}{bK} \sum_{i=1}^{bK} (\phi(F_d^{-1}(l_i), F_d^{-1}(l_i) + r_i) - \phi(G_d^{-1}(l_i), G_d^{-1}(l_i) + r_i) \right|$$

$$\leq \frac{1}{\sqrt{bK}} + \mathbb{E}^T \left| F_d^{-1}(l_i) - G_d^{-1}(l_i) \right|,$$

where we used the Cauchy-Schwarz inequality for the first summand, as in the proof of Theorem 6.10.

The difficulty lies again in bounding

$$\mathbb{E}^T \left| F_d^{-1}(l_i) - G_d^{-1}(l_i) \right|.$$

In contrast to the proof of Theorem 6.10, we now differentiate, giving that

$$
\begin{aligned}
|F_d^{-1}(l_i) - G_d^{-1}(l_i)| &= |G_d^{-1}(G_d(F_d^{-1}(l_i))) - G_d^{-1}(F_d(F_d^{-1}(l_i)))| \\
&= |(G_d^{-1})'(l_i + \theta)(G_d(F_d^{-1}(l_i))) - F_d(F_d^{-1}(l_i)))|,
\end{aligned}
$$

for some θ. As we assumed that $G_d' = I(t_0 + t) \geq \rho$ on the interval considered, we may bound

$$\mathbb{E}^T \left| F_d^{-1}(l_i) - G_d^{-1}(l_i) \right| \leq \frac{1}{\rho} \mathbb{E}^T |G_d(F_d^{-1}(l_i))) - l_i|.$$

Observe that, if we omit individual 1 from the population, the course of the epidemic up to time $F_d^{-1}(l_1)$ is not affected. Let $F_{K,1}$ be as in Lemma 6.12. Then $F_K(t) = F_{K,1}(t)$ for all $t \leq l_1$ by construction. Moreover, put

$$F_{d,1}(t) = F_{K,1}(\tau + t).$$

Then

$$
\begin{aligned}
F_{d,1}^{-1}(l_1) &= \inf\{t : F_{d,1} = l_1\} \\
&= \inf\{t : F_{K,1}(d + t) = l_1\} \\
&= \inf\{t : F_K(d + t) = l_1\} \\
&= F_d^{-1}(l_1).
\end{aligned}
$$

Similarly, $F_{d,1}^{-1}(s) = F_d^{-1}(s)$ for all $s \leq l_1$. Furthermore, for any t we have

$$
\begin{aligned}
F_{d,1}(t) &- G_d(t) \\
&= L_{K,1}F_{K,1}(\tau + t) - LG_T(t_0 + t) \\
&= L_{K,1}F_{K,1}(\tau + t) - LF_{K,1}(\tau + t) + LF_{K,1}(\tau + t) - LG_T(t_0 + t) \\
&= L_{K,1}F_{K,1}(\tau + t) - LF_{K,1}(\tau + t) + LF_{d,1}(t) - LG_d(t).
\end{aligned}
$$

Hence, for all $s \leq F_d^{-1}(l_1)$,

$$
\begin{aligned}
G_d(s) &- F_d(s) \\
&= L_{K,1}F_{K,1}(\tau + s) - LF_{K,1}(\tau + s) + LF_{d,1}(s) - LG_d(s) \\
&= L_{K,1}F_{K,1}(\tau + s) - LF_{K,1}(\tau + s) + LF_d(s) - LG_d(s).
\end{aligned}
$$

Thus

$$\sup_{s \leq F_d^{-1}(l_i)} |G_d(F_d^{-1}(l_i)) - l_i|$$

$$\leq \sup_{h \in \mathcal{H}_t} \| L_{K,1}h - Lh \|_T + \sup_{s \leq F_d^{-1}(l_i)} |LF_d(t) - LG_d(t)|.$$

As before, using Inequality (6.17),

$$\mathbb{E} \sup_{h \in \mathcal{H}_t} \| L_{K,1}h - Lh \|_T \leq \alpha T \left\{ (1+b)\sqrt{\frac{1}{K}} + \frac{2}{K} \right\} = S(K).$$

Now we bound $|LF_d(F_d^{-1}(l_1)) - LG_d(F_d^{-1}(l_1))|$. Recall that, for any function f, as $\lambda(t, x) = \alpha x(t)$,

$$Lf(t)$$
$$= \alpha \int_0^t (1 - \hat{\Phi}(s))ds + \alpha b \int_0^t \left(\Psi(f(s)) - \int_0^s \Psi(f(s-u))\phi(u)du \right) ds.$$

For the last integral, interchanging the order of integration gives

$$\int_0^t \int_0^s \Psi(f(s-u))\phi(u)duds$$
$$= \int_0^t \int_0^s \Psi(f(y))\phi(s-y)dyds$$
$$= \int_0^t \Psi(f(y)) \int_y^t \phi(s-y)dsdy$$
$$= \int_0^t \Psi(f(y)) \int_0^{t-y} \phi(s)dsdy$$
$$= \int_0^t \Psi(f(y))\Phi(t-y)dy.$$

Thus

$$Lf(t)$$
$$= \alpha \int_0^t (1 - \hat{\Phi}(s))ds + \alpha b \int_0^t \Psi(f(s))(1 - \Phi(t-s))ds.$$

Hence,

$$|LF_d(F_d^{-1}(l_1)) - LG_d(F_d^{-1}(l_1))|$$
$$\leq \alpha b \left| \int_0^{F_d^{-1}(l_1)} (\Psi(F_d(s)) - \Psi(G_d(s)))(1 - \Phi(F_d^{-1}(l_1) - s))ds \right|$$
$$\leq \psi(d) \| F_d - G_d \|_{F_d^{-1}(l_1)} \int_0^{F_d^{-1}(l_1)} (1 - \Phi(F_d^{-1}(l_1) - s))ds$$
$$= \psi(d)\mathbb{E}^T(r_1) \| F_d - G_d \|_{F_d^{-1}(l_1)},$$

noting that $\psi(t) < \psi(d)$ on the interval considered, as $F_d \geq d$ and $G_d \geq d$. This yields

$$\mathbb{E}^T \sup_{s \leq F_d^{-1}(l_1)} |LF_d(s) - LG_d(s)| \leq \alpha b \psi(d)\mathbb{E}^T(r_1) \| G_d - F_d \|_{F_d^{-1}(l_1)} \cdot$$

By Assumption (6.22), we may apply the contraction argument without having to dissect the target interval. Thus

$$\mathbb{E}^T \parallel F_d - G_d \parallel_{F_d^{-1}(l_1)} \leq S(K) + \alpha b \psi(d) \mathbb{E}^T(r_1) \mathbb{E}^T \parallel G_d - F_d \parallel_{F_d^{-1}(l_1)},$$

yielding that

$$\mathbb{E}^T \parallel G_d - F_d \parallel_{F_d^{-1}(l_1)}$$
$$\leq \frac{1}{1 - \alpha b \psi(d) \mathbb{E}^T(r_1)} S(K)$$
$$= \frac{1}{1 - \alpha b \psi(d) \mathbb{E}^T(r_1)} \alpha b T \left\{ (1 + b) \sqrt{\frac{1}{K}} + \frac{2}{K} \right\}.$$

This completes the proof. $\square$

Typically, $I(t)$ would be unimodal on a large interval, as would be $I_K(t)$, so that the restriction on being at least as large as ρ would be natural. To see this, in the next section we show some plots.

6.3.6 Some plots of the limiting expression

For simplicity, here we consider Bartlett's GSE. Here we have the case that $\Phi(x) = 1 - e^{-\beta x}$, and that $\psi(x) = e^{-x}$; if $\alpha = 1$ and $b \approx 1$, say, then, for $\beta \geq 1$ any $d > 0$ would satisfy Assumption (6.22).

Here, β can be interpreted as the relative removal rate. It is well-known (see Bailey (1975)) that, if $\beta < 1$, then the chance of ultimate extinction of the epidemic is less than unity, whereas for $\beta \geq 1$, the chance of ultimate extinction of the epidemic is unity. In the latter case, only a minor outbreak of the epidemic would be expected, whereas in the first case, a major build-up may occur. For the case of a minor outbreak, Proposition 6.13 is suitable. (Often, instead of the relative removal rate, the basic reproductive rator $R_0 = \frac{1}{\beta}$ is studied; see, e.g., Ball et al. (1997).)

Firstly we choose as parameters $a = 0.01$, $\alpha = 1$, as usual infection rate 1, and removal rate $\beta = 2$. The first plot, Figure 6.1, shows the proportion of infectives. The proportion of infectives first decreases, corresponding to the initially infected individuals in the population, until enough infectivity in the population has accumulated; then it increases to a peak, and then decreases until it reaches 0; then the epidemic dies out. Thus the unimodality is valid for a large time interval. Figure 6.2 shows the cumulative infectivity function G. As mentioned above, any value $d > 0$ would be admissible. The bound in Proposition 6.13 improves with larger ρ, which corresponds to a shorter time interval. For $\rho = .0001$, for example, dependent on the observations, the time interval could be chosen as $[17, 134]$. As $G(17) = 0.00459$, choosing $d = 0.00459$, Proposition 6.13 provides as bound, for any $t \in [17, 134]$, $\frac{1}{\sqrt{K}}\left(1.006 + 19{,}517(t - 17) + 2 + \frac{2}{\sqrt{K}}\right)$. For $\rho = .001$, in comparison, the time interval could be chosen as $[50, 100]$; as $G(50) = 0.009451$, Proposition 6.13 provides as bound, for any $t \in [50, 100]$, $\frac{1}{\sqrt{K}}\left(1.006 + 1{,}942(t - 50) + 2 + \frac{2}{\sqrt{K}}\right)$. So, in practice, K has to be very large to make this bound useful.

The next series of plots illustrate the "critical" case $\beta = 1$ all the other parameters are as above. For $\rho = .01$, for example, dependent on the observations, the time interval could be chosen as $[31, 40]$. As $G(16) = 0.040931$, Proposition 6.13 provides as bound, for any $t \in [31, 40]$, the value $\frac{1}{\sqrt{K}}\left(1.006 + 200(t - 40) + 2 + \frac{2}{\sqrt{K}}\right)$. Still, K has to be very large to make this bound useful in practice.

Lastly, Figures 6.5 and 6.6 show the corresponding plots with $\beta = 0.3$. Here, the proportion removals and the cumulative infectivity behave almost identically. In this case, Assumption (6.22) translates to $e^{-d} < 0.3$, or $d > \ln(10/2.97) \approx 1.21$, a case that is clearly not of interest, as the function G never reaches that level. Thus the scope of Proposition 6.13 is limited. However, it would be possible to refine the contraction approach, by separating the time interval into only a few intervals, for which the contraction property then would hold again. This would extend the scope of Proposition 6.13, but would yield bounds that are higher order polynomials in T.

6.4 Discussion

In the above we have derived an explicit bound on the distance of the time course of the GSE over a finite time interval $[0, T]$ to its mean field limit. This is the first explicit bound known for this problem. Of course normal approximations, as given by Barbour (1974) provide the order of the distance, but only in the Markovian case, and do not give an explicit expression. The results above not only give an explicit bound, but also a numerically fast procedure (using a contraction construction) to derive the approximating deterministic system. Moreover, we confirmed a heuristic by Metz by showing that the deterministic approximation is linear in time if the approximation is started at a random time τ where the epidemic has taken off, and if the approximation is only derived on an interval where the epidemic grows strictly.

Note that considering only a finite time interval is not a strong restriction, as all our observations from an epidemic process will necessarily be over a finite time interval. For Bartlett's GSE, from Barbour (1975) it is known that the duration of the epidemic is a nontrivial random quantity

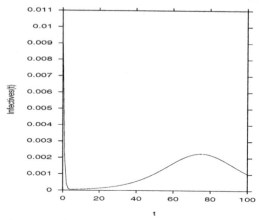

Figure 6.1 *Proportion infectives in Bartlett's GSE; a = 0.01, α = 1, and β = 2.*

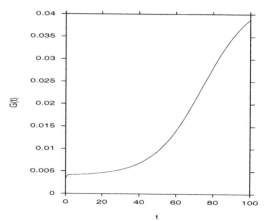

Figure 6.2 *Cumulative infectivity function G in Bartlett's GSE; a = 0.01, α = 1, and β = 2.*

with center of order $\log K$, so that our bounds will be expected to be satisfactory for a relatively long time period. This fact also illustrates why the approximation breaks down when used for a very large time period; as the duration of the epidemic is truly random, no good deterministic approximation for it exists. Moreover, in Bartlett's GSE the parameter $\alpha\beta$ corresponds to the transmission rate. As a rule of thumb, the faster the transmission rate, the shorter the duration of the epidemic; in this sense $\alpha\beta$ scales with T.

The deterministic approximation agrees, for Bartlett's GSE, with the ones obtained in the literature. Isham (1991) shows how to derive the approximation using moment closure methods (although her focus is on stochastic approximations). Moment closure methods have been proven a

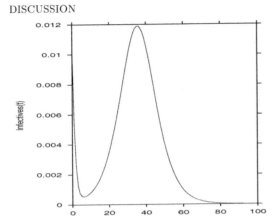

Figure 6.3 *Proportion infectives in Bartlett's GSE;* $a = 0.01$, $\alpha = 1$, *and* $\beta = 1$.

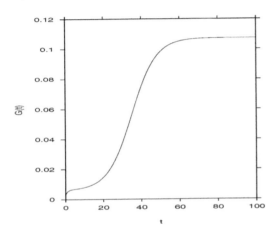

Figure 6.4 *Cumulative infectivity function G in Bartlett's GSE;* $a = 0.01$, $\alpha = 1$, *and* $\beta = 1$.

powerful approach for Markovian models, in particular for spatial models. However, I am not aware of any extension to non-Markovian settings. Furthermore, the moment closure approach has heuristic character, to determine the covariance, for example, when a Gaussian approximation is known to be valid. It does not give a bound on the distance to the approximation.

The model discussed here is a closed epidemic, without reinfection, and not admitting birth of individuals, nor removals not due to the infection. These assumptions exclude many biologically relevant cases. However, independent birth events and independent death events would only result in a slightly modified empirical measure, for which the above arguments, in particular the contraction argument, should still be valid. In this extension

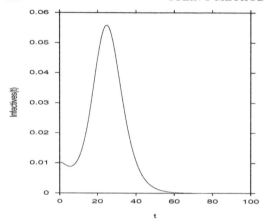

Figure 6.5 *Proportion infectives in Bartlett's GSE;* $a = 0.01$, $\alpha = 1$, *and* $\beta = 0.3$.

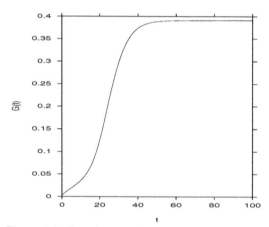

Figure 6.6 *Cumulative infectivity function G in Bartlett's GSE;* $a = 0.01$, $\alpha = 1$, *and* $\beta = 0.3$.

we would expect a recurrent epidemic, due to the possibility of births after the epidemic has taken off; some complex dynamics could then occur, see, e.g., Earn et al. (2000) and Andersson and Britton (2000). Moreover, it would be possible to treat a stratified population, with different types of individuals (the (l_i, r_i)'s may come from different distributions); for a motivation, see, e.g., May (2000).

Also it would be desirable to generalize the above for accumulation functions λ that might not involve just the proportion of infectives over the time course, but also some neighborhood structure. This would allow to treat spatial epidemic models in a similar way.

The above bounds have only been derived for smooth test functions. Deriving bounds for nonsmooth test functions is not a fundamental problem but a rather tedious enterprise; see, for example, Rinott and Rotar (1996), Götze (1991), or Bolthausen (1984). Lastly, a Gaussian approximation together with error bounds would be a natural next step to investigate.

Acknowledgments

This work has been motivated by many fruitful discussions with Andrew Barbour. John Einmahl was very helpful in connection with the Glivenko-Cantelli bound; Guy Gielis and Nico Stollenwerk made many useful comments on the manuscript. Frank Ball drew my attention to Metz' work and intuition. Finally, I would like to thank the organizers for the invitation, and the referees for their comments, which yielded a substantial improvement of the paper.

6.5 References

Aldous, D. (1989). Stein's method in a two-dimensional coverage problem. *Statist. Probab. Lett.* **8**, 307–314.

Andersson, H. and Britton, T. (2000). Stochastic epidemics in dynamic populations: quasi-stationarity and extinction. Preprint, Uppsala University. Submitted to *J. Math. Biol.*

Arratia, R., Goldstein, L., and Gordon, L. (1989). Two moments suffice for Poisson approximations: the Chen-Stein method. *Ann. Probab.* **17**, 9–25.

Bailey, N.T.J. (1975). *The Mathematical Theory of Infectious Diseases and Its Applications.* (2nd ed.) Griffin, London.

Baldi, P., Rinott, Y., and Stein, C. (1989). A normal approximation for the number of local maxima of a random function on a graph. In *Probability, Statistics and Mathematics, Papers in Honor of Samuel Karlin.* T.W. Anderson, K.B. Athreya and D.L. Iglehart, eds., Academic Press, 59–81.

Ball, F., Mollison, D., and Scalia-Tomba, G. (1997). Epidemics with two levels of mixing. *Ann. Appl. Probab.* **7**, 46–89.

Barbour, A.D. (1974). On a functional central limit theorem for Markov population processes. *Adv. Appl. Probab.* **6**, 21–39.

Barbour, A.D. (1975). The duration of the closed stochastic epidemic. *Biometrika* **62**, 477–482.

Barbour, A.D. (1990). Stein's method for diffusion approximations. *Probab. Theory Relat. Fields* **84**, 297–322.

Barbour, A.D., Chen, L.H.Y., and Loh, W.-L. (1992a). Compound Poisson approximation for nonnegative random variables via Stein's method. *Ann. Probab.* **20**, 1843–1866.

Barbour, A.D., Holst, L., and Janson, S. (1992b). *Poisson Approximation.* Oxford Science Publications.

Barbour, A.D., Karoński, M., and Ruciński, A. (1989). A central limit theorem for decomposable random variables with applications to random graphs. *J. Comb. Theory, Ser. B*, **47**, 125–145.

Barbour, A.D., and Utev, S. (1998). Solving the Stein equation in compound Poisson approximation. *Adv. Appl. Prob.* **30**, 449–475.

Bartlett, M.S. (1949) Some evolutionary stochastic processes. *J. R. Statist. Soc., Ser. B*, **11**, 211–229.

Berard, D.R., Burgess, J.W., and Coss, R.G. (1981). Plasticity of dendritic spine formation: a state-dependent stochastic process. *Intern. J. Neuroscience* **13**, 93–98.

Bolthausen, E. (1984). An estimate of the remainder in a combinatorial central limit theorem. *Z. Wahrscheinlichkeitstheor. Verw. Geb.* **66**, 379–386.

Chen, L.H.Y. (1975). Poisson approximation for dependent trials. *Ann. Probab.* **3**, 534–545.

Dawson, D.A. (1993). Measure-valued Markov processes. In *Ecole d'Eté de Probabilité de Saint-Flour 1991*, LNM 1541, Hennequin, P.L., eds., Springer-Verlag, 1–260.

Diaconis, P. (1989). An example for Stein's method. Technical Report, Stanford University. Stat. Dept.

Diaconis, P., and Zabell, S. (1991). Closed form summation for classical distributions: variations on a theme of de Moivre. *Statistical Science* **6**, 284-302.

Ethier, S.N. and Kurtz, T.G. (1986). *Markov Processes: Characterization and Convergence*. Wiley, New York.

Earn, D., Rohani, P., Bolker, B.M., and Grenfell, B. (2000). A simple model for complex dynamical transitions in epidemics. *Science* **287**, 667–670.

Ehm, W. (1991). Binomial approximation to the Poisson binomial distribution. *Statist. Probab. Lett.* **11**, 7–16.

Götze, F. (1991). On the rate of convergence in the multivariate CLT. *Ann. Probab.* **19**, 724–739.

Hall, P., and Barbour, A.D. (1984). Reversing the Berry-Esséen theorem. *Proc. AMS* **90**, 107–110.

Isham, V. (1991). Assessing the variability of stochastic epidemics. *Math. Biosci.* **107**, 209–224.

Keeling, M., and Grenfell, B.Y. (1998). Effect of variability in infection period on the persistence and spatial spread of infectious diseases. *Math. Biosciences* **147**, 207–226.

Loh, W.-L. (1992). Stein's method and multinomial approximation. *Ann. Appl. Probab.* **2**, 536–554.

Loh, W.-L. (1997). Stein's method and Pearson Type 4 approximations. Research report, Dept. of Statistics and Applied Probability, National University of Singapore.

Luk, H.M. (1994). Stein's method for the gamma distribution and related statistical applications. Ph.D. thesis. University of Southern California, Los Angeles.

Mann, B. (1997). Stein's method for χ^2 of a multinomial. Preprint, Harvard.

Massart, P. (1990). The tight constant in the Dvoretzky-Kiefer-Wolfowitz inequality. *Ann. Probab.* **18**, 1269–1283.

May, R. (2000). Simple rules with complex dynamics. *Science* **287**, 601–602.

Metz, J.A.J. (1978). The epidemic in a closed population with all susceptibles equally vulnerable; some results for large susceptible populations and small initial infections. *Acta Biotheoretica.* **27**, 75–123.

Peköz, E. (1995). Stein's method for geometric approximation. Technical Report 225, Stat. Dept. University of California, Berkeley.

Picard, P. and Lefevre, C. (1991). The dimension of Reed-Frost epidemic models with randomized suscpetibility levels. *Math. Biosci.* **107**, 225–233.

Reinert, G. (1994). A weak law of large numbers for empirical measures via Stein's method. *Ann. Probab.* **23**, 334–354.

Reinert, G. (1995) The asymptotic evolution of the General Stochastic Epidemic. *Ann. Appl. Probab.* **5**, 1061–1086.

Reinert, G. (1998a). Couplings for normal approximations with Stein's method. In *DIMACS Series in Discrete Mathematics and Theoretical Computer Science* **41**, 193–207.

Reinert, G. (1998b). Stein's method in application to empirical measures. In *Modelos Estocasticos.* J.M. Gonzales Barrios and L.G.Gorostiza, eds. Sociedad Matematica Mexicana, 65–120.

Rinott, Y., and Rotar, V. (1996). A multivariate CLT for local dependence with $n^{-1/2} \log n$ rate and applications to multivariate graph related statistics. *J. Multivariate Analysis.* **56**, 333–350.

Roos, M. (1993). Stein-Chen method for compound Poisson approximation. Ph.D. thesis, University of Zürich.

Sellke, T. (1983). On the asymptotic distribution of the size of a stochastic epidemic. *J. Appl. Probab.* **20**, 390–394.

Solomon. W. (1987). Representation and approximation of large population age distributions using Poisson random measures. *Stoch. Proc. Appl.* **26**, 237–255.

Stein, C. (1972). A bound for the error in the normal approximation to the distribution of a sum of dependent random variables. *Proc. Sixth Berkeley Symp. Math. Statist. Probab.* **2**, 583–602. Univ. of California Press, Berkeley.

Stein, C. (1986). *Approximate computation of expectations.* Institute of Mathematical Statistics Lecture Notes – Monograph Series, Vol. **7**, Hayward, California.

Swinton, J., Harwood, J., Grenfell, B.T., and Gilligan, C. (1998). Persistence thresholds for phocine distember virus infection in harbour seal *Phoca vitulina* metapopulations. *Journ. Animal Ecology* **67**, 54–68.

Wang, F.S.J. (1975). Limit theorems for age and density dependent stochastic population models. *J. Math. Bio.* **2**, 373–400.

Wang, F.S.J. (1977). A central limit theorem for age-and-density-dependent population processes. *Stoch. Proc. Appl.* **5**, 173–193.

Index

Milton Keynes UK
Ingram Content Group UK Ltd.
UKHW020022071024
449327UK00032B/2888